Frontiers in Statistical Quality Control 7

Hans-Joachim Lenz
Peter-Theodor Wilrich
Editors

Frontiers in Statistical Quality Control 7

With 70 Figures
and 64 Tables

Physica-Verlag
A Springer-Verlag Company

Professor Dr. Hans-Joachim Lenz
lenz@wiwiss.fu-berlin.de

Professor Dr. Peter-Theodor Wilrich
wilrich@wiwiss.fu-berlin.de

Freie Universität Berlin
Institut für Statistik und Ökonometrie
Garystraße 21
14195 Berlin, Germany

ISBN 3-7908-0145-3 Physica-Verlag Heidelberg New York

Cataloging-in-Publication Data applied for

A catalog record for this book is available from the Library of Congress.

Bibliographic information published by Die Deutsche Bibliothek
Die Deutsche Bibliothek lists this publication in the Deutsche Nationalbibliografie; detailed bibliographic data available in the internet at *http.//dnb.ddb.de*

Physica-Verlag is a part of Springer Science+Business Media

springeronline.com

Cover design: Erich Kirchner, Heidelberg

SPIN 10972977 88/3130 – 5 4 3 2 1 0 – Printed on acid-free paper

Editorial

The VIIth International Workshop on "Intelligent Statistical Quality Control" took place in Waterloo, Canada, and was hosted by Professor St. H. Steiner, Institute for Improvement in Quality and Productivity, University of Waterloo.

The papers collected in this volume fall into four main categories: General SQC Methodology, On-line Control including Sampling Inspection and Statistical Process Control, Off-line Control with Data Analysis and Experimental Design, and, finally, fields related to Reliability.

In the first group there are three papers on **"General Aspects of Statistical Quality Control"**.
v. Collani in "A Note on the Quality of Statistical Quality Control Procedures" proposes substituting classical designs of statistical procedures by a unified approach called the "Neyman procedure". The problem is to minimise the weighted volume of the measurement procedure and the prediction procedure constrained by the confidence level and the sample size. ***Hryniewicz*** in "Statistical Decisions in Quality Audits - a Possibilistic Interpretation of Single Statistical Tests" uses possibilistic and necessity measures defined on the null hypothesis space H and the alternative hypothesis space K. This approach enables the user, for example in quality audits, to substitute p-value based decisions by degrees of indication that a specific alternative hypothesis is true. ***Takeuchi, Suenaga, Yadohisa, Yamaguchi, Watanabe and Asano*** in "Dynamic Link Library and Electronic Book for Statistical Quality Control" present a statistical software library system called DLLQC that is particularly useful for SQC problems, and an e-book system on the Web named EBQC.

In the second group there are twelve papers on **"On-line Control"**.
Wilrich in "Sampling Plans for Inspection by Variables under a Variance Component Situation" extends single sampling plans for inspection by variables to applications where the variance of the quality characteristic X consists of two components. One component describes the variation of X due to repeated use of a specific item, the other one the variation between mean values of X of different items. ***Lam and Rahim*** in "Joint Determination of Economic Design of $\bar{X}$ -Control Charts, Economic Production Quantity, and Production Run Length for a Deteriorating Production System" simultaneously determine the economic production rate and the economic production run length. ***Stoumbos and Reynolds*** in "The Robustness and Performance of CUSUM Control Charts Based on the Double-Exponential and Normal Distributions" investigate the effects of non-normality on the statistical performance of CUSUM control charts for monitoring the process mean μ. They consider the standard CUSUM chart, which is based on the normal distribution, and a second CUSUM chart, which is based on the heavy-tailed double-exponential distribution. ***Ohta and Kusukawa*** in "An Application of Confirmation Sample Control Chart in the High-Yield Processes and Its Economic Design" deal with control chart design for detecting small and moderate changes of the fraction defective. The four chart parameters are the reaction point, the sampling interval and the lower and upper control limits LCSL and UCSL. The parameters are determined from a cost model due to Collani and Dräger (1995). ***Saniga, Davis and Lucas*** in "Statistical Design of Attribute Charts for Monitoring and Continuous Improvement When Count Levels are Low" present a procedure to find a

design of a control chart for count data when the lower Shewhart control limit is zero. The approach is based on a proposal of Lucas, Davis and Saniga (2001). ***Nishina and Matsubara*** in "A Markov Approach to the Average Run Length of CUSUM Charts for an AR(1) Process" derive analytically the item-dependent distribution of the CUSUM statistic and the Average Run Length (ARL). Thus, one obtains the conditional out-of-control ARL for any change point, and, of course, the steady-state ARL. ***Steiner and MacKay*** in "Effective Monitoring of Processes with Parts Per Million Defective. A Hard Problem!" analyse the monitoring of the defective rate where a process produces defectives measured in parts per million. As sample sizes tend to be huge in such cases, the authors propose the use of related (auxiliary) variables that are identified in a pre-processing step. ***Thyregod, Melgaard and Iwersen*** in "On Monitoring Processes and Assessing Process Capability under a Hierarchical Model" discuss sub-grouping in control charting and suggest charting on a hierarchical model that reflects variation between subgroups. In the second part of their paper, the authors develop a simple hierarchical model of "between subgroup" variation based on historical data, in order to specify the long-term distribution of subgroup averages. ***Alexopoulos, Goldsman and Tsui*** in "SPC Monitoring and Variance Estimation" review common monitoring methods for SPC in the case of auto-correlated data. The underlying characteristics and relationships are investigated. ***Knoth and Schmid*** in "Control Charts for Time Series: A Review" classify charts into the two classes 'residual charts' and 'modified control charts'. For each type, charts with memory like EWMA or CUSUM exist. By using the ARL as the performance measure, the authors show that there is no clear preference when there is no auto-correlation. They recommend a modified EWMA chart for positive auto-correlation, and the classical EWMA or CUSUM chart applied to residuals for negative auto-correlation. ***Morais and Pacheco*** in "A Note on Stochastic Ordering Criteria in SPC for the Case of Correlated Output" establish stochastic ordering results concerning the in-control and out-of-control run length of residual schemes for the mean of stationary auto-regressive processes of first and second order.

In the third group entitled "**Off-line Control**" there are five contributions.
Yamaguchi, Kono and Asano in "A New Statistical Tool for Finding Causal Conditions for Small Data Sets and Its Software" propose a sub-grouping method for small data sets based on algorithms due to Quine (1952) and Miller (1965), and illustrate the user interface of the corresponding software. ***Sparks*** in "What has Data Mining got to do with Statistical Process Monitoring? A Case Study" gives best-practise recommendations to monitor complex multivariate systems. He advocates simple charting in combination with a stepwise refinement involving the study of corresponding reports by the management. For example, slicing and dicing of SPC data cubes helps them to focus on important issues and avoids looking at too many low level charts. ***Hirotsu and Ohta*** in "Profile Analysis Based on Repeated Measurements – A New Approach to the Analysis of Interaction" give their own view on the analysis of interactions within multiple comparison techniques when there exists a natural ordering of the categories in two-way contingency tables. ***Lenz*** in "Trouble-Shooting by Graphical Models" applies cause-and-effect modelling based on Bayesian Network models to problems of troubleshooting. As call-backs, accidents etc. are isolated events, no-data driven procedures can be applied. Instead of estimation, marginal and conditional probabilities have to be settled by experts. Risk assessment and a sensitivity study within a case study show the potential of the approach, which dominates the classic FMEA set-up. ***Ojima, Suzuki and Yasui*** in "An Alternative Expression of the Fractional Factorial Designs for Two-level and Three-level Factors" present an alternative approach to two- and three-level factor designs by using the Hadamard product from group theory based on the columns of the corresponding design matrix. As is well established in the

theory of experimental design, this operator helps to solve many related design problems like fold-over designs in an elegant way.

In the final group " **Studies in Areas Related to Reliability**" there are three papers. ***Prendergast and Murphy*** in "Building in Reliability. Is Statistical Process Control Statistical Reliability Control?" demonstrate that, although many nodes (units) are controlled in an integrated circuit manufacturing process, only few have an impact on reliability. If these nodes are identified and controlled carefully in a real environment, reliability can be controlled and predicted. ***Ladany and Shore*** in "Optimal Warranty Period when Sale-Price Increases with the Lower Specification Limit" address the optimal warranty period. It is assumed that the increase of the warranty period causes proportionality of sale price and the cost of non-conforming items. The problem is solved analytically in the case of an exponentially distributed quality characteristic, other distributional assumptions requiring numerical solution. ***Solojentsev and v. Collani*** in "Evaluation of Quality and Market Risk of Complex Industrial Products" study risk assessment of complex products with several quality characteristics, where each of them may fail or be in one of several operating states, and initiating events leading to object failure define logical connections between the characteristics.

The impact of such a workshop is mainly shaped by the quality of papers, which are read at the meeting, revised later and finally submitted. We would like to express our deep gratitude to the following members of the scientific programme committee, who did an excellent job with respect to the recruiting of invited speakers as well as refereeing all the submitted papers:

Prof. Choichiro Asano, Japan
Mr David Baillie, United Kingdom
Prof. Elart von Collani, Germany
Prof. Hans-J. Lenz, Germany
Prof. Hiroshi Ohta, Japan
Prof. Poul Thyregod, Denmark
Prof. Peter.-Th. Wilrich, Germany
Prof. William H. Woodall, United States of America.

We would like to close with our cordial thanks to *Mrs. M. Eckardt,* who managed to clean up and to integrate WINWORD and LATEX papers. A financial grant of the Institute of Statistics and Econometrics, Free University of Berlin, which finally made it possible to get this volume to press, is gratefully acknowledged. Moreover, we kindly acknowledge the smooth collaboration with Physica-Verlag, Heidelberg.

On behalf of all participants, the editors would like to thank Professor Steiner and his staff for their superb hospitality, the perfect organisation, and the stimulating atmosphere. We are happy and proud to announce that the International Workshop on Intelligent Statistical Quality Control will be continued in Warszawa, Poland, in the near future.

Berlin, August 2003

Hans - J. Lenz
Peter - Th. Wilrich

Contents

PART 3: OFF-LINE CONTROL

PART 4: STUDIES IN AREAS RELATED TO RELIABILITY

Author Index

Alexopoulos, Christos., Ph.D., Prof., Georgia Institute of Technology, School of Industrial and Systems Engineering, Atlanta, GA 30332-0205, U.S.A
e-mail:christos.alexopoulos@isye.gatech.edu

Asano, Chooichiro, Prof. Dr., Soka University, Faculty of Engineering, Tokyo, Japan
e-mail: QYG04340@niftyserve.ne.jp

Collani, Elart von, Prof. Dr., Universität Würzburg, Volkswirtschaftliches Institut, Sanderring 2, D-97070 Würzburg, Germany
e-mail: collani@mathematik.uni-wuerzburg.de

Davis, Darwin J., Ph.D., Prof., Department of Business Administration, College of Business and Economics, University of Delaware, 204 MBNA America Building, Newark, DE 19716, U.S.A
e-mail: davisd@lerner.udel.edu

Goldsman, David, Ph.D., Georgia Institute of Technology, School of Industrial and Systems Engineering, Atlanta, GA 30332-0205, U.S.A
dave.goldsman@isyse.gatech.edu

Hirotsu, Chihiro, Prof. Dr., Meisei University, Faculty of Science and Technology, 2-1-1- Hodokubo, Hino City, Tokyo 191-8506, Japan
e-mail: hirotsu@ge.mesei-u.ac.jp

Hryniewicz, Olgierd, Prof. Dr., Polish Academy of Sciences and University of Applied Informatics and Management, Systems Research Institute, Newelska 6, 01-447 Warsaw, Poland
e-mail: hryniewi@ibspan.waw.pl

Iwersen, Jørgen, Dr., Novo Nordisk A/S, Novo Alle, DK 2880 Bagsvaerd, Denmark
e-mail: jiw@novo.dk

Jiang, W., INSIGHT, AT&T Labs, Short Hills, NJ 07078, U.S.A

Knoth, Sven, Dr., Europe University Viadrina, Department of Statistics, Postfach 1786, D-15207 Frankfurt(Oder), Germany
e-mail: knoth@euv-frankfurt-o.de

Kono, Yasunari, Rikkyo University, Graduate School of Social Relations, Tokyo, Japan

Kusukawa, Etsuko, Dr., Osaka Prefecture University, Department of Industrial Engineering, Sakai, Osaka 599-8531, Japan
e-mail: kusukawa@ie.osakafu-u.ac.jp

Ladany, Shaul P., Prof. Dr., Ben-Gurion University of the Negev, Department of Industrial Engineering and Management, P.O. Box 653, Beer Sheva 84105, Israel
e-mail: ladany@bgumail.bgu.ac.il

Lam, Kian Keon, University of Brunswick, Faculty of Administration, P.O. Box 4400, Fredericton, New Brunswick, E3B 5A3, Canada
e-mail: a89kh@unb.ca

Lenz, Hans-J., Prof. Dr., Freie Universität Berlin, Institut für Statistik und Ökonometrie, Garystr. 21, 14915 Berlin, Germany
e-mail: hjlenz@wiwiss.fu-berlin.de

Lucas, James M., Dr., J. M. Lucas and Associates
e-mail: jamesm.lucas@worldnet.att.net

MacKay, R. Jock, University of Waterloo, Department of Statistics and Actuarial Science, Waterloo, Ontario, N2L 3G1, Canada
e-mail: rjmackay@uwaterloo.ca

Matsubara, Shin'ichiro, Nagoya Institute of Technology, Department of Systems Engineering, Gokiso-cho, Showa-ku, Nagoya 466-8555, Japan

Melgaard, Henrik, Dr., Novo Nordisk A/S, Novo Alle, DK 2880 Bagsvaerd, Denmark
e-mail: hmel@novo.dk

Morais, Manuel C., Technical University of Lisbon, Department of Mathematics and Centre for Mathematics and its Applications, Instituto Superior Técnico, Av. Rovisco Pais, 1049-001 Lisboa, Portugal
e-mail: maj@math.inst.utl.pt

Murphy, Eamonn, Prof., University of Limerick, Plassey Technology Park, Limerick, Ireland
e-mail: eamonn.murphy@ul.ie

Nishina, Ken, Prof. Dr., Nagoya Institute of Technology, Department of Systems Engineering, Gokiso-cho, Showa-ku, Nagoya 466-8555, Japan
e-mail: nishina@system.nitech.ac.jp

Ohta, Eri, Prof. Dr., Meisei University, Faculty of Science and Technology, 2-1-1-Hodokubo, Hino City, Tokyo 191-8506, Japan
e-mail: otha@ge.meisei-u.ac.jp

Ohta, Hiroshi, Prof. Dr., Osaka Prefecture University, Department of Industrial Engineering, Sakai, Osaka 599-8531, Japan
e-mail: ohta@ie.osakafu-u.ac.jp

Ojima, Yoshikazu, Prof. Dr., Science University of Tokyo, Department of Industrial Administration, 2641, Yamazaki, Noda, Chiba, 278-8510, Japan
e-mail: ojima@ia.noda.tus.ac.jp

Pacheco, António, Technical University of Lisbon, Department of Mathematics and Centre for Mathematics and its Applications, Instituto Superior Técnico, Av. Rovisco Pais, 1049-001 Lisboa, Portugal
e-mail: apacheco@math.inst.utl.pt

Prendergast, James, Dr., Institute of Technology, Clash, Tralee, Co. Kerry, Ireland
email: james.prendergats@ittralee.ie

Rahim, M.A., Prof. Dr., University of New Brunswick, Faculty of Administration, P.O. Box 4400, Fredericton, New Brunswick, E3B 5A3, Canada
e-mail: rahim@unb.ca

Reynolds Jr., Marion, Prof. Dr., Virginia Polytechnic Institute and State University, Department of Statistics, Blacksburg, VA 24061-0439, U.S.A.
e-mail: mrr@vt.edu

Saniga, Erwin M., Prof. Dr., University of Delaware, Department of Business Administration, Newark, DE 19716, U.S.A.
e-mail: sanigae@be.udel.edu

Schmid, Wolfgang, Prof. Dr., Europe University Viadrina, Department of Statistics, Postfach 1786, D-15207 Frankfurt(Oder), Germany
e-mail: schmid@euv-frankfurt-o.de

Shore, Haim, Prof. Dr., Ben Gurion University of the Negev, Department of Industrial Engineering and Management, P.O. Box 653, Beer-Shewa 84105, Israel
e-mail: shor@bgumail.bgu.ac.il

Solojentsev, Evgueni, Prof. Dr., Russian Academy of Science, Institute of Mechanical Engineering Problems, Intelligent Integrated Systems Laboratory, St. Petersburg, Russia
e-mail: sol@sapr.ipme.ru

Sparks, D. Ross, Dr., CSIRO Mathematical and Information Sciences, Locked Bag 17, North Ryde, NSW 2113, Australia
e-mail: ross.sparks@cmis.csiro.au

Steiner, Stefan, Prof. Dr., University of Waterloo, Department of Statistics and Actuarial Science, Waterloo, Ontario, N2L 3G1, Canada
e-mail: shsteine@uwaterloo.ca

Stoumbos, Zachary, Prof. Dr., Rutgers, University of New Jersey, Piscataway, NJ 08854-8054, U.S.A
e-mail: stoumbos@andromeda.rutgers.edu

Suenaga, Katsuyuki, Prof. Dr., Kagoshima Immaculate Heart College, Japan,
e-mail: sue@k-junshin.ac.jp

Suzuki, Tomomichi, Tokyo University of Science, 278-8510, Japan
e-mail: suzuki@tqm.t.u-tokyo.ac.jp

Takeuchi, Akinobu, Ph.D., Rikkyo University, College of Social Relations, Nishi-Ikebukuro 3-34-1, Tokyo 171-8501, Japan
e-mail: akitake@rikkyo.ac.jp

Thyregod, Poul, Prof. Dr., The Technical University of Denmark, Informatics and Mathematical Modelling (IMM), Richard Petersen Plads 321, DK-2800 Lyngby, Denmark
e-mail: pt@imm.dtu.dk

Tsui, Kwok-Leung, Prof. Dr., Georgia Institute of Technology, School of Industrial and Systems Engineering, Atlanta, GA 30332-0205, U.S.A.
e-mail: ktsui@isye.gatech.edu

Watanabe, Michiko, Prof. Dr., Toyo University, Faculty of Economics, Japan
e-mail: watanabe_michiko@nifty.com

Wilrich, Peter-Th., Prof. Dr., Freie Universität Berlin, Institut für Statistik und Ökonometrie, Garystr. 21, 14195 Berlin, Germany
e-mail: wilrich@wiwiss.fu-berlin.de

Yadohisa, Hiroshi, Prof. Dr., Kagoshima University, Faculty of Science, Department of Mathematics and Computer Science, Japan
e-mail: yado@sci.kagoshima-u.ac.jp

Yamaguchi, Kazunori, Prof. Dr., Rikkyo University, Department of Industrial Relations, Nishi-Ikebukuro 3-34-1 Tokyo 171-8501, Japan
e-mail: kyamagu@rikkyo.ac.jp

Yasui, Seiichi, Tokyo University of Science, 278-8510, Japan

Part 1

General Aspects of Statistical Quality Control

A Note on the Quality of Statistical Quality Control Procedures

Elart von Collani

University of Würzburg, Sanderring 2, D-97070 Würzburg, Germany

Abstract. Statistical Quality Control (SQC) deals with the questions how to monitor, control and improve the quality of products and manufacturing processes by means of statistical methods. To this end the quality of products and processes is defined and statistical methods are used to determine the actual quality value. As there are many different statistical methods which can be used in a given situation, the question arises about the quality of statistical methods. Searching statistical literature for an answer reveals that statistical methods are evaluated rather by means of auxiliary criteria like, for instance, unbiasedness and consistency than by a unified measure reflecting quality. Hence, this paper is concerned with filling the existing gap by developing a measure for the quality of statistical procedures.

1 Quality

Almost everything is at one time or another evaluated. The result of the evaluation reflects its actual quality and is often the basis of far-reaching decisions. Clearly, any meaningful evaluation must necessarily be based on a definition of quality, and, therefore, "quality" as a concept has universal importance for individuals, enterprises and nations. Hence, it is not surprising that "quality" became an important catchword particularly in modern industries.

There are many meanings connected with the word "quality", some of them are listed in Juran's Quality Control Handbook [12]. The most important definition is the one given in the International Standard ISO 3534-2:1993, which reads as follows:

- The totality of features and characteristics of a product or service that bear on its ability to satisfy stated or implied needs.

According to this "official" definition of quality, three tasks have to be accomplished in order to determine the actual value of quality of an object.

- The needs to be satisfied, i.e., the object's usage has to be defined.
- Those features and characteristics have to be identified that contribute to reaching the defined objective.
- A measure defined on the relevant features must be developed reflecting the extent to which they satisfy the defined needs.

2 Stochastic Procedures

Statistics may be looked upon something being close to the "science of stochastics"[1], which deals with uncertainty either caused by ignorance with respect to a historical development or by randomness with respect to a future development. The science of stochastics quantifies and utilizes the existing relation between past and future. The modeling task may be assigned to 'theoretical stochastics' and the methods linking model and reality by experiments build the body of 'empirical stochastics'.

There are stochastic procedures satisfying different needs, but any of these methods may be reduced to one of two types of procedures. In order to show that this statement holds, one has to consider the general situation in which stochastic procedures are used, where a procedure is called "stochastic", if it is based on a probability distribution, i.e., if it takes into account the uncertainty about the future development.

Table 1. Past and future connected by the law of randomness

Past	Presence	Future
the historical event determines the present situation	the present situation determines the law of randomness	the law of randomness determines the future event

The law of randomness given by a probability distribution connects the past and the future. The past event determines the present conditions, which are fixed, but often known only partially. Assume a process of interest, then clearly its outcome depends on the initial conditions. The variable representing that part of the determinate, initial conditions which is relevant to the process of interest, is called deterministic variable, denoted by D with actual value d.

The variable representing the indeterminate, random outcome of the process is called random variable denoted by X.

The presence, represented by the actual value d of the deterministic variable D, determines the probability distribution of the random variable, which is therefore denoted by $X|d$ with probability distribution $P_{X|d}$. The probability distribution contains the actual value d of the deterministic variable D as distribution parameter and gives for each possible future event of $X|d$ on a scale from zero to one the degree of certainty (i.e. probability) of its occurrence. The connection of the past and future by $P_{X|d}$ enables to draw

[1] The "science of stochastics" was inaugurated by Jakob Bernoulli about 300 years ago as "science of prediction" or in Latin "Ars conjectandi".

inference from the past to the future and vice versa. Thus, the basic objectives of stochastic procedures are

- to predict a future event, i.e., the outcome x of X and
- to determine[2] a past event, i.e., the actual value d of D.

The past event is fixed by the actual value d of the deterministic variable and, therefore, procedures for determining d are called *measurement procedures* in the remainder. The future event of the random variable X is uncertain and, therefore, procedures for determining it are called prediction procedures. Any stochastic procedure makes either statements about a past event or a future event and, therefore, may be reduced to a prediction procedure or to a measurement procedure. Thus, the general problem of defining the quality of statistical procedures is basically solved by defining the quality of measurement and prediction procedures, respectively.

3 Measurement and Prediction Procedures

A measurement procedure refers to a deterministic variable D with bounded image set $\mathcal{D}$ and aims at measuring its fixed actual value $d \in \mathcal{D}$. A prediction procedure refers to a random variable $X|d$ with image set $\mathcal{X}(d)$ and aims at predicting the future event $E \subset \mathcal{X}(d)$. The situation may be described in the following way:

Table 2. Schematic representation of the situation dealt with in classical statistics

variable	symbol	image space	model
deterministic variable	D	$\mathcal{D}$	$P_D : \mathcal{B} \to [0,1]$ $P_D(\{d\}) = 1$
random variable	$X\|d$	$\mathcal{X}(d)$	$P_{X\|d} : \mathcal{B} \to [0,1]$ $d \in \mathcal{D}$

The measurement problem is solved by a mapping

$$C_D : \mathcal{X} \to \mathcal{B}(\mathcal{D}) \tag{1}$$

[2] In statistics "to determine" is called "to estimate".

where $\mathcal{X} = \cup_{d \in \mathcal{D}} \mathcal{X}(d)$ contains all possible realizations of X and $\mathcal{B}(\mathcal{D})$ is the σ-field of Borel-sets over $\mathcal{D}$.

The set $\mathcal{D}$ is called measurement space and each subset $C_D(x)$ is called a measurement.

The prediction problem is solved by a mapping

$$A_X : \mathcal{D} \to \mathcal{B}(\mathcal{X}) \tag{2}$$

where $\mathcal{B}(\mathcal{X})$ is the σ-field of Borel-sets over $\mathcal{X}$.

The set

$$\mathcal{X} = \bigcup_{d \in \mathcal{D}} A_X(d) \tag{3}$$

is called prediction space and each $A_X(d)$ is called a prediction.

A mapping C_D or, equivalently, the respective system of sets

$$\mathcal{C}_D = \left\{ C_D(x) \,\middle|\, x \in \mathcal{X} \right\} \tag{4}$$

is called a measurement procedure. The mapping A_X or, equivalently, the respective system of sets

$$\mathcal{A}_X = \left\{ A_X(d) \,\middle|\, d \in \mathcal{D} \right\} \tag{5}$$

is called prediction procedure.

Any measurement and prediction procedure can be represented graphically by a subset of

$$\mathcal{M}_{D,X} = \left\{ (d, x) \,\middle|\, x \in \mathcal{X}(d), d \in \mathcal{D} \right\} \tag{6}$$

which therefore is called *measurement and prediction space*[3].

---

[3] The measurement and prediction space was already considered by Jerzy Neyman in [16], who called it "general space" and assumed it to be unbounded in general.

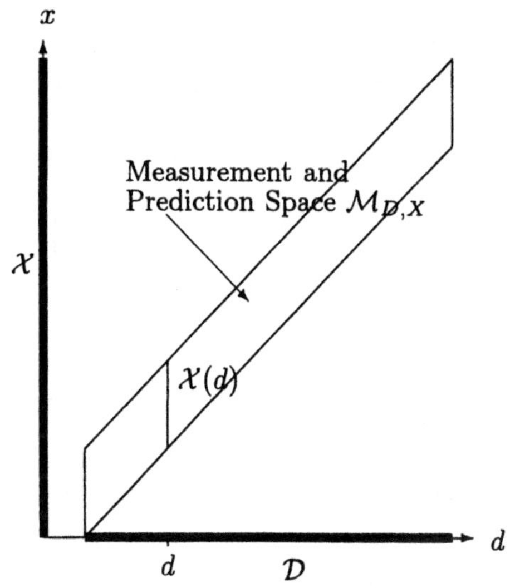

Fig. 1. Schematic representation of a measurement and prediction space $\mathcal{M}_{D,X}$

4 Reliability of a Stochastic Procedure

The reliability of a stochastic procedure refers to the frequency of the occurrence of an error: A measurement error occurs if the actual value d is not an element of the obtained measurement $C_D(x)$, and a prediction error occurs if the outcome x of the random variable X is not an element of the prediction $A_X(d)$.

In order to have a reliable prediction procedure, the probability of the occurrence of an error must be bounded by a small value, say α, which is called the *significance level* of the procedure:

$$P_{X|d}\Big(\mathcal{X} \setminus A_X(d)\Big) < \alpha \qquad \text{for any } d \in \mathcal{D} \tag{7}$$

where $\mathcal{X} \setminus A_X(d)$ denotes the set difference of $\mathcal{X}$ and $A_X(d)$.

Clearly, the requirement (7) is equivalent to the condition

$$P_{X|d}\Big(A_X(d)\Big) \geq 1 - \alpha \qquad \text{for any } d \in \mathcal{D} \tag{8}$$

where $(1 - \alpha)$ is called the confidence level of the procedure.

A prediction procedure meeting the requirement (7) is denoted by $A_X^{(1-\alpha)}$. A prediction $A_X^{(1-\alpha)}(d)$ is called an $(1-\alpha)$-prediction.

Let $A_X^{(1-\alpha)}$ be a prediction procedure with reliability $1-\alpha$. Setting

$$C_D^{(1-\alpha)}(x) = \left\{ d \,\middle|\, x \in A_X^{(1-\alpha)}(d) \right\} \tag{9}$$

yields a measurement procedure with

$$P_{X|d}\Big(\{x \,|\, d \in C_D^{(1-\alpha)}(x)\}\Big) \geq 1-\alpha \quad \text{for any } d \in \mathcal{D} \tag{10}$$

Hence, the measurement procedure $C_D^{(1-\alpha)}$ defined by (9) has a confidence level 1-α.

The relations (9) and (10) reveal that the problem of determining a prediction procedure with given reliability is dual to the problem of determining a measurement procedure with the same reliability. Any prediction procedure with reliability 1-α determines uniquely a measurement procedure with reliability 1-α and vice versa. The advantage of this duality is the fact that for the derivation of a prediction procedure complete knowledge of the underlying probability measure is available.

A prediction procedure aims at predicting the future outcome x, a measurement procedure aims at measuring the actual value d. The features relevant for reaching these aims are

- procedure's reliability
- procedure's expense
- procedure's accuracy

The reliability of a procedure is taken care by the reliability requirement given by the significance level α or equivalently by the confidence level $\beta = 1-\alpha$ which is assumed to be fixed here. The expense of a procedures is determined in general by the sample size n, which is also assumed to be fixed. Thus, only the accuracy of the procedure has to be taken into account for defining the procedure's quality.

As shown, any stochastic procedure may be represented by a system of subsets of the measurement and prediction space $\mathcal{M}_{D,X}$ implying that a serious evaluation of a statistical procedure consists of an evaluation of subsets of the measurement and prediction space. Therefore, a necessary condition for an adequate evaluation of a statistical method is the availability of suitable measures for different types of sets related to the measurement and prediction space $\mathcal{M}_{D,X}$.

5 Measures Related to $\mathcal{M}_{D,X}$

The necessary measures refer to the following types of sets:

1. Subsets of $\mathcal{M}_{D,X}$ which correspond to the pair of variables (D, X). The corresponding measure denoted by $P_{D,X}$ is called the joint measure of (D, X).
2. Subsets of the projection of $\mathcal{M}_{D,X}$ on the d-axes which correspond to the marginal variable D and, therefore, the corresponding measure denoted by P_D is called marginal measure of the deterministic variable D.
3. Subsets of the projection of $\mathcal{M}_{D,X}$ on the x-axes which correspond to the marginal variable X and, therefore, the corresponding measure denoted by P_X is called marginal measure of the random variable X.
4. Subsets of the cross-section of $\mathcal{M}_{D,X}$ for given value d of the deterministic variable D. The corresponding measure denoted by $P_{X|d}$ is called conditional measure of X under the condition $\{d\}$.
5. Subsets of the cross-section of $\mathcal{M}_{D,X}$ for given value x of the random variable X. The corresponding measure denoted by $P_{D|x}$ is called conditional measure of D under the condition $\{x\}$.

There is a natural measure on the measurement space $\mathcal{D}$, which represents the existing ignorance about the initial conditions. In contrast to uncertainty caused by randomness, uncertainty caused by ignorance is not structured implying that any value of $\mathcal{D}$ has to be weighted equally, thus, leading to the geometric or uniform measure.

Assuming that any relevant set is measurable and introducing

$$|\mathcal{D}| = \int\limits_{\mathcal{D}} dy$$

as the geometric volume of $\mathcal{D}$, a normed measure for subsets of $\mathcal{D}$ is immediately obtained.

$$P_D\left(\mathcal{D}_0\right) = \int\limits_{\mathcal{D}_0} p_{\mathcal{D}}(y)dy \quad \text{for any } \mathcal{D}_0 \subset \mathcal{D} \tag{11}$$

where

$$p_{\mathcal{D}}(y) = \begin{cases} \frac{1}{|\mathcal{D}|} & \text{for } y \in \mathcal{D} \\ 0 & \text{elsewhere} \end{cases} \tag{12}$$

Subsets on the cross-section of the measurement and prediction space for given value d of the deterministic variable D

$$\left\{(d,x)\,\middle|\, x \in \mathcal{X}(d)\right\} \quad \text{for fixed } d \in \mathcal{D} \tag{13}$$

are appropriately measured by the conditional probability measure $P_{X|d}$ which is assumed to be known.

The product measure $P_D \otimes P_{X|d}$ is appropriate for subsets of the measurement and prediction space $\mathcal{M}_{D,X}$ resulting in the joint measure $P_{D,X}$ of the two-dimensional variable (D, X).

The joint measure immediately yields the measure on the prediction space $\mathcal{X}$, i.e., the marginal measure of the random variable X denoted by P_X.

The marginal measure P_X together with the joint measure $P_{D,X}$ results in a measure on the cross-sections of the measurement and prediction space for given value x of the random variable X:

$$\left\{(d, x) \,\middle|\, d \in \mathcal{D}\right\} \quad \text{for fixed } x \in \mathcal{X} \tag{14}$$

i.e., the conditional measure of D denoted by $P_{D|x}$.

By means of the five measures P_D, $P_{X|d}$, $P_{D,X}$, P_X and $P_{D|x}$ any relevant set may be evaluated taken into account the accuracy (represented by the geometric size of the measurements) on the one hand and the probability structure (represented by the given probability measure) on the other hand.

The above approach is based on the realistic assumption that there are always information available about the initial conditions (represented by D) and the process of interest (represented by X) which eliminate the possibility of a unbounded parameter space on the one hand and an unbounded support of the probability distribution on the other hand. Thus, it becomes possible to utilize the complete existing knowledge in order to arrive at procedures guaranteeing with certainty meaningful results. This is important particularly in quality control, because the variables considered in quality control have often a comparatively small range of variability and neglecting this information may lead to severe difficulties. Assume, for instance, automatic sorting of product which leads to a truncation of the values of the characteristic of interest. If this truncation is not taken into account by the underlying model, wrong decisions are almost certain and as a matter of fact none of the relevant industrial standards allows for including truncation into the model.

6 Procedure's Quality

Let $C_D^{(\beta)}$ be a measurement procedure with reliability β called β-measurement procedure. The quality of the procedure $C_D^{(\beta)}$ is determined by the accuracy of the resulting measurements $C_D^{(\beta)}(x)$ given by their geometrical size. Let

$\mathcal{S}(x)$ be the geometrical size of $C_D^{(\beta)}(x)$ measured by P_D, then

$$\mathcal{S}(x) = \int\limits_{C_D^{(\beta)}(x)} dP_D(y) \tag{15}$$

Let $\mathcal{S}(X)$ be the random variable defined on $\mathcal{X}$ by its realizations (15). Then the mean accuracy of the measurement procedure and, hence, the procedure's quality is obtained by the expectation of $\mathcal{S}(X)$.

The expectation $E[S]$ denoted by $V_w\left(C_D^{(\beta)}\right)$ is called weighted volume of the β-measurement procedure and is given by:

$$V_w\left(C_D^{(\beta)}\right) = \int\limits_{\mathcal{X}} \left(\int\limits_{C_D^{(\beta)}(x)} dP_D(y) \right) dP_X(x) \tag{16}$$

Clearly, the smaller the weighted volume of a measurement procedure the better its quality, i.e., its accuracy.

Assuming realistically a bounded measurement and prediction space and utilizing the duality relation between measurement and prediction procedures yields a second representation of the weighted volume of a β-measurement procedure.

$$V_w\left(C_D^{(\beta)}\right) = \int\limits_{\mathcal{D}} \left(\int\limits_{A_X^{(\beta)}(d)} dP_X(x) \right) dP_D(y) \tag{17}$$

which only depends on the predictions $A_X^{(\beta)}(d)$ and not any more on the measurements $C_D^{(\beta)}(x)$.

With the weighted volume V_w a quality indicator for the basic stochastic procedures has been derived enabling to compare different procedures with respect to their quality, i.e., their "ability to satisfy the stated needs". This quality indicator replaces the rather meaningless quality criteria used in classical statistics, as for instance "unbiasedness", "consistency" or the criterion of a "uniformly minimum variance unbiased estimator" (UMVUE).

7 Optimal Statistical Procedures

The objective function for measurement procedures (16) or equivalently (17) enables the definition of optimal procedures.

Definition:
A β-measurement procedure ${}^*C_D^{(\beta)}$ and the dual β-prediction procedure ${}^*A_X^{(\beta)}$ are called optimal or most accurate if their common weighted volume V_w^* is minimum. Most accurate procedures are also called *Neyman procedures.* •

Remark 1. What is called here measurement and prediction space has been considered already by Jerzy Neyman, who named it *general space.* As, the here developed approach is a straightforward extension of Neyman's ideas [16] we call the optimal procedures after him. •

The problem of determining Neyman procedures is to minimize the weighted volume V_w under the side-condition of the reliability requirement given by the confidence level β and the procedure's expense given by the sample size n.

This problem may be formulated in the following way:

- For any $d \in \mathcal{D}$ select an appropriate subset $A_X^{(\beta)}(d)$ of $\mathcal{X}(d)$ in such a way that
  - its measure with respect to P_X is minimum, and
  - its measure with respect to $P_{X|d}$ is not less than β.

In the continuous case with respect to the measures P_D and $P_{X|d}$, respectively, the solution of the problem is to determine the subsets according to a rule which is based on an analogous algorithm as the maximum likelihood ratio principle of the Neyman-Pearson theory.

- For any $d \in \mathcal{D}$ order the elements $x \in \mathcal{X}(d)$ according to the *likelihood ratio* of the two measures concerned:

$$\frac{f_{X|d}(x)}{f_X(x)} \tag{18}$$

  where $f_{X|d}$ and f_X are the density functions of $P_{X|d}$ and P_X, respectively.
- The optimal β-predictions are obtained by composing $A_X^{(\beta)}(d)$ by those $x \in \mathcal{X}(d)$ which have the highest likelihood-ratio (18) until the reliability requirement given by the confidence level β is fulfilled. •

In case of discrete measures $P_{X|d}$ it is generally not possible to meet exactly the confidence level β. Therefore, proceeding as described above does not yield an optimal procedure. However, the result may be used as a starting point. An optimal solution may be found by a subsequent search utilizing the given discrete structure of the involved measure.

In [4] the approach outlined above has been applied to the case that the deterministic variable D represents a proportion or probability, the random variable X is given by the number of successes among n random experiments, and both are connected by the binomial law. The resulting procedures are more accurate and flexible than the traditional ones. As the binomial case combined with small sample sizes is frequently used in quality control, it should be worthwhile to apply the new and more accurate procedure for the purpose of industrial quality control. In [9] the approach is used to determine the variance assuming a normal distribution and similar as in the binomial case the procedure yields more accurate results and avoids meaningless measurements.

8 Conclusion

The above derived quality indicator for stochastic procedures results in any case in new methods which are superior to the traditional ones with respect to accuracy. This advantage has to be paid by the fact that the determination of the procedures is numerically more difficult. The greater complexity of the new methodology demands new ways of presenting the results to the users. In [4] the presentation of the methods is done using a CD-ROM by means of which the user does not need to perform any mathematical transformation of the data, but immediately obtains the result either as a measurement or a prediction.

The quality indicator is founded on an approach which is neither classical nor Bayesian. It has some characteristic features.

- It is completely based on the issue of probability and thus resembles to some extent the Bayesian approach.
- It always leads to a meaningful result in contrast to the classical approach.
- It allows a geometric interpretation and a graphical representation of any procedure.
- It allows the definition of objective-oriented quality indicators and, thus, a rational comparison of any two stochastic procedures serving the same aim.

The new approach is beneficial in particular in case of small sample sizes as are characteristic for statistical quality control. With increasing sample sizes the differences between the new and the classical methods are getting smaller, and asymptotically they are in many instances equivalent. The last fact shows that classical statistics is basically an asymptotic theory which does not necessarily be valid in case of a finite sample size.

References

1. Bernoulli, J. (1713). *Ars Conjectandi.* Thurnisorium, Basel.

2. Cantor, M. (1877). *Das Gesetz im Zufall.* Berlin SW., Verlag von Carl Habel.
3. Clopper, C., J., and Pearson, E. S. (1934). The Use of Confidence or Fiducial Limits Illustrated in the Case of the Binomial. *Biometrika*, 26, 404-413.
4. v. Collani, E., Dräger, K. (2001). *Binomial Distribution Handbook for Scientists and Engineers.* Birkhäuser, Boston.
5. v. Collani, E., Dumitrescu, M., and Lepenis, R.. (2001). An Optimal Measurement and Prediction Procedure. Economic Quality Control 16, 109-132.
6. v. Collani, E., and Dumitrescu, M. (2001). Complete Neyman Measurement Procedure. METRIKA 54, 111-130.
7. v. Collani, E., and Dumitrescu, M. (2000). Neyman Exclusion Procedure. *Economic Quality Control* 15, 15-34.
8. v. Collani, E., and Dumitrescu, M. (2000). Neyman Comparison Procedure. *Economic Quality Control* 15, 35-53.
9. v. Collani, E., Dumitrescu, M. and Victorina Panaite (2002). Measurement Procedures for the Variance of a Normal Distribution. *Economic Quality Control* 17, 155-176.
10. Crow, E. L. (1956). Confidence Intervals for a Proportion. *Biometrika*, 43, 423-435.
11. Hauser, W. (1997). *Die Wurzeln der Wahrscheinlichkeitsrechnung.* Franz Steiner Verlag, Stuttgart.
12. Juran, J.M. (Ed.) (1974): Quality Control Handbook. 3rd ed., McGraw-Hill, New York.
13. Laplace, P.-S. (1814). *Essai philosophique sur les probabilités*, Christian Burgois, Paris.
14. Nelder, J. (1986). Statistics, Science and Technology. *J. Roy. Statist. Soc. A* 149, 109-121.
15. Neyman, J. (1935). On the Problem of Confidence Intervals. *Ann. Math. Statist.*, 6, 111-116.
16. Neyman, J. (1937). Outline of a Theory of Statistical Estimation Based on the Classical Theory of Probability. *Phil. Trans. Roy. Soc., Ser. A*, 236, 333-380.
17. About, P.-J., Boy, M. (1982). *La Correspondance de Blaise Pascal et de Pierre de Fermat. - La Géométrie du Hasard ou le deébut du Calcul des Probabilités.* Les Cahiers de Fontanay. Fontenay aux Roses.
18. Schmitz, Ph. S.J. (1990) Probabilismus - das jesuitischste der Moralsysteme. In: *Ignatianische Eigenart und Methode der Gesellschaft Jesu.* Freiburg.
19. Sterne, T.E. (1954). Some Remarks on Confidence or Fiducial Limits. *Biometrika*, 41, 275 -278.

Statistical Decisions in Quality Audits - a Possibilistic Interpretation of Single Statistical Tests

Olgierd Hryniewicz

Systems Research Institute of the Polish Academy of Sciences
and University of Applied Informatics and Management
Newelska 6, 01-447 Warsaw, Poland

Abstract. Properties of statistical tests which are used in statistical quality control have very good interpretation in the case of sequences of similar tests, for example in the acceptance sampling of series of production lots. However, there is no convincing practical interpretation of test results when tests are performed only once, as in the case of the acceptance sampling of isolated lots. This happens especially in quality auditing, when auditors have to verify claims about declared quality levels using only the available statistical data. For such a case we propose a simple interpretation of tests results in terms of the theory of possibility [10]. Alternative hypotheses concerning the declared quality levels are validated using Possibility of Dominance (PD), Possibility of Strict Dominance (PSD), Necessity of Dominance (ND), and Necessity of Strict Dominance (NSD) indices.

1 Introduction

In statistical quality control decisions are made after the evaluation of existing statistical data describing a certain quality characteristic. In a classical setting we assume that the quality characteristics of interest is described by a random variable Z distributed according to a certain probability distribution P_θ belonging to a family of probability distributions $\mathcal{P} = \{P_\theta : \theta \in \Theta\}$ indexed by a parameter θ. In the case of inspection by attributes Z may describe a number of nonconforming items (or nonconformities) in a lot, and in the case of inspection by variables Z may be interpreted as any continuous quality characteristic. We assume that decisions which are to be made depend entirely on the value of the parameter θ. If the value of θ were known we might take an appropriate decision without any problem. However, the value of θ is not known, and we could only formulate a respective hypothesis about it. Therefore, our original decision is equivalent to the acceptance (or rejection) of a null hypothesis $H : \theta \in \Theta_H$ (where $\Theta_H \subset \Theta$) upon a value of the parameter θ. According to the Neyman-Pearson theory of statistical tests we should also formulate an alternative hypothesis $K : \theta \in \Theta_K$, where $\Theta_K \subset \Theta$, and $\Theta_H \cap \Theta_K = \emptyset$.

Suppose that we observe a random sample $Z_1, Z_2, ..., Z_n$, and the following decisions are to be made: either to reject H (and to accept K) or not to reject H (usually identified with the acceptance of H). Let's denote by 0 the acceptance, and by 1, the rejection of H. Hence, the decision rule, called a statistical test, can be defined as a function $\varphi : \mathcal{R}^n \to [0, 1]$. According to the theory of statistical tests each nonrandomised statistical test divides the whole space of possible observations of the random variable Z into two exclusive subspaces: $\{(z_1, z_2, ..., z_n) \in \mathcal{R}^n : \varphi(z_1, z_2, ..., z_n) = 0\}$, and $\{(z_1, z_2, ..., z_n) \in \mathcal{R}^n : \varphi(z_1, z_2, ..., z_n) = 1\}$. First of these subspaces is called *an acceptance region*, and the second is called *a critical region*. In the majority of practical cases we deal with a certain *test statistic* $T = T(Z_1, Z_2, ..., Z_n)$, and we reject the considered null hypothesis H when the value of T belongs to a certain critical region $\mathcal{K}$, i.e. if $T = T(Z_1, Z_2, ..., Z_n) \in \mathcal{K}$. In such a case the decision rule looks like this

$$\varphi(Z_1, Z_2, ..., Z_n) = \begin{cases} 1 \text{ if } T(Z_1, Z_2, ..., Z_n) \in \mathcal{K}, \\ 0 \text{ otherwise.} \end{cases} \tag{1}$$

To define the critical region we usually set an upper value for the probability of a wrong rejection of the null hypothesis H (the so called probability of type I error). This probability, denoted by α, is called *a significance level* of the test. Thus, we have

$$P(\varphi(Z_1, Z_2, ..., Z_n) = 1|H) \leq \alpha. \tag{2}$$

In general, for a given sample number n there may exist many statistical tests which fulfil this condition. However, only some of them may have additional desirable properties, and only those are used in practice. For more detailed description of the problem we refer the reader to the excellent book of Lehmann[9].

Despite their clear mathematical description, statistical tests are very often difficult to understand for a general user. The reason of this is the asymmetry between the null and the alternative hypotheses. It is clear, that only the probability of type I error α (α risk) is under control. The probability of wrong acceptance of the null hypothesis H (called the probability of type II error β, or β risk) is generally either not explicitly defined or even not possible to define.

In a special case of testing simple statistical hypotheses, i.e. when we test the null hypothesis $H : \theta = \theta_0$ against the alternative $K : \theta = \theta_1$, we may introduce an additional requirement for the probability of type II error β, defined as

$$P(\varphi(Z_1, Z_2, ..., Z_n) = 0|K) \leq \beta. \tag{3}$$

This enables us to have both risks, α and β, under control. In many practical cases the setting of α, β, θ_0, and θ_1 is sufficient for the design of a statistical test, i.e. setting the sample number n, and defining the critical region $\mathcal{K}$ for the values of the test statistic T. In the case of acceptance sampling a user may analyse the OC-curves of different acceptance sampling plans that are presented in international standards from the ISO 2859 and ISO 3951 series in order to find the procedure with appropriate α and β risks. However, even in this case there still exist some problems with practical interpretation of the results of statistical tests, especially for an unexperienced user. There exist many explanations of these difficulties. First, in many practical cases the choice of θ_0, and θ_1 seems to be rather artificial. Even if a limit for "good quality" given by θ_0 is easily established, quality practitioners are very often not able to set a particular value of θ_1, as for many of them any value of the quality characteristic θ that is worse than the value of θ_0 is unacceptable. Second, the notions of risks are easily interpreted in the case of series of repeated statistical tests. In such a case a practitioner knows that the risk has its interpretation in terms of the expected fraction of correct (or false) decisions. This way of understanding fails when users are faced with results of single statistical tests. Nowadays, practitioners are facing these problems much often than previously. Acceptance procedures are much often than previously applied to short series of delivered production lots. Moreover, due to an improved logistics and global markets products are submitted for inspection in lots which should be considered as isolated. The situation is even worse in quality audits. The statistical data presented to auditors have usually a unique character. In such a case every statistical test that can be used for the interpretation of such data should be considered as a single one. Therefore, the interpretation of the results of statistical tests that is rather obvious in the case of series of repeated identical tests in the case of single test is not convincing for the majority of practitioners who are usually not well trained in statistics. To overcome these problems we need a new simple interpretation of the results of statistical tests. This interpretation should be presented using a language that is easy to understand for a person who is not well trained in statistics. Thus, statistical terms should be rather avoided, and the results of tests should be presented using a language that is close to a common language. In this paper we propose a new method of interpretation in which we apply a methodology that is used in computer science for building man-machine interfaces. In the second section of the paper we present some methods which statisticians traditionally use to interpret their test results. In many practical cases these methods do not provide the user with sufficient information about consequences of his/her decisions. Therefore, there is a need to provide the user with some additional information. In the third section of the paper we propose a new decision support tool which could be helpful in such cases. The proposed tool uses notions of the theory of possibility, and provides the user with a possibility measure of the correctness of his/her decision.

2 Statistical interpretation of the results of tests

Statistical tests designed according to the Neyman-Pearson theory can be well interpreted if the same test is repeatedly applied to new data. When we repeat the test many times, then the percentage of cases when the null hypothesis is wrongly rejected is not greater than $100\alpha\%$. When the probability of type II error β is well defined we can also say that the percentage of cases when the null hypothesis is wrongly accepted (or, the alternative hypothesis is wrongly rejected) is smaller than $100\beta\%$. This interpretation is generally accepted in statistical quality control, especially in the case of acceptance sampling of continuous series of lots. However, this type of interpretation generally fails if we have to analyse the result of a single test. To present this problem let us consider the following example.

Example 1. Suppose that we want to inspect a lot of $N = 1000$ items that is purchased from a supplier who uses a system of acceptance sampling plans from the international standard ISO 2859-1 [6]. If it is known that the supplier uses a sampling scheme indexed by the AQL (Acceptance Quality Limit) equal to 1%, and we want to have a good protection against bad quality defined as 5% of nonconforming items in the inspected lot, then we can use a sampling plan from the international standard ISO 2859-2 [7] with the sample size $n = 125$ items, and the critical number of nonconforming items in the sample equal to $c = 3$ items. From the investigation of the properties of the ISO 2859-1 sampling system we know that in the case of a long series of accepted lots the *average* level of supplier's quality must be better than the stated value of AQL. Therefore, we may use the chosen sampling plan to verify if the quality of the lot (in percent of nonconforming items) is not worse than 1%, i.e. that the lot does not contain more than 10 nonconforming items. If we were to apply this procedure to a series of purchased lots we would be assured that *on average* the quality of all inspected lots is not worse than required. However, in our case we deal with *a single* lot, and we have to accept it even if the fraction of nonconforming items in a sample is equal to $3/125 * 100 = 2,4\%$, i.e. much more than the required value of 1%. In such a case a statistician says that the available statistical data is not sufficient to reject the null hypothesis.

The situation may be even more difficult in the case of quality auditing where statistical tests are usually performed in isolation, i.e. in circumstances which are rather unlikely to be repeated. Despite the fact that there exists an international standard ISO 2859-4 [8] that provides an auditor with sampling plans for assesssment of stated (declared) quality levels, an auditor very often is not able to use such a sampling plan, and has to use the only available quality data. It may happen that the estimated value of quality level is significantly worse than the declared one, however from the statistical point of view there is still no sufficient evidence to reject the verified hypothesis. To illustrate this problem let us consider additional examples.

Example 2. Suppose that a quality auditor has to verify that the quality level of a certain production process (measured in the percentage of nonconforming items) is not worse than 1%. The auditor is interested in a small probability of falsely contradicting a correctly declared quality level $DQL = 1\%$, and with a possibly small probability of accepting the declared quality level when the actual quality level is worse. The auditor consults the international standard ISO 2859-4 [8], takes the value of $DQL = 1\%$, and selects the limiting quality ratio (LQR) level III sampling plan. In such a case the sample size is $n = 125$ items, and the limiting number of nonconforming items in the sample is $L = 3$. Again, as in the previous example, the auditor has to accept the declared quality level even if the estimated quality level is significantly worse than the declared one.

Example 3. Suppose that a quality auditor does not have a possibility to apply the sampling plan from the standard. He has to verify the correctness of the declaration that the quality level is equal to $DQL = 1\%$ having a sample of $n = 75$ items with $d = 2$ nonconforming items. His estimated quality level is obviously worse than the declared value of 1%. What should be his decision? Is he entitled do reject the quality level?

These situations, when encountered in practice, could create some doubts among many quality practitioners, especially those who are not well trained in statistics. Therefore, there is a need to offer a simple measure of the correctness of possible decisions when we deal with isolated lots or singular sets of statistical data.

To provide the user with a simple evaluation method for the result of a statistical test many statisticians use the notion of the so called *p-value* (which is also called the observed test size, or the *significance* of the observed test result). The observed test size p^* is the minimum value of the significance level α for which the null hypothesis H has to be rejected for the given observed value of the test statistics T. In other words, if the chosen value of the significance level α is greater than the value of p^*, then the null hypothesis H is rejected. Otherwise, we do not have any important reason to reject it. In practice, we reject the null hypothesis H, if p^* is smaller than 5% (or even 1%). Large values of p^* (for example, larger than 10%) indicate that the results of the test strongly support the null hypothesis. Moreover, a larger value of p^* is interpreted as a stronger support of the null hypothesis H. Another loose interpretation of p^* is the following: it is the measure to what extent the difference between the observed phenomenon (estimated value of a parameter, for example) and the hypothetical one is due only to the random character of the sample. In this interpretation, small values of p^* mean that we cannot neglect the observed difference between the hypothetical value of the parameter and its estimated value as purely haphazard, and therefore we shall reject the null hypothesis.

The value of p^* can be easily computed if we know how to calculate the limits of the confidence interval for the parameter of interest θ. It is equal to the confidence level of the widest confidence interval (one- or two-sided) that does not contain the value of θ_0.

Example 1 (continued). In the considered of inspection by attributes of a single lot, the null hypothesis can be stated as $H : D \leq D_0$, where D is the number of nonconforming items in a lot, and D_0 is its maximum allowable value, the number of nonconforming items in the sample is described by the hypergeometric distribution and the value of p^* can be calculated as

$$p^* = \sum_{i=d}^{n} \frac{\binom{D_0}{i}\binom{N-D_0}{n-i}}{\binom{N}{n}} \tag{4}$$

where d is the observed number of nonconforming items in the sample of n items taken from a lot of N items.

Example 2 (continued). In the case of the problem of quality auditing, the null hypothesis can be stated as $H : \Theta \leq \Theta_0$, where Θ is the fraction of nonconforming items in the audited process, and Θ_0 is its declared value, the number of nonconforming items in the sample is described by the binomial distribution and the value of p^* can be calculated (see Bickel and Doksum [1]) as

$$p^* = \sum_{i=d}^{n} \binom{n}{i} {\Theta_0}^i (1 - \Theta_0)^{n-i} \tag{5}$$

where d is the observed number of nonconforming items in the sample of n items

Example 3 (continued). In this case the calculation of p^* is exactly the same as in *Example 2*.

It is worthy to note that many statistical packages such as STATISTICA, SPSS, STATGRAPHICS, etc. provide the information about the value of p^* for all statistical tests.

Example 1 (continued). In the considered case of the inspection of a single lot we have $D_0 = 10$, $N = 1000$, and $n = 125$. Suppose now, that exactly $d = 3$ nonconforming items have been found in that sample. The observed test size in such a case equals $p^* = 0,12$, and indicates that the test results rather strongly support the null hypothesis, and we should not reject the lot, as it is recommended by a sampling plan. If the number of nonconforming items is $d = 2$ the observed test size is much greater, and is approximately equal to $p^* = 0,36$. In both cases, according to all statistical recommendations, we have to accept the null hypothesis $H : D \leq 10$ despite the fact that the estimated fraction nonconforming in the lot is visibly greater than 1%.

Example 2 (continued). In the case of the quality audit sampling the situation is similar. When exactly $d = 3$ nonconforming items are found in the sample the observed test size approximately equals $p^* = 0,13$. If the number of nonconforming items is $d = 2$ the observed test size is approximately equal to $p^* = 0,36$. Again, there is no reason to reject the hypothesis about the declared quality level despite the fact that the estimated fraction nonconforming in this process is significantly greater than 1%.

Example 3 (continued). In this case it is easy to find that $p^* = 0,17$. According to all known recommendations that can be found in statistical textbooks and manuals of statistical packages there is no reason to reject the declaration that the the actual quality level is not worse than 1% despite the fact that in the sample its value is significantly worse than DQL.

The concept of *p-value* is not accepted by many statisticians as it does not have any frequency interpretation. In the case of continuous probability distributions it can be related to the so called fiducial probability distributions of the parameters of probability distributions that were intoduced by R.Fisher in the early 1930's. For a composite null hypothesis *p-value* can be computed as an appropriate fiducial probability. This approach, however, is also strongly criticised by many statisticians for the same reasons as the *p-value*. Controversies between different schools of statisticians that concern the notion of *p-value* make us believe that there is a need to look for its another interpretation using a different approach.

3 Possibilistic interpretation of the results of single statistical tests

Suppose that the quality auditor has to verify the truth of the claim that the actual quality level is *not worse* than a certain stated (declared) value. Additionally, he may try to verify that the actual quality level is *worse* than a certain limiting (disqualifying) value. In the setting of the statistical decisions this problem may be described by setting two alternative hypotheses $H : \theta \in \Theta_H$ and $K : \theta \in \Theta_K$. Without loosing generality we assume that $H : \theta \leq \Theta_0$, and $K : \theta > \Theta_1$, where $\Theta_0 \leq \Theta_1$. Now, let us consider these two hypotheses *separately*. First, let us analyse only the null hypothesis H. For thc observed value of the test statistics $t = T(z_1, z_2, ..., z_n)$ we can find the observed test size p_H^* for this hypothesis. The value of observed test size p_H^* shows how the observed data support the null hypothesis. When this value is relatively large we may say that the observed data strongly support H, i.e. that the actual quality level is not worse than the declared one. Otherwise, we should say that the data do not sufficiently support H, and that the verified declaration about the quality level is probably not true. It is worthwhile to note that in the latter case we do not claim that the data support K. The same can be done for the alternative hypothesis K, so we can find for this hypothesis the observed test size p_K^*. When $\Theta_H \cup \Theta_K = \Theta$ we have $p_K^* = 1 - p_H^*$. Thus, to find the value of p_K we have to find the *p-value* for the test of the hypothesis $\theta \leq \Theta_1$, and to subtract this value from one.

Now, let us denote by $x = 1$ a situation when we decide that the data do not support the considered hypothesis, and by $x = 0$ a situation when we decide to accept the hypothesis. We propose to evaluate the results of testing of both considered hypotheses in terms of the theory of possibility.

Theory of possibility has been proposed by L.Zadeh [10] and is used in modern computer science, especially in expert systems, approximate reasoning, data mining, etc. The basic notion of the theory of possibility is a possibility distribution π on a set of possible states S. According to the definition given in [5] the possibility distribution is a mapping $\pi : S \to L$ from S to a bounded, linearily ordered valuation set $(L, >)$. In practical terms a possibility distribution represents a state of knowledge distinguishing what is plausible (not surprising) from what is less plausible (to some extend surprising). The value of the possibility function $\pi(s) = 1$ means that the state $s \in S$ is totally possible (plausible). On the other hand, $\pi(s) = 0$ means that the state $s \in S$ is impossible. The values of $\pi(s)$ that belong to the interval $(0, 1)$ mean that the state s is only partially possible. There exists a link between the theory of possibility and the theory of fuzzy sets. In general, possibility distributions can be regarded as membership function of fuzzy sets describing the set of possible states S. It is worth noting that the notion of possibility is not the same as the notion of probability which is related to the notion of the frequency of occurence. The degree of possibility of a certain state is not equivalent to the probability of its ocuurence. For a more detailed discussion of the principles of the theory of possibility see, for example, [4] or [5].

Let us describe the result of a single statistical test in terms of a possibility distribution. In order to do it we propose to evaluate the acceptance of the null hypothesis H by a fuzzy set $\widetilde{H}$ defined by the following membership function

$$\mu_H(x) = \begin{cases} \min\left[1, 2p_H^*\right] & , x = 0 \\ \min\left[1, 2\left(1 - p_H^*\right)\right] & , x = 1 \end{cases} \tag{6}$$

which may be interpreted as a *possibility distribution* $\pi_H(x, p_H^*) = \mu_H(x)$ of the acceptance of H.

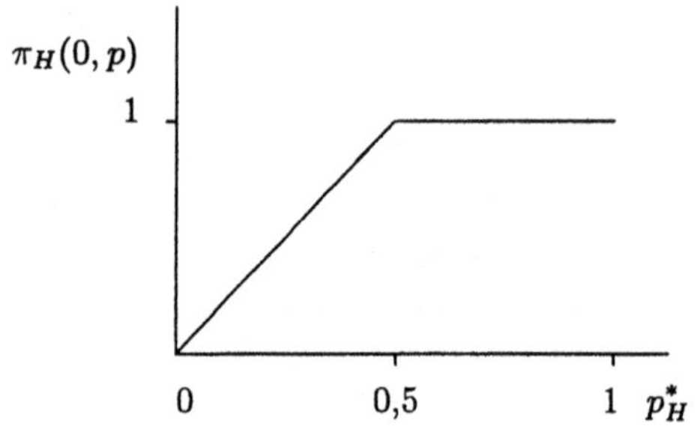

Fig. 1. Possibility of the acceptance of H as the function of the observed test size

It is worthy to note that $\sup(\mu_H(0), \mu_H(1)) = 1$, and $\mu_H(1) = 1$ indicates that it is ***plausible*** that the hypothesis H is not true. On the other hand, when $\mu_H(0) = 1$ we wouldn't be surprised if H was true. It is necessary to stress here that the values of $\mu_H(x)$ do not represent the probability that H is true, but only a *possibility* of a correct decision.

The same can be done for the alternative hypothesis K. We propose to evaluate the alternative hypothesis K by a fuzzy set $\widetilde{K}$ with the following membership function

$$\mu_K(x) = \begin{cases} \min[1, 2p_K^*] & , x = 0 \\ \min[1, 2(1 - p_K^*)] & , x = 1 \end{cases} \tag{7}$$

which may be interpreted as a *possibility distribution*$\pi_K(x, p_K^*) = \mu_K(x)$ of K.

To choose an appropriate decision, i.e. to choose either H or K, we propose to use four measures of possibility defined by Dubois and Prade [3]. First measure is named the *Possibility of Dominance*, and for two fuzzy sets A and B is defined [3] as

$$PD = Poss(A \succeq B) = \sup_{x,y;x \geq y} \min\{\mu_A(x), \mu_B(y)\} \tag{8}$$

where $\mu_A(x)$ and $\mu_B(y)$ are the membership functions of A and B, respectively. PD is the measure for a possibility that the set A is not dominated by the set B. In the considered problem of testing statistical hypotheses we have

$$\begin{aligned} &PD = Poss\left(\widetilde{H} \succeq \widetilde{K}\right) = \\ &= \sup[\min\{\mu_H(1), \mu_K(1)\}, \min\{\mu_H(0), \mu_K(1)\}, \min\{\mu_H(0), \mu_K(0)\}] = \\ &= \max\{\mu_H(0), \mu_K(1)\}, \end{aligned} \tag{9}$$

and PD represents a *possibility* that choosing H over K is not a worse solution.

Second measure is named the *Possibility of Strict Dominance*, and for two fuzzy sets A and B is defined [3] as

$$PSD = Poss(A \succ B) = \sup_x \inf_{y;y \geq x} \min\{\mu_A(x), 1 - \mu_B(y)\} \tag{10}$$

PSD is the measure for a possibility that the set A strictly dominates the set B. In the considered problem of testing statistical hypotheses we have

$$\begin{aligned} &PSD = Poss\left(\widetilde{H} \succ \widetilde{K}\right) = \\ &= \sup[\inf[\min\{\mu_H(1), 1 - \mu_K(1)\}, \min\{\mu_H(1), 1 - \mu_K(0)\}] = \\ &= \min\{\mu_H(0), 1 - \mu_K(0)\}, \end{aligned} \tag{11}$$

and PSD represents a *possibility* that choosing H over K is a correct decision.

Third measure is named the *Necessity of Dominance*, and for two fuzzy sets A and B is defined [3] as

$$ND = Ness(A \succeq B) = \inf_x \sup_{y,y \leq x} \max\{1 - \mu_A(x), \mu_B(y)\} \tag{12}$$

and ND represents a *necessity* that the set A is not dominated by the set B. In the considered problem of testing hypotheses we have

$$\begin{aligned} &ND = \inf\{\sup[\max(1-\mu_H(0),\mu_K(1)),\\ &\max(1-\mu_H(0),\mu_K(0)),\ \max(1-\mu_H(1),\mu_K(1))]\} =\\ &= \max\{1-\mu_H(1),\mu_K(1)\}, \end{aligned} \tag{13}$$

and ND represents a *necessity* of choosing H over K.

Fourth measure is named the *Necessity of Strict Dominance*, and for two fuzzy sets A and B is defined [3] as

$$\begin{aligned} &NSD = Ness(A \succ B) = 1 - \sup_{x,y;x\leq y} \min\{\mu_A(x),\mu_B(y)\} =\\ &= 1 - Poss(B \geq A) \end{aligned} \tag{14}$$

and NSD represents a *necessity* that the set A dominates the set B. In the considered problem of testing hypotheses we have

$$\begin{aligned} &NSD = 1 - \sup[\min\{\mu_H(1),\mu_K(1)\},\\ &\min\{\mu_H(1),\mu_K(0)\},\min\{\mu_H(0),\mu_K(0)\}] =\\ &= 1 - \max\{\mu_H(1),\mu_K(0)\}, \end{aligned} \tag{15}$$

and NSD represents a strict *necessity* of choosing H over K.

Close examination of the proposed measures reveals that

$$PD \geq ND \geq PSD \geq NSD. \tag{16}$$

It means that according to the practical situation we can choose the appropriate measure of the correctness of our decision. If the choice between H and K leads to serious consequences we should choose the NSD measure. In such a case $p_H^* > 0,5$ is required to have $NSD > 0$. When these consequences are not so serious we may choose the PSD measure. In that case $PSD > 0$ when $p_K^* < 0,5$, i.e. when there is no strong evidence that the alternative hypothesis is true. Finally, the PD measure, which may be equal to zero only for sampling from finite populations, gives us the information of the possibility that choosing H over K is not a wrong decision. It is also possible to use the ND measure instead of PD, especially when we expect a slightly stronger evidence that choosing H is allowable.

Another interpretation of the possibility and necessity indices can be found in the framework of the theory of preference relations. Let $\nu : R \times R \rightarrow [0,1]$ be a measure of preference such that $\nu(x,y)$ be a measure of the preference of x over y. The preference relation is complete (see [2] for further references) when

$$\nu(x,y) + \nu(y,x) \geq 1\ \forall x,y \tag{17}$$

For a complete set of preference relations we may define the measure of the indifference between the alternatives x and y

$$\nu_I(x,y) = \nu(x,y) + \nu(y,x) - 1, \tag{18}$$

the measure that the alternative x is better than y

$$\nu_B(x,y) = \nu(x,y) - \nu_I(x,y), \tag{19}$$

and the measure that the alternative x is worse than y

$$\nu_W(x,y) = \nu(y,x) - \nu_I(x,y). \tag{20}$$

For the possibility and necessity measures defined above it is easy to show that

$$\nu_I(x,y) = PD - NSD \tag{21}$$

$$\nu_B(x,y) = NSD \tag{22}$$

$$\nu_W(x,y) = 1 - PD \tag{23}$$

Thus, the knowledge of the indices PD and NSD is sufficient for the complete description of preference relations for two alternatives: to choose the null hypothesis H or to choose the alternative hypothesis K.

Example 1 (concluded). Continuing the example from previous section for $d = 3$ we have $p_H^* = 0,12$, and we find that for the alternative hypothesis $K : D > D_1$, where $D_1 = 50$, the observed test size equals $p_K^* = 0,038$. Thus, the NSD measure of the correctness of the acceptance of H is equal to 0. However, its possibility measures are non-negative. The PD measure is equal to 1, and the PSD measure is equal to $\min\{0,24\ ,\ 1 - 0,076\} = 0,24$. This result may be interpreted as follows. There is rather no evidence that there is 50 or more nonconforming items in the inspected lot. However, there is also a rather slight possibility that there is only 10 nonconforming items in that lot. Moreover, the indifference measure introduced above indicates that for the given results of sampling both alternatives are rather indifferent $(\nu_I(x,y) = 1)$. With such an interpretation of test results a decision maker might be able to decide whether to accept that lot or not to accept it.

When there are only $d = 2$ nonconforming items in this sample we have $p_H^* = 0,36$, and and for the alternative hypothesis the observed test size equals $p_K^* = 0,009$. In such a case we have: $PD = 1$, $PSD = 0,72$, $ND = 1$, and $NSD = 0$. It seems that there is rather a strong possibility that the null hypothesis is better confirmed by the data than the alternative one $(PSD = 0,72)$. However, the necessity of this is equal to zero. This result is not surprising as the estimated quality level is worse than 1% but significantly better than 5%.

Example 2 (concluded). In this case the results are different as the alternative hypothesis $K : \Theta_1 > 0,01$ is much more restrictive. When $d = 3$ nonconforming items are found in the sample the observed test size for the alternative hypothesis equals $p_K^* = 0,87$. Thus, we have: $PD = 0,26$, $PSD = 0$, $ND = 0,26$, and $NSD = 0$. It means that there is only small possibility $(PD = 0,26)$ that the actual quality level is not worse than its declared value. The measure of indifference is rather low $(\nu_I(x,y) = 0,26)$, and there is rather strong indication ($\nu_W(x, y) = 0,74$) that the alternative hypothesis is true.

When there are only $d = 2$ nonconforming items in the sample we have: $PD = 0,72$, $PSD = 0$, $ND = 0,72$, and $NSD = 0$. These results rather support the alternative hypothesis ($\nu_W(x, y) = 0,28$). However, the level of infifference is rather high ($\nu_I(x, y) = 0,72$), so we should look for additional evidence in order to make a correct decision.

Example 3 (concluded). In this case the observed test size for the alternative hypothesis equals $p_K^* = 0,83$. Hence, we have: $PD = 0,34$, $PSD = 0$, $ND = 0,34$, and $NSD = 0$. As in the first part of the previous example we see, that the alternative hypothesis should be rather preferred ($\nu_W(x, y) = 0,66$). However, there are still some doubts ($\nu_I(x, y) = 0,34$) suggesting that additional evidence should be considered when a final decision has to be made.

4 Conclusions

In this paper we propose another way of looking at the results of single statistical tests for attributes. The proposed method allows a practitioner to evaluate the test results using intuitive concepts of possibility, necessity or indifference. It has to be stressed, however, that by the application of the proposed methodology we do not provide the user with any *new* information. The only difference is in a language that is used for the interpretation of test results. The proposed method allows to use a language that is closer to a common language used by practitioners than the specialised language of statistics. This might be very valuable, especialy for computerised decision support systems that are oriented on users who are not well trained in statistics.

The proposed method may be straightforwardly generalised for the majority of statistical tests used in practice. It allows to assign possibility and necessity measures for the results of classical statistical tests. An appropriate index can be used for given circumstances. The PD (or ND) index shows even the weakest sign of dominance of the null hypothesis over an alternative one. If there is a need to reveal a strict domination of the null hypothesis over the alternative one (or to give a strict discrimination measure for these two hypotheses) we should use the NSD index. The PSD measure may be used in intermediate cases.

The proposed method may be also generalised for the case of imprecisely defined statistical hypotheses and vague statistical data. In these cases the fuzzy sets methodology is used to describe both hypotheses and data. As the proposed method is based on the possibility distributions that are closely related to fuzzy sets such a generalisation is not so difficult.

References

1. Bickel, P. J., Doksum, K. A. (1977) Mathematical Statistics. Basic Ideas and Selected Topics. Holden Day, Inc., San Francisco.
2. Cutell, V., Montero, J. (1999) An Extension of the Axioms of Utility Theory Based on Fuzzy Rationality Measures. In: Preference and Decisions under Incomplete Knowledge, J.Fodor, B.De Baets, P.Perny (Eds.), Physica-Verlag, Heidelberg, 33–50.
3. Dubois, D., Prade, H. (1983) Ranking fuzzy numbers in the setting of possibility theory, Information Sciences, **30** , 184–244.
4. Dubois, D., Prade, H. (1997) Qualitative possibility theory and its applications to reasoning and decision under uncertainty, Belgian Journal of Operations Research, Statistics and Computer Science, **37**, 5–28.
5. Dubois, D., Prade, H., Sabbadin, R.. (2001) Decision-theoretic foundations of qualitative possibility theory. European Journal of Operational Research, bf128, 459–478.
6. ISO 2859-1: 1989, Sampling procedures for inspection by attributes - Part 1: Sampling plans indexed by acceptable quality level (AQL) for lot-by-lot inspection.
7. ISO 2859-2: 1985, Sampling procedures for inspection by attributes - Part 2: Sampling plans indexed by limiting quality (LQ) for isolated lot inspection.
8. ISO 2859-4: 1999, Sampling procedures for inspection by attributes - Part 4: Procedures for assessment of stated quality levels.
9. Lehmann, E. L. (1986) Testing Statistical Hypotheses, 2nd ed., J.Wiley, New York, 1986.
10. Zadeh, L. A. (1978) Fuzzy sets as a basis for a theory of possibility, Fuzzy Sets and Systems, **1** , 3–28.

Dynamic Link Library and Electronic Book for Statistical Quality Control

Takeuchi, A.[1], Suenaga, K.[2], Yadohisa, H.[3], Yamaguchi, K.[1], Watanabe, M.[4], and Asano, Ch.[5]

[1] College of Social Relations, Rikkyo University, Nishi-Ikebukuro 3-34-1, Tokyo 171-8501, Japan
[2] Kagoshima Immaculate Heart College, Japan
[3] Department of Mathematics and Computer Science, Kagoshima University, Japan
[4] Faculty of Economics, Toyo University, Japan
[5] Kyushu University, Japan

Abstract. In this paper, we produce two projects and their results for statistical quality control on the web. The first system named DLLQC: Dynamic Link Library for Statistical Quality Control, is a statistical software library system with the DLLs. The second system named EBQC: Electronic Books for Quality Control, is an e-book system on the Web. Both systems, which can be used freely through the internet, have many contents with the DLLs of statistical analysis and PDF files of e-books.

1 Introduction

It may not be too much to say that statistical techniques have been mainly introduced under the guidance of consultants, e.g. W. E. Deming, at the Headquarter (GHQ) of the Allied Forces after the Second World War, in order to investigate the cultivated area and the amount of foods in occupied Japan. Although this primarily involved sample survey techniques was soon introduced to industry, e.g. several kinds of control charts, in order to improve manufacturing processes and to export such products. In connection with such activities, statistical sampling inspection plans also have been applied, e.g. Dodge-Romig, A. Wald, MIL-STD etc., and jointly statistical methods to improve manufacturing processes have been developed on experimental designs and other statistical methods, e.g. W. E. Deming, J. M. Juran, A. V. Feigenbaum and others. Subsequently, enthusiastic SQC activities have developed into QC circles and so-called TQC, originating in Japan. By this means, industries in Japan were modernized and recovered miraculously. Thus some Japanese industries had been called model students of SQC for about ten years, and showed high prosperity. However, after that, the Japanese economy broke down into a bubble situation in the 1990s. Thus the previous superiority of Japanese quality has been lost, and economic conditions in Japan are stagnant. With the passage of time, in the recent high quality age with IT, it is now emphasized for every businessman in industries and students to be aware of the need for solid and sober education in SQC. Previously, when the quality of Japanese products and the QC management systems was applauded, many books were published on SQC by influential authors. But nowadays, such books are rare, and it is difficult for us to study the background of those days, philosophical and conceptual directions, and methodological developments, because such books have now gone out of print. It will be a big loss and inconvenience at present for junior engineers and students to lose such authors' intelligence, experience and contributions. Additionally, the software for studying SQC methods in these books are usually expensive and has many options that may confuse the novice user. They desire software that dispenses with unnecessary options and has a simplified interface.

It is the present purpose to prevent such losses by collecting such rare books related to SQC, and to put these books with full pages on display through the internet. For this are developing, we propose an e-book system on the Web, called EBQC: Electronic Books for Quality Control. We are also developing a new statistical software library for SQC, called DLLQC: Dynamic Link Library for Statistical Quality Control, that is free and more user-friendly. Thus, anyone would be able to study those books and use the library from anywhere at any time. All staff of this project hope that it will be contribute socially to SQC education in industries and schools.

2 DLLQC

Total quality control (TQC) refers to procedures in broad fields at the respective companies. But actually these procedures in each fields may be defined concretely by Deming circle, consisting of i) design and planning, ii) doing/manufacturing, iii) checking for outgoing, iv) market response and improvement of quality of actions. Namely, at required stages of the circle, statistical methods, like experimental designs, estimation and testing hypothesis, analysis of variances, sampling inspection, sampling survey and so on, are able to apply.

In this section, essential softwares of such techniques are provided in Table 2, although still they are not enough and now preparing in haste. In ordinary condition for data management, several techniques of control charts are applicable, and also randomness tests are used in view of time-series variation. For characteristic diagram, exploratory multivariate techniques may be effective, like principal component analysis, methods of multidimensional scaling, of cluster analysis, factor analysis and so on.

As an example of the interface of the DLLQC, analysis of variances (ANOVA) is presented to explain features of the DLLQC.

2.1 Dynamic link library of the DLLQC

At first, we will briefly introduce the dynamic link library (DLL) of the DLLQC in this section.

A DLL can be considered part of a library for applications within the Windows operating system as well as for programming. A DLL contains one or more functions that are compiled, linked, and stored separately from the applications using them. One advantage of a DLL is that multiple applications can access the same DLL. The DLLs in our library can be accessed from any software if such software has the facility to call functions or subroutines in DLLs. Since the source code for DLLs can be written in several programming languages, such as Basic, C/C++, Fortran, and Pascal, we think it will be significant to construct a library using DLLs.

Users can DLLize source programs with commercial software suitable for each programming language. For example, when source programs are written in Fortran, users could make a DLL using Visual Fortran. Table 1 lists commercial software that can create DLLs. For further details, users can refer to manuals for their software according to need.

Table 1. Software to make a DLL

Programing Language	Software
Basic	Visual Basic
C/C++	Visual C++
Fortran	Visual Fortran
Pascal	Delphi

The process of DLLizing an existing source program is as follows:

1. Rewrite the source program of the subroutine or function-type program, removing any output-to-screen commands.
2. Add descriptions for input/output suited to the environment creating the DLL. For example, if using Visual Fortran, like Figure 1, the user must describe the following.
   i) "`!DEC$ ATTRIBUTES STDCALL, REFERENCE, ALIAS:'Subroutine or Function name' :: Subroutine or Function name`"
   ii) Each variables for input/output, "`!DEC$ ATTRIBUTES REFERENCE ::Argument name`"
   iii) "`!DEC$ ATTRIBUTES DLLEXPORT ::Subroutine name or Function name`"
3. Build the above file to create a DLL file.

The procedures vary a little, depending on the software used to create the DLL and the software used to call it. The following are the procedures used to call a DLL created in Visual Fortran from Excel.

```
subroutine anova(arg1,arg2,arg3,arg4,arg5,arg6,arg7)
!dec$ attributes stdcall, reference, alias:'anova' ::anova
!dec$ attributes reference ::arg1
!dec$ attributes reference ::arg2
!dec$ attributes reference ::arg3
!dec$ attributes reference ::arg4
!dec$ attributes reference ::arg5
!dec$ attributes reference ::arg6
!dec$ attributes reference ::arg7
!dec$ attributes dllexport ::anova
```

Figure 1. Sample (anova.dll) input/output description with Visual Fortran

The descriptions necessary to call the DLL must be typed into the header of the macro file. As shown in Figure 2, these descriptions indicate the names of the procedures, the locations of the DLL file, and the names and type of variables to transport the values. Then the `call` command is used at the location for the variables, in order to call the procedure defined in the DLL file.

It is impossible for an independent creator or the small group to develop and publish a software package that contains a wide variety of SQC methods. It is not easy for the users to use a variety of software packages with different interfaces and varying procedures. The DLLQC project could offer a solution to this problem. DLLizing existing program sources cuts down the time needed to develop such softwares, and makes it possible to create a library with procedures that can be used for any software. Currently, the Fortran sources of the NISAN System (Asano and Kimura, 1990) are most often used in the DLLQC.

The DLLs in this library consist of different numerical calculations and statistical methods. Users can choose the DLL from interface software. Users, for instance, could feed data into Excel and then let the DLL perform multivariate analysis, which Excel cannot do. Separating the interface from the statistical engine allows users to perform the entire process in a familiar environment, from data input to analysis to output of results. At the same time, the library frees software authors from interface problems, allowing them to concentrate on developing what is most needed.

Each SQC method included in the library includes not only the DLL file, but also the manual and sample files for using the DLL. We propose the following items for each method.

- DLL
- Format of the DLL (Figure 3)
- Sample data (Comma Separated Value (CSV) or Excel)
- Results of analysis of sample data (Text or Excel)
- Information on interface software (macro of Excel and so on)
- Reference

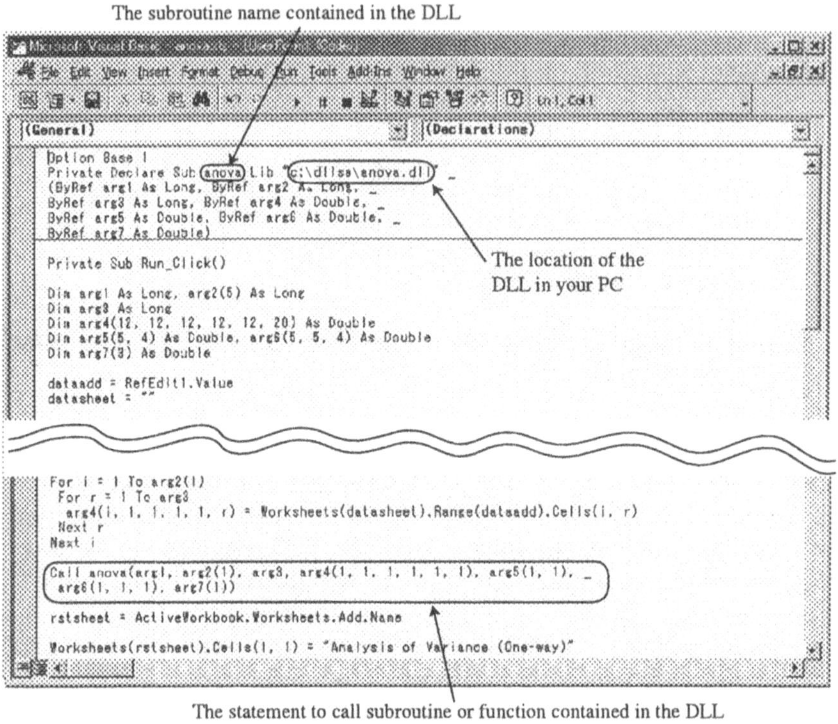

Figure 2. Call the DLL in Excel

The contents of our library can be downloaded and used freely and we show the list of DLLs of DLLQC in Table 2. Details for the library and published DLLs can be obtained from our web site (Figure 4). The URL is http://www.sci.kagoshima-u.ac.jp/~dllqc/. On this web site, we also provide examples of interfaces for several applications, including, Microsoft Excel, S-PLUS, STATISTICA, etc. Examples of specifications for some DLLs are also available.

The requirements for the software vary according to the user's goals. In the field of education, in which the level of computer skill varies widely amongst users, it would be more effective if the interface could be changed according to the users' level of computer proficiency. For novice users, complicated input and output options are confusing and often result in users missing the objective of a lesson. They may also cause errors in operation. Dispensing with unnecessary options and simplifying the interface, therefore, would be of greater benefit to this type of user. Using graphics, such as charts, figures, and diagrams, to represent the results of analysis is often more effective than numerical description.

Table 2. List of contents of DLLQC

Name	Content
Mathematical computing	
UPPE	Statistical Distribution Tables (Upper Probability ↔ Percentile) 1. N(0,1)-dist. 2. Chi-sq. dist. 3. t-dist. 4. F-dist. 5. Beta dist. 6. Gamma dist. 7. exponen. dist. 8. t non-central dist., F non-central dist., and Chi-sq. non-central dist.
Generating random numbers	
RAND	One million rectangular digital random numbers[0,9], which are available to create m-digit random numbers ($m < 10$)
RANDOMS	Random numbers of 1. Uniform [0,m] dist., 2. N(0,1) dist., 3. Exponential dist., 4. Chi-square dist., 5. Binomial dist., 6. Poisson dist., 7. Multivariate N(0, σ^2) dist., 8. Autocorrelated N(0,1) dist.
Data investigation (Data cleaning)	
CMDIS	Complete multivariate data investigation system (basic statistics, regression, histogram, correlation diagram)
CONTI	Contingency table ($m \times n$ Chi-square test)
Data transformation	
RIDIT	Ridit transformation
PROBIT	Probit analysis
Basic test	
MEANTEST	Mean-vector
VARCOVARTEST	Variance-covariance matrix
Design of experiment and analysis of variances	
LATIN	Generating latin squares of reduced type
ANOVA	Analysis of variances of m factors, where $m < 6$, each level size < 13, replication size < 21. Every parameter is estimated for main factor and 2-factor interaction effects.
GANOVA2W	Generalized two way ANOVA table is given in case of unequal replication size and missing values.
Multivariate analysis	
PCA	Principal component analysis
CORR	Multivariate correlation analysis
REG	Multivariate regression analysis
DISCRIMINANT	Discriminant analysis
CACORR2	Canonical correlation analysis for 2-vectors
CACORRM	Canonical correlation analysis for m-vectors
CAANL2	Canonical analysis for 2-groups
CAANLM	Canonical analysis for m-groups
Cluster analysis	
LCA	Latent class analysis
HCLUSTER	Hierarchical cluster analysis with dendrogram based on Lance-Williams method
CRICLUS	New criteria for hierarchical clustering in order to evaluate the result
Metric multidimensional scaling	
TOR	Multidimensional scaling method 1 by variables (Torgerson)
ADD	Multidimensional scaling method 2 by variables (ADDSCAL)
Factor analysis	
BCOEF	Grouping of variables by B-coefficient method
VARIMAX	Varimax solution
KDA	Kaiser-Dickman solution (oblique rotation)
QMAX	Quartimax solution (orthogonal rotation)
FACT	Complete process of multiple factor analysis
Statistical quality control	
QCG	QC-game
Multivariate nonparametric statistics	
MNONPARA	Multivariate nonparametric statistics
Biomedical analysis	
KMTEST	Kaplan-Meier test
WILCOXON	Wilcoxon test
GPW-SIDK	Sequential selection procedure based on play-the-winner sampling with successive success-, inverse-, and difference of success- stopping rules

```
Name of a library:
  [Name of a DLL file (must)]: anova.dll
  [Name of a subroutine (must)]: anova

 Arguments
  [Number of arguments (must)]: 7

  Argument 1
   [Type (must)]: integer
   [Size (must)]: 1
   [Input or Output (must)]: Input
   [Explanation]: Number of Factors (less than 6)

  Argument 2
   [Type (must)]: integer
   [Size (must)]: 1
   [Input or Output (must)]: Input
   [Explanation]: Number of each Levels the Factor (less than 13)

  ......
```

Figure 3. The sample format of the DLL

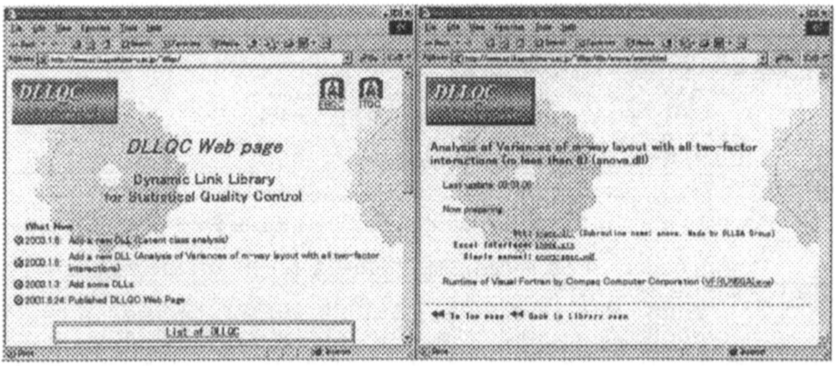

Figure 4. Top page of DLLQC and the distribution of the DLL (anova.dll) page

There are also diversified demands for using the software for practical purposes. Those users who only do routine work only need the interface necessary for that work, so unused options should be discarded. On the other hand, users who actively use statistical methods for analysis of every kind would need an interface that is applicable to a variety of cases. Such users demand numerical accuracy for the output of results, in a format that often must be adaptable to further complicated analysis.

The DLLQC proposes separation of the interface and the statistical engine to meet such demands. It provides a statistical engine that satisfies the general user, while the interface can be custom-made according to the users' purposes. As an illustration, the following section examines a sample of the interface for general use of the DLLQC for ANOVA, which is a very useful analysis for SQC.

2.2 How to use the DLLQC

We state the concrete utilization of the DLLQC published by our web site in this section.

Here, we explain how to use an Excel Macro to call the DLL of the ANOVA (anova.dll) from Excel. The system requirement of this DLL is Windows95 and Excel97, or later versions of both.

On the web site (right in Figure 4), the files needed in the analysis are packed in one self-expanding file per analysis. For individual specifications of the interfaces, please refer to the manuals and sample files included. Note that since anova.dll is created with Visual Fortran, users need to install Visual Fortran or the runtime of Visual Fortran when they use this DLL. The Visual Fortran runtime files are published freely as a runtime distribution kit on the Hewlett-Packard Company (COMPAQ Computer Corporation) web site (http://h18009.www1.hp.com/fortran/visual/redist.html).

First, open the anova.xls, because this file includes an example of how to use the anova.dll. Users are then advised that if using a Macro from Excel, they should click [Enable Macros]. There are sample data and results of analysis on each sheet. We describe the how to input and analyze new data in this section.

Users only need to perform steps i) and ii) once. For subsequent analyses, they begin with step I) in II] Analysis.

I] Preparations

i) Create a directory (folder) named "dllsa" at the top of the C drive (e.g. C:\dllsa).
ii) Copy or move anova.dll to dllsa directory (folder) (e.g. C:\dllsa\anova.dll).

II] Analysis

I) Input the data in the appropriate cells on the sheet.
II) Select [ANOVA (One-way)] or [ANOVA (Two-way)] or [ANOVA (m-way) (m <= 5)] from [dllsa] on the Excel menu bar according to data format (Figure 5).
III) If user selected [ANOVA (Two-way)], the input dialog box appears.
IV) Click the right button on the input box for input range. Then, the window that assigns the data range of the sheet is activated. Select the data range that you have already inputted by dragging the range (Figure 6).

V) When you click the right button in the input dialog box, the screen backs up one step. Input number of replications in text box (Figure 7). Finally, Click the [Run] button.

When the analysis is completed, the results (table of analysis of variances) are displayed in a new sheet (Figure 8). Thus, users can make calculations that do not exist in Excel by using the DLL. In addition, users can input the data and output the results using Excel. Users can also get future results using the new statistical analysis when they call another DLL on the same Excel macro.

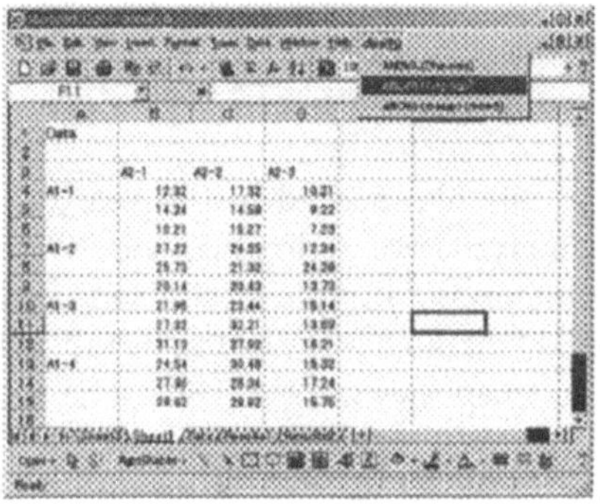

Figure 5. Menu (anova.xls)

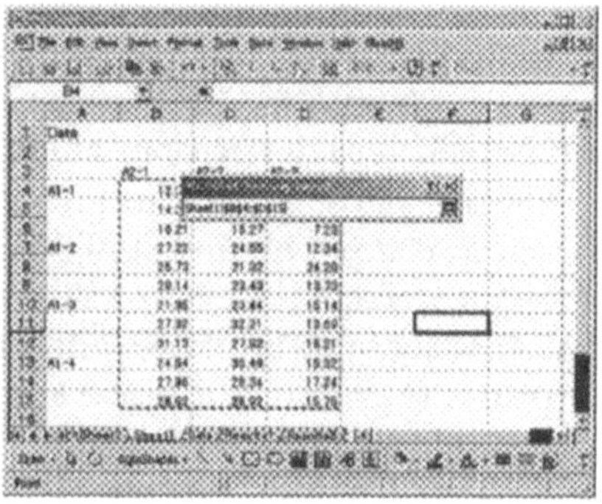

Figure 6. Select the data range

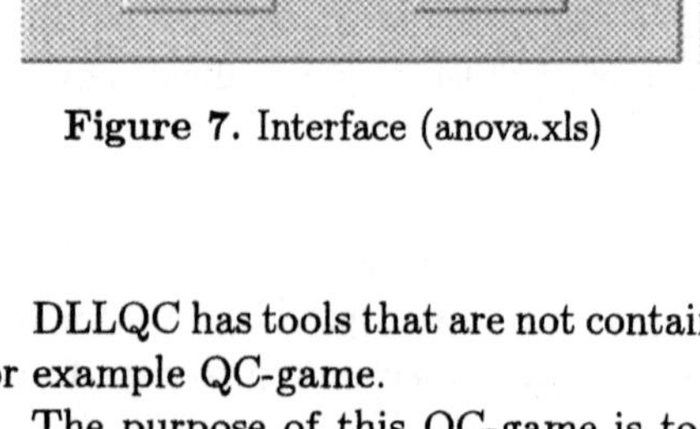

Figure 7. Interface (anova.xls)

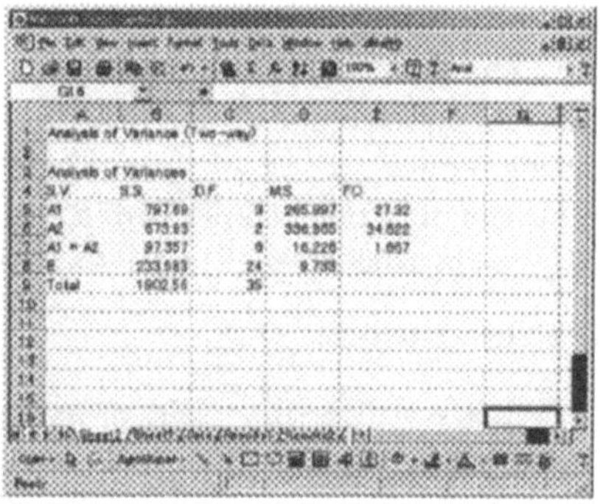

Figure 8. The output by anova.dll

DLLQC has tools that are not contained the popular commercial software, for example QC-game.

The purpose of this QC-game is to compete with another player(s) for the maximal yield by finding optimal levels of manufacturing factors under a restricted number of trials given in advance, where the yields are shown by a value between zero and one hundred, i.e. the worst and the best yields, and each yield has a random error distributed as N(0, $k\sigma^2$), where $\sigma^2 = 1$.

In case of default, k is one. Factors may or may not have a main effect and interactions of order two with other factors.

The referee determines the numbers of factors, the main effects, the interactions, and these effects are given by contours of yields. The magnitude of random normal deviation is also changeable. By default, the standard deviation is one. The maximal number of trials for each player has to be given in advance.

According to the referee's planning, each player is given only these levels, so that the player can obtain the maximal yield. Depending on the arrangement of factor levels, the yield is automatically shown on monitor, disk or print. Usually the number of players will be less than ten.

3 EBQC

Many valuable books formally used as scientific books had been published for SQC, especially theoretical and technical texts and manuals, in the various fields of QC in industries. However, such texts and manuals are already out of print and no stocks at publishers at the present time. Therefore, the beginners of engineers in industries and students are difficult to find such books of SQC as the references. Actually these books include original and historical ideas and conceptions to solve problems in these days. It will be useful for them to reconsider new development of methodologies at the present situations. To tackle this these problems, a site named EBQC is created and digitized on the web, by receiving official approval of such books from these copyright holders, i.e. the authors and publishers.

In this section, the system EBQC is introduced with main features.

3.1 Features of the EBQC

Main representative and important books are provided on the Web site. Every page of the original books is converted to PDF,and published on the web for browsing and printing. Figure 9 shows the web site of EBQC (http://www.sci.kagoshima-u.ac.jp/~ebqc/). Every person is able to read such books by accessing on Web site. These electronic books give public to open under the permission of the respective copyright holders and publishers. Since these electronic books are scanned and transformed into PDF format files, everybody can download and print out all selected contents in the books. Table 3 shows the list of involved books of EBQC, and Figure 10 shows an example of the top page of an electronic book. Every top page consists of four parts. The first and the second parts are links to the respective contents and index pages. The third part is used for keywords search, and the fourth part is print image of actual book.

Figure 11 shows a content of an illustrative electronic book, and everyone can read relevant part of the book by checking the name of the section or subsection.

Table 3. List of contents of EBQC

Title and Authors
Introduction to Statistical Quality Control
General Remarks of Quality Control (by Susumu Sonobe, 1955)
Promotion for Quality Control (by QCC Committee, 1955)
What a manufacturing standard should be and how to make it (by Seikichi Miyagi, 1955)
Total Quality Control
Explanation of Quality Control Terminology (by Hidehiko Higashi, 1955)
Experimental Designs
Experimental Designs (by Motosaburo Masuyama, 1956)
Lecture I of Experimental Designs, Basic Course(1) (by Tosio Kitagawa, 1955)
Lecture II of Experimental Designs, Basic Course(2) (by Tosio Kitagawa, 1956)
Experimental Designs (by Chooichiro Asano, 1958)
Sampling Inspection Plans
JIS-Z Series (JSA)
Sampling Inspection Plans (by Hiroshi Matsumoto, 1955)
Theories and Their Applications of Sampling Inspection Plans (by Chooichiro Asano, 1981)
Methods of Sampling Survey
Sampling Survey in Manufacturing (by Kaoru Ishikawa, 1952)
Techniques of Survey (by Akira Asai, 1992)
An Introduction to Statistical Survey, - mainly Randomized Block Design (by Yoshiakira Funazu, 1977)
An Introduction to Statistical Survey,(conti.),- mainly One Way Layout (by Yoshiakira Funazu, 1986)
Experimental Designs for Observations (by Chooichiro Asano, 1992)
Graphic Representations
Methods of Control Charts (by Shi Miura, 1955)
QC Games
QC game (by Yuukou Fujishiro, 1957)
Simulations
Monte Carlo Method (by Osamu Miyatake and Takashi Nakagawa, 1960)
Analytical Techniques
Statistical Analysis (by Gen'ichi Taguchi, 1972)
Statistical Methods (by Tetsuichi Asaka and Yoshitsugu Ohmae, 1955)
An Introduction to Correlation and Regression Analysis (by Ryuichi Kobayashi, 1980)
An Introduction to Quantification Methods (by Ryuichi Kobayashi, 1981)
Applied Statistics (by Ryuichi Kobayashi, 1984)
Multivariate Analysis
Theories of Multivariate Analysis (by Minoru Siotani and Chooichiro Asano, 1967)
Multivariate Analysis of Qualitative Informations (by Noriyuki Matsuda, 1988)
Theories and Applications of Factor Analysis (by Chooichiro Asano, 1972)
Menus of Lecture Course
Seminar Course on Quality Control and Standardization
TQM Seminar Course for Section Directors
Seminar course of Industrial Standardization/Quality Control for responsible promoters
Miscellaneous Books
Brief History of Agricultural Experimental Designs (by Chuochi Okuno, 1994)
Food and Agriculture in 21st Century (ed. by Chuochi Okuno, 1975)
Settlement of medical small samples (by Motosaburo Masuyama, 1953)
Stochastics and its Contribution to the present Society (Jour.Science, Spec.issue, 1951)
Mathematical Methods for Demand Estimation (by Ryuichi Kobayashi, 1967)

3.2 Search system of the EBQC

All index words are collected and integrated in all books and a glossary database is created. Each word is referred in all the applicable pages. Therefore if a user enters a key word of SQC, the engine searches the word across all the registered books. It often excludes heavy image files and enables users to see and print desired pages only. The system also has text-based index pages so that it can be often retrieved by the prevailing search engines. The opportunities for the original books to be read by people will thus be increased.

Figure 12 represents the search page. This search system can 'and/or' search by multiple keywords. The regular expression (such as * or ^) can also be used in this system and the search can be applied to a selected book or all books in our system.

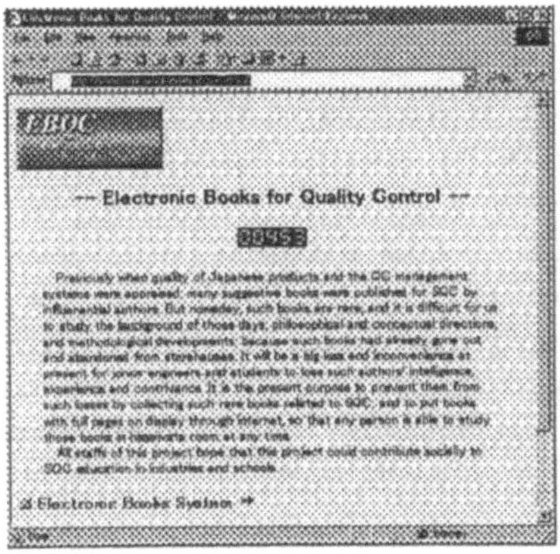

Figure 9. Top page of EBQC

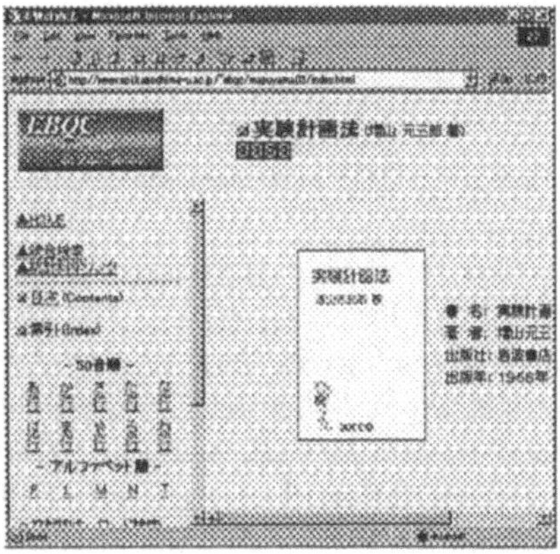

Figure 10. Top page of an electronic book

Figure 11. Contents of an electronic book

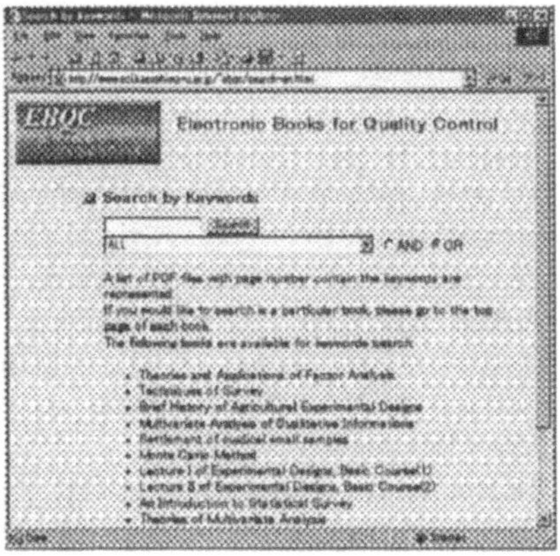

Figure 12. Keywords search system

4 Concluding remarks

The DLLQC can satisfy diverse demands of users with one DLL by changing its interface. Changing the interface, however, requires a certain level of computer proficiency on the user's part and those who are inexperienced with programming would find it difficult to understand or perform this task. To help such users, an Excel macro is published with the DLL as an example of the interface.

We are planning to propose a new interface which can call the library by using the Web browser. By using the browser, we can use popular and free tools such as HTML, Form, and CGI (Common Gateway Interface). By using HTML especially, we can provide the information appropriate for the user's level and purpose together with analysis tools.

The EBQC has a powerful search engine and a user friendly interface. When you want to obtain the knowledge of particular methods or particular words of SQC, you may get them easily by using the EBQC as an online dictionary.

Our research group is also undertaking similar projects with respect to statistics (Takeuchi, et al., 2001; Suenaga, et al., 2000). Since we use the same technology in these projects to make the DLLQC and EBQC systems, we can expand them easily and quickly. We also plan to develop an educational system for SQC using the DLLQC and EBQC. By using these systems, engineers and students can learn SQC more easily wherever they are, at the lecture room, PC practicing room, home or any other place with an internet environment.

The authors would like to thank the editor(s) and the referee(s) for providing helpful comments which led to the improvement of this paper.

References

1. Asano, Ch. and Kimura, H. (1990) A study on system construction of general purpose statistical package for microcomputers — Micro-NISAN system, Computational Statistics Quarterly, **6**, 65–77.
2. Suenaga, K., Yadohisa, H., Sakurai, N., Yamaguchi, K., Watanabe, M., and Asano, Ch. (2000) WWW-supported environments for learning and teaching statistics, Proceedings of The 8th International Conference on Computers in Education/International Conference on Computer Assisted Instruction 2000, 1, 724–730.
3. Takeuchi, A., Yadohisa, H., Yamaguchi, K., Watanabe, M., and Asano, Ch. (2001) DLLSA: Dynamic link library for statistical analysis, Bulletin of The Computational Statistics Society of Japan, **13(2)**, 115-125, (in Japanese).

Part 2

On-line Control

Single Sampling Plans for Inspection by Variables under a Variance Component Situation

Peter-Theodor Wilrich

Freie Universität Berlin, Institut für Statistik und Ökonometrie, Garystrasse 21, 14195 Berlin, Germany

Abstract. International Standard ISO 3951 (1989), being based on the Military Standard MIL-STD-414 (1957), specifies an acceptance sampling system of single sampling plans for inspection by variables, under the assumptions that X, the quality characteristic of items in a lot, is (approximately) normally distributed with unknown mean and known or unknown variance and that the measurement error in measuring X is negligible . The acceptance sampling procedure determines the acceptability of a lot from an estimate of the fraction of nonconformimg items in the lot, i.e. the fraction of items with $x \leq UL$ where UL is an upper specification limit (or $x \geq LL$ where LL is a lower specification limit). In this paper the acceptance procedure is extended to the case where the quality characteristic X is a performance characteristic of the items, and these items can be used repeatedly so that X takes a value for each repetition of use. In this situation the variance of X consists of two components, one describing the variation of X under repeated use of an item, the other one the variation between the mean values of the performance characteristic of different items. The paper considers the cases of known and of unknown components of variance, and gives an example.

1 Introduction

International Standard ISO 3951 (1989), being based on the Military Standard MIL-STD-414 (1957), specifies an acceptance sampling system of single sampling plans for inspection by variables, under the following conditions: the inspection procedure is to be applied to a continuous series of lots of discrete items, all coming from one and the same production process,

- only a single quality characteristic X of the items is taken into consideration, which must be measurable on a continuous scale,
- the measurement error in measuring X is negligible, i.e., its standard deviation is no more than 10% of the process (or within-lot) standard deviation,
- the production process is stable and the quality characteristic X is (approximately) distributed according to a normal distribution,

- an upper specification limit UL (or a lower specification limit LL) is specified: an item is qualified as conforming if and only if its value x of the quality characteristic satisfies $x \leq UL$ (or $x \geq LL$),
- an acceptance quality level (AQL) is specified: This is the worst tolerable fraction of nonconforming items in the process and, hence, in the lots, i.e., the acceptance inspection shall accept lots in which the fraction of nonconforming items, p, is less or equal to AQL and reject for $p >$ AQL.

The acceptance sampling procedure determines the acceptability of a lot from an estimate of the fraction of nonconforming items in the process, based on the inspection of a random sample of items from the lot.

In this paper we extend the acceptance procedure to the case where the quality characteristic X is a performance characteristic of the items, and these items can be used repeatedly so that X takes a value for each repetition of use. In this situation the variance of X consists of two components, one describing the variation of X under repeated use of an item, the other one the variation between the mean values of the performance characteristic of different items.

Section 2 introduces the underlying model, sections 3 and 4 are dealing with the cases of known and of unknown components of variance and section 5 gives an example.

2 The underlying model

X is a performance characteristic of a discrete item in a lot which can be used infinitely often; it takes a value x whenever the item is used. X can be modelled as

$$X = \mu + X_{between} + X_{within} \tag{1}$$

where μ is the lot mean, $X_{within} \sim N(0, \sigma_w^2)$ is the random variable describing the variation between the values of the performance characteristic when an item is used repeatedly (and it is assumed that the standard deviation of X_{within}, σ_w, is equal for all items in the lot) and $X_{between} \sim N(0, \sigma_b^2)$ is the random variable, independent of X_{within}, describing the variation between the mean values of the performance characteristic of different items of the lot.

Since $X_{between}$ and X_{within} are assumed to be normally distributed with zero means and variances σ_b^2 and σ_w^2, resp., the random variable X follows the normal distribution with mean μ and variance

$$\sigma^2 = \sigma_b^2 + \sigma_w^2 , \tag{2}$$

i.e. it consists of two components σ_b^2 and σ_w^2. The fraction of nonconforming values of the performance characteristic, i.e. values *x* being larger than *UL*, in a lot is

$$p = P(X > UL) \;=\; 1 - \Phi\left(\frac{UL - \mu}{\sigma}\right), \tag{3}$$

where $\Phi(\cdot)$ is the cumulative distribution function of the standardised normal distribution. A single sampling plan for inspection by variables is a statistical test of the null hypothesis

$$H_0 : p \le AQL := p_1 \tag{4}$$

against the alternative hypothesis

$$H_1 : p < p_1 . \tag{5}$$

3 The case of known components of variance

In the case of unknown μ and known components of variance, σ_b^2 and σ_w^2, an estimator of the fraction of nonconforming items, *p*, based on $n = n_{\sigma 1} n_{\sigma 2}$ measurements obtained at $n_{\sigma 1}$ items, each of them measured $n_{\sigma 2}$ times repeatedly, is

$$\hat{p} = 1 - \Phi\left(\frac{UL - \bar{\bar{X}}}{\sigma}\right), \tag{6}$$

where

$$\bar{\bar{X}} = \frac{1}{n_{\sigma 1} n_{\sigma 2}} \sum_{i=1}^{n_{\sigma 1}} \sum_{j=1}^{n_{\sigma 2}} X_{ij} \tag{7}$$

is the mean of the $n_{\sigma 1} n_{\sigma 2}$ measurements X_{ij}; $i = 1, \ldots, n_{\sigma 1}$; $j = 1, \ldots, n_{\sigma 2}$. If p_c is the critical value of the test, the lot will be accepted for

$$\hat{p} \le p_c \tag{8}$$

or, equivalently,

$$\frac{UL - \bar{\bar{X}}}{\sigma} \ge u_{1-p_c} := k_\sigma , \tag{9}$$

$$Z_\sigma := \bar{\bar{X}} + k_\sigma \sigma \le UL , \tag{10}$$

where u_{1-p_c} is the (1-p_c)-quantile of the standardised normal distribution. Since the test statistic Z_σ is normally distributed with

$$E(Z_\sigma) = \mu + k_\sigma \sigma , \tag{11}$$

$$V(Z_\sigma) = V(\bar{\bar{X}}) = \frac{\sigma_b^2}{n_{\sigma 1}} + \frac{\sigma_w^2}{n_{\sigma 1} n_{\sigma 2}} , \tag{12}$$

the operating characteristic function of the single sampling plan with given parameters $(n_{\sigma 1}, n_{\sigma 2}, k_\sigma)$ is

$$L(p) = P(Z_\sigma \leq UL) = \Phi\left(\sqrt{\frac{n_{\sigma 1}\, n_{\sigma 2}}{\gamma(n_{\sigma 2} - 1) + 1}} \left(u_{1-p} - k_\sigma\right)\right) \tag{13}$$

where

$$\gamma = \frac{\sigma_b^2}{\sigma_w^2 + \sigma_b^2} \tag{14}$$

is the ratio of the within to the between component of variance. In the limiting case $\gamma = 0$, i.e. if the between component of variance, σ_b^2, does not exist, the formula reduces to

$$L(p) = \Phi\left(\sqrt{n_{\sigma 1}\, n_{\sigma 2}}\ (u_{1-p} - k_\sigma)\right) ; \tag{15}$$

the OC depends only on the total number of measurements, $n_{\sigma 1} n_{\sigma 2}$, because the expected values of the performance characteristic *X* of different items are equal so that measuring another item is equivalent to measuring the same item again. In the limiting case $\gamma = 1$, i.e. if the within component of variance, σ_w^2, does not exist, the formula reduces to

$$L(p) = \Phi\left(\sqrt{n_{\sigma 1}}\ (u_{1-p} - k_\sigma)\right) ; \tag{16}$$

$n_{2\sigma}$ drops out of the expression because repeated measurements at one and the same item would be identical.

A single sampling plan for which

$$L(p_1) \geq 1 - \alpha, \qquad L(p_2) \leq \beta , \tag{17}$$

where $P_1(p_1; 1-\alpha)$ and $P_2(p_2; \beta)$ with $0 < p_1 < p_2 < 1$ and $0 < \beta < 1 - \alpha < 1$ are two given points on the OC curve (see Figure 1) is given by

$$k_\sigma = \frac{u_{1-\alpha}\ u_{1-p_2} + u_{1-\beta} u_{1-p_1}}{u_{1-\alpha} + u_{1-\beta}} \tag{18}$$

and

$$\frac{n_{\sigma 1}\ n_{\sigma 2}}{\gamma(n_{\sigma 2}-1)+1}=\left(\frac{u_{1-\alpha}+u_{1-\beta}}{u_{1-p_1}-u_{1-p_2}}\right)^2 ; \tag{19}$$

for $n_{\sigma 2}=1$ the value of $n_{\sigma 1}$ does not depend on γ.

The values of the pairs $(n_{\sigma 1}, n_{\sigma 2})$ which fulfil equation (19) have to be rounded to the next largest integers. The solution with smallest value $n_{\sigma 2}=1$ needs the largest value for $n_{\sigma 1}$,

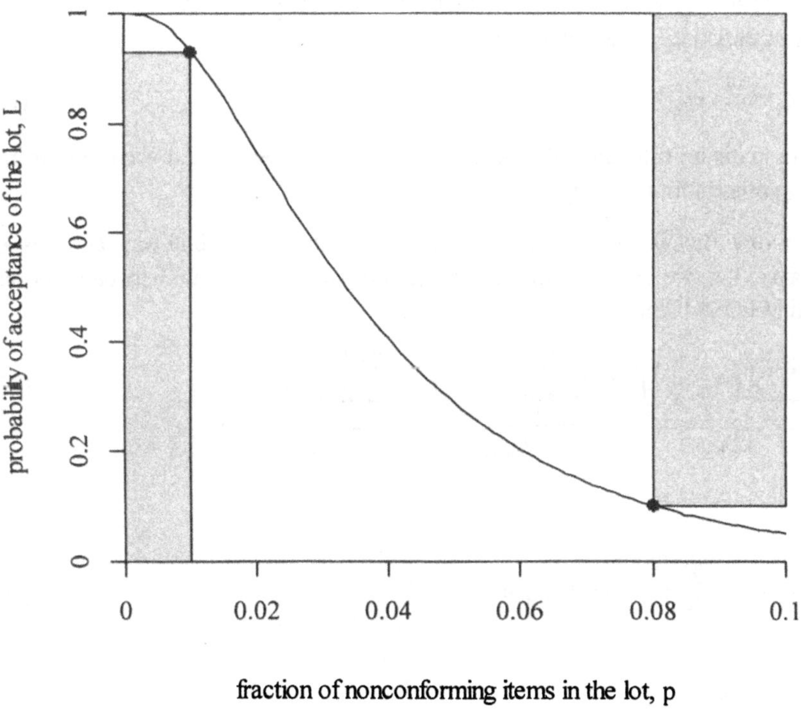

Fig. 1. The OC curve of the single sampling plan *($n_{\sigma 1}=9$, $n_{\sigma 2}=1$, $k=1.83$)* which fufils the requirements L 1%() $\geq 1-\alpha=93\%$ and L (8%) $\leq \beta=10\%$. The forbidden areas for the OC curve are shaded.

$$n_{\sigma 1,\max}=\left(\frac{u_{1-\alpha}+u_{1-\beta}}{u_{1-p_1}+u_{1-p_2}}\right)^2 ; \tag{20}$$

on the other hand, for $n_{\sigma 2} \to \infty$, i.e. when each sampled item is measured infinitely often, $n_{\sigma 1}$ takes its smallest value,

$$n_{\sigma 1,\min} = \gamma \left(\frac{u_{1-\alpha} + u_{1-\beta}}{u_{1-p_1} + u_{1-p_2}} \right)^2 . \tag{21}$$

See the example in section 5.

4 The case of unknown components of variance

In the case of unknown μ and unknown components of variance, σ_b^2 and σ_w^2, the test statistic Z_σ according to (9) is modified to

$$Z_s := \bar{\bar{X}} + k_s \hat{\sigma} \tag{22}$$

where $\hat{\sigma}$ is an estimator of σ and k_s a factor to be determined according to the design criteria for the sampling plan.

In order to get an estimator $\hat{\sigma}$ for σ we perform a one-way analysis of variance, i.e., we break down the total sum of squares into one between and one within sampled items:

$$\underbrace{\sum_{i=1}^{n_{s1}} \sum_{j=1}^{n_{s2}} \left(X_{ij} - \bar{\bar{X}}\right)^2}_{Q_{total}} = \underbrace{n_{s2} \sum_{i=1}^{n_{s1}} \left(\bar{X}_i - \bar{\bar{X}}\right)^2}_{Q_{between}} + \underbrace{\sum_{i=1}^{n_{s1}} \sum_{j=1}^{n_{s2}} \left(X_{ij} - X_i\right)^2}_{Q_{within}} , \tag{23}$$

where

$$\bar{X}_i = \frac{1}{n_{s2}} \sum_{j=1}^{n_{s2}} X_{ij} \quad ; \; i = 1, \ldots, n_{s1} , \tag{24}$$

$$\bar{\bar{X}} = \frac{1}{n_{s1} n_{s2}} \sum_{i=1}^{n_{s1}} \sum_{j=1}^{n_{s2}} X_{ij} \; ; \tag{25}$$

$\bar{X}_i$ is the mean of the n_{s2} performance measurements X_{ij} ; $j = 1,\ldots, n_{s2}$ of the sampled item is and $\bar{\bar{X}}$ is the total mean of the measurements X_{ij} ; $i = 1,\ldots, n_{s1}$, $j = 1,\ldots, n_{s2}$. The variances

$$S_I^2 = \frac{Q_{within}}{n_{s1}(n_{s2}-1)}, \qquad S_{II}^2 = \frac{Q_{beween}}{n_{s1}-1} \tag{26}$$

have expectations

$$E(S_I^2) = \sigma_w^2, \qquad E(S_{II}^2) = \sigma_w^2 + n_{s2}\sigma_b^2, \tag{27}$$

see Graf et. al. (1987). Hence, an unbiased estimator of $\sigma^2 = \sigma_w^2 + \sigma_b^2$ is given by the weighted sum

$$\hat{\sigma}^2 = \frac{1}{n_{s2}} S_{II}^2 + \frac{n_{s2}-1}{n_{s2}} S_I^2 = \frac{Q_{between}}{(n_{s1}-1)n_{s2}} + \frac{Q_{within}}{n_{s1}n_{s2}} \tag{28}$$

of $Q_{between}$ and Q_{within}, where $Q_{between}$ and Q_{within}, divided by their expectations, are independently chisquared distributed. According to Satterthwaite (1946) the ratio $\hat{\sigma}^2 / E(\hat{\sigma}^2) = \hat{\sigma}^2 / \sigma^2$ follows approximately a χ_f^2 / f-distribution with degrees of freedom determined by equating the variance $V(\chi_f^2 / f) = 2/f$ and the variance of $\hat{\sigma}^2 / \sigma^2$,

$$\begin{aligned} V\left(\frac{\hat{\sigma}^2}{\sigma^2}\right) &= \frac{1}{\sigma^4} V\left(\hat{\sigma}^2\right) \\ &= \frac{1}{\sigma^4}\left(\frac{2}{n_{s2}^2(n_{s1}-1)}\left(\sigma_w^2 + n_{s2}\sigma_b^2\right)^2 + \frac{2(n_{s2}-1)^2}{n_{s2}^2 n_{s1}(n_{s2}-1)}\sigma_w^4\right). \end{aligned} \tag{29}$$

From $2/f = V\left(\hat{\sigma}^2/\sigma^2\right)$ follows

$$f = \frac{n_{s2}^2\sigma^4}{\frac{\left(\sigma_w^2 + n_{s2}\sigma_b^2\right)^2}{n_{s1}-1} + \frac{(n_{s2}-1)\sigma_w^4}{n_{s1}}} = \frac{n_{s2}^2}{\frac{\left(1-\gamma+n_{s2}\gamma\right)^2}{n_{s1}-1} + \frac{(n_{s2}-1)(1-\gamma)^2}{n_{s1}}}. \tag{30}$$

For given parameters $(n_{s1}^*, n_{s2}^*, k_s)$ of the single sampling plan the OC function is

$$\begin{aligned} L(p) &= P(Z_s \le UL) \\ &= P\left(\bar{\bar{X}} + k_s\hat{\sigma} \le UL\right) = P\left(\frac{UL - \bar{\bar{X}}}{\hat{\sigma}} \ge k_s\right) \end{aligned}$$

$$= P\left(\frac{\frac{UL-\mu}{\sigma_{\bar{\bar{X}}}} - \frac{\bar{\bar{X}}-\mu}{\sigma_{\bar{\bar{X}}}}}{\hat{\sigma}/\sigma} \geq \frac{\sigma}{\sigma_{\bar{\bar{X}}}} k_s \right) \tag{31}$$

$$\approx P\left(\frac{-U+\delta}{\sqrt{\chi_f^2/f}} \geq \frac{\sigma}{\sigma_{\bar{\bar{X}}}} k_s \right) = P\left(\frac{U+\delta}{\sqrt{\chi_f^2/f}} \geq \frac{\sigma}{\sigma_{\bar{\bar{X}}}} k_s \right)$$

where U denotes the standardised normal random variable and

$$\delta = \frac{UL-\mu}{\sigma_{\bar{\bar{X}}}} = \frac{UL-\mu}{\sigma} \cdot \frac{\sigma}{\sigma_{\bar{\bar{X}}}} = u_{1-p} \cdot \frac{\sigma}{\sigma_{\bar{\bar{X}}}}. \tag{32}$$

$$T_{f,\delta} = \frac{U+\delta}{\sqrt{\chi_f^2/f}} \tag{33}$$

is noncentral t-distributed with f degrees of freedom, where f is given by (30), and noncentrality parameter δ, and since

$$\left(\frac{\sigma_{\bar{\bar{X}}}}{\sigma} \right)^2 = \frac{\frac{\sigma_b^2}{n_{s1}} + \frac{\sigma_w^2}{n_{s1}n_{s2}}}{\sigma_b^2 + \sigma_w^2} = \frac{n_{s2}\gamma + 1 - \gamma}{n_{s1}n_{s2}} \tag{34}$$

it follows

$$\begin{aligned} L(p) &\approx P\left(T_{f,\delta} \geq \sqrt{\frac{n_{s1}n_{s2}}{n_{s2}\gamma + 1 - \gamma}} \cdot u_{1-p} \right) \\ &= 1 - F_{T_{f,\delta}}\left(\sqrt{\frac{n_{s1}n_{s2}}{n_{s2}\gamma + 1 - \gamma}} \cdot u_{1-p} \right); \end{aligned} \tag{35}$$

$F_{T_{f,\delta}}(\cdot)$ denotes the cumulative distribution function of the random variable $T_{f,\delta}$.

Two sampling plans are equivalent with regard to their operating characteristic function if they have identical factor k_s and if, for the ratio γ given, their pairs (n_{s1}, n_{s2}) are chosen so that

$$\frac{n_{s1}}{\gamma + \dfrac{1-\gamma}{n_{s2}}} = const. \tag{36}$$

For all γ with $0 \le \gamma \le 1$ the value of n_{s1} is largest for the pair with $n_{s2} = 1$. The largest value is denoted as $n_{\max,1}$. Figure 2 shows

$$\frac{n_{s1}}{n_{\max,1}} = \gamma + \frac{1-\gamma}{n_{s2}} \tag{37}$$

as a function of n_{s2} for $\gamma = 0$ ($\sigma_b^2 = 0$), $\gamma = 0.5$ ($\sigma_b^2 = \sigma_w^2$) and $\gamma = 1$ ($\sigma_w^2 = 0$).

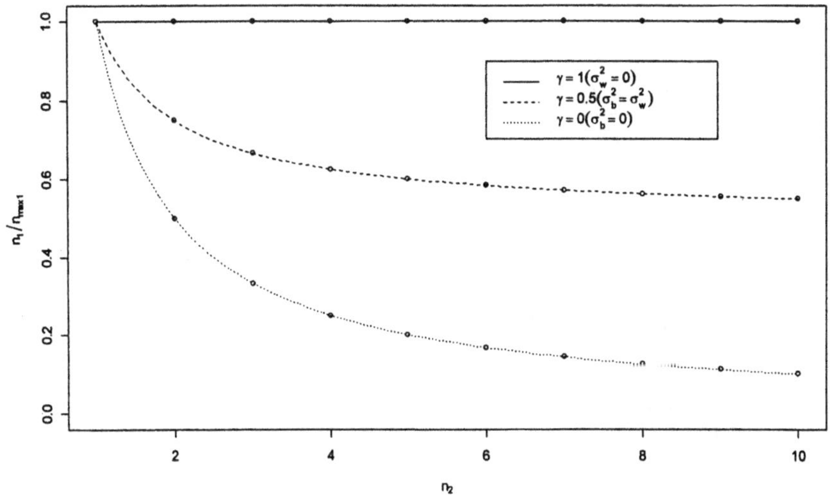

Fig. 2. The ratio $n_1/n_{\max,1}$ as a function of n_2 for $\gamma = 0$, 0.5 and 1. Only for $\gamma = 0$ resp. $\sigma_b^2 = 0$ the ratio is equal to $1/n_2$ so that the OC functions of all plans with $n_1 n_2 =$ const. are equivalent.

If the variance component σ_w^2 does not exist ($\sigma_w^2 = 0$ resp. $\gamma = 1$), measurement results obtained repeatedly at a particular item are identical. Hence, the operating characteristic curves are identical for any number n_{s2} of repetitions and an alteration of n_{s2} does not result in an alteration of $n_{s1} \Rightarrow n_{s1}/\, n_{\max,1} = const.$

On the other hand, if the variance component σ_b^2 does not exist ($\sigma_b^2 = 0$ resp. $\gamma = 0$), repeated measurements at one and the same item are equivalent to measurements at different items. Hence, operating characteristic curves are identical for all plans with identical measurement effort $n_{s1} n_{s2} = n_{\max,1}$ $\Rightarrow n_{s1}/\, n_{\max,1} = 1/n_{s2}$.

In practice, γ is unkown, and therefore, neither the operating characteristic curve nor the equivalent pairs (n_{s1}, n_{s2}) can be calculated. However, if $n_{s2} = 1$ (and hence, $n_{s1} = n_{\max,1}$) the operating characteristic curve does not depend on γ.

A numerical calculation of the OC curve is based on the fact that the statistic Z_s according to (22) is approximately normally distributed with

$$E(Z_s) \approx \mu + k_s\sigma, \tag{38}$$

$$V(Z_s) \approx \sigma^2_{\overline{X}} + k_s^2 V(\hat{\sigma}).$$

With

$$E(\hat{\sigma}^2) = \sigma^2 \; ; \; E(\hat{\sigma}) \approx \sigma \tag{39}$$

and

$$V(\hat{\sigma}^2) = \frac{2}{n_{s2}^2}\left(\frac{\left(\sigma_w^2 + n_{s2}\sigma_b^2\right)^2}{n_{s2}\left(n_{s1}-1\right)} + \frac{\left(n_{s2}-1\right)\sigma_w^4}{n_{s1}}\right) \tag{40}$$

the variance $V(\hat{\sigma})$ is

$$V(\hat{\sigma}) \approx \frac{1}{4}\cdot\frac{E^2(\hat{\sigma})}{E^2\left(\hat{\sigma}^2\right)}\cdot V\left(\hat{\sigma}^2\right)$$

$$\approx \frac{1}{4\sigma^2} V\left(\hat{\sigma}^2\right)$$

$$= \frac{\sigma^2}{2n_{s2}^2}\left(\frac{(1-\gamma+n_{s2}\gamma)^2}{n_{s1}-1} + \frac{(n_{s2}-1)(1-\gamma)^2}{n_{s1}}\right) \tag{41}$$

and therefore

$$V(Z_s) \approx \sigma^2\left[\frac{\gamma}{n_{s1}} + \frac{1-\gamma}{n_{s1}n_{s2}} + \frac{k_s^2}{2n_{s2}^2}\left(\frac{(1-\gamma+n_{s2}\gamma)^2}{n_{s1}-1} + \frac{(n_{s2}-1)(1-\gamma)^2}{n_{s1}}\right)\right] := A^2\sigma^2 \tag{42}$$

The operating characteristic function, approximated by the normal distribution of the test statistic, is

$$L(p) = P(Z_s \le UL) \approx \Phi\left(\frac{UL - E(Z_s)}{\sqrt{V(Z_s)}}\right) = \Phi\left(\frac{1}{A}\left(u_{1-p} - k_s\right)\right) \tag{43}$$

with A according to (42) and $\Phi(\cdot)$ denoting the cumulative distribution function of the standardised normal distribution.

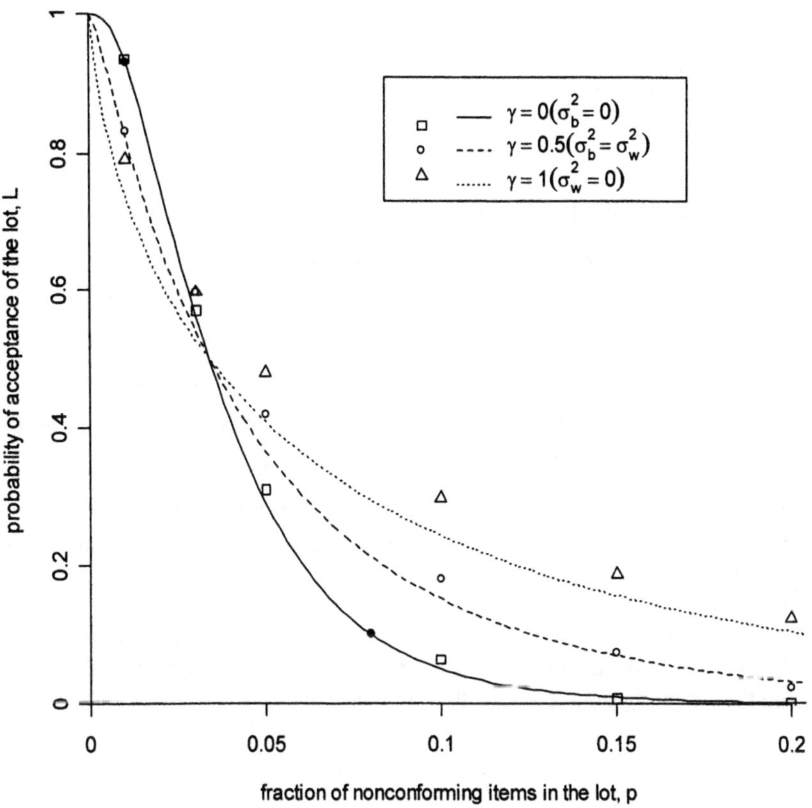

Fig. 3. The OC of the single sampling plan ($n_{s1} = 5$, $n_{s2} = 5$, $k = 1.83$) for $\gamma = 0$, 0.5 and 1. For $\gamma = 0$ the plan fulfils the reqirement $L(1\%) \geq 1 - \alpha = 93\%$ and $L(8\%) \leq \beta = 10\%$, however, it fails if it is used in case of $\gamma \neq 0$. Each symbol represents the acceptance frequencly on the basis of 1000 simulation runs.

For the case $\sigma_w^2 = 0$ resp. $\gamma = 1$, A^2 reduces to

$$A_{\gamma=1}^2 = \frac{1}{n_{s1}} + \frac{k_s^2}{2(n_{s1} - 1)} \approx \frac{1}{n_{s1}}\left(1 + \frac{k_s^2}{2}\right). \tag{44}$$

Sampling plans with largest A and hence, smallest sampling effort, for which

$$L(p_1) \geq 1 - \alpha \ , \ L(p_2) \leq \beta \tag{45}$$

where $P_1(p_1;1-\alpha)$ and $P_2(p_2;\beta)$ with $0<p_1<p_2<1$ and $0<\beta<1-\alpha<1$ are two points on the OC curve (see Figure 1), are given by

$$\frac{1}{A}=\left(\frac{u_{1-\alpha}+u_{1-\beta}}{u_{1-p_1}-u_{1-p_2}}\right)^2, \tag{46}$$

$$k_s=\frac{u_{1-\beta}u_{1-p_1}+u_{1-\alpha}u_{1-p_2}}{u_{1-\alpha}+u_{1-\beta}}; \tag{47}$$

u_p is the p-quantile of the standardised normal distribution.

Unfortunately, the OC function and, hence, (n_{s1}, n_{s2}) depend on γ. However, since k_s is determined only by $P_1(p_1;1-\alpha)$ and $P_2(p_2;\beta)$ for all values of γ, the OC functions of all sampling plans with identical k_s have the common point with p given by $u_{1-p}=k_s$ and $L(p)=\Phi(0)=0.5$. For $\gamma\to 0$ they become steeper, for $\gamma\to 1$ they become more flat. If the sampling plan is designed for $\gamma=1$ and the actual unknown γ is smaller than one, the actual OC fulfils the requirement $L(p_1)\geq 1-\alpha$ and $L(p_2)\leq\beta$, however, it is steeper than necessary, i.e. the actual risks α of the producer and β of the consumer would both be smaller than the required values; see Figure 3.

In order to be on the safe side one could start with a sampling plan with $n_{s2}=1$, repeat some of the measurements in order to estimate γ and use this estimate to redesign the sampling plan to one with a smaller value n_{s1}.

5 An example

A pen-injector is a medical device intended for parenteral administration by injection of medicinal products from a multi-dose cartrigde.

Figure 4 shows the dose of 4 repeated applications of each of 15 pen-injectors. The dose varies with repeated application of one and the same pen-injector (variance component σ_w^2) and between different pen-injectors (variance component σ_b^2); $\sigma_b\approx 0.8\sigma_w$, i.e. $\gamma\approx 0.5$.

According to the International Standard ISO 11608-1.2 (1998), section 4.2., the absolute error of the pre-set dose shall not be larger than 0.01 ml. In order to verify this dose accuracy requirement $n_{s2}=4$ repeated tests with each of $n_{s1}=5$ selected pen-injectors (at three different dose levels) shall be performed, mean $\bar{x}$ and standard deviation s of the $n_{s1}n_{s2}=20$ measurements and the two-sided statistical tolerance interval $\bar{x} \pm k \cdot s$ computed and the lot accepted if this tolerance interval lies completely inside the dosage tolerance interval (dosage setting ±0.01 ml). The rationale behind this procedure is the requirement „To pass the dose accuracy requirement, there shall be a 95% confidence that at least p of all doses delivered will fall within the proposed upper and lower specification limits for the 3 dose settings", where p is fixed at 95% (section 5.4 of the Standard). Expressed in terms of sampling by variables the probability of acceptance of a lot of pen-injectors shall be L = 5% if 5% „of all doses" fall outside the dosage tolerance interval. The test statistic is computed with the total standard deviation s instead of the unbiased estimator $\hat{\sigma}$ according to (28) and the factor k is chosen as k = 2.760 (for $n = n_{s1}n_{s2} = 20$, 1 - α = 95% 5 and p = 9%), erroneously neglecting the variance component situation.

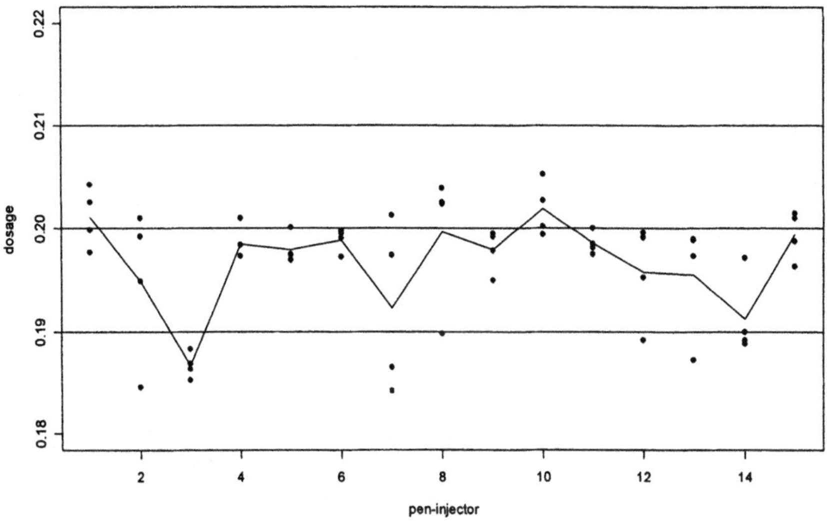

Fig. 4. Each of 15 pen-injectors selected from a lot have been tested with 4 replications each. The pre-set dose is 0.2 ml, lower and upper specification limit 0.19 ml and 0.21 ml, resp. The polygonial line connects the mean values.

In order to demonstrate the effect of this misapplication we look at the proposed procedure as if being one sided, i.e. for lot acceptance if $\bar{x} + k \cdot s$ is not larger than the upper specification limit (dosage setting + 0.01 ml) resp. $\bar{x} - k \cdot s$ is not smaller than the lower specification limit (dosage setting – 0.01 ml). This changes the $0C$ curve , however, the comparison between the various cases for n_{s1} and n_{s2} is not affected. The effect is highly dependent on γ. For $\gamma = 0$, i.e. $\sigma_b^2 = 0$, sampling plans with all pairs (n_{s1}, n_{s2}) where $n_{s1} \cdot n_{s2} = n = 20$, have almost the same OC function (they differ only slightly because of the different estimator for σ^2 according to (28)). Figure 5 shows the OC functions for

$k = 2.760$, $n_{s1} = 5$, $n_{s2} = 4$ and $\gamma = 0$, $\gamma = 0.5$ and $\gamma = 1$, if the unbiased $\hat{\sigma}$ of σ according to (28) is used. For the percent nonconforming equal to 5% the probablility of acceptance is 0.013 for $\gamma = 0$ (this value is different to the required value 0.05 because we look at the procedure as if being one sided), 0.063 for $\gamma = 0.5$ and 0.149 for $\gamma = 1$, i.e. for the realistic value $\gamma = 0.5$ the acceptance probability is five times larger than the required value. This required value holds only if each of $n_1 = 20$ pen-injectors is applied once ($n_{s2} = 1$). Increasing the number of repetitions, n_{s2}, decreases the number of pen-injectors, n_{s1}, neccesary for the test, however not to $n_{s1} = 20/\, n_{s2}$, but to $n_{s1} = 20, 13, 11, 10, 9$ for $n_{s2} = 1, 2, 3, 4, 5$, resp.; $n_{s1} = 7$ for $n_{s2} \to \infty$. For $\gamma = 5$ a sampling plan with $n_{s1} = 5$ (as required in the standard) being equivalent to ($n_{s1} = 20$; $n_{s2} = 1$) does not exist.

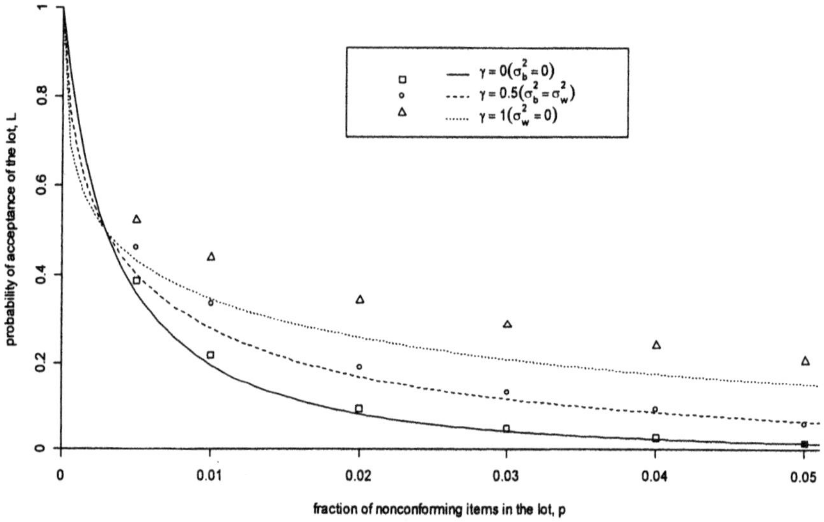

Fig. 5. The OC curves of the single sampling plan ($n_{s1} = 5$, $n_{s2} = 4$, $k = 2.76$) for $\gamma = 0, 0.5$, and 1. For $\gamma = 0$ the plan fulfils the requirement $L(0.05) = 0.013$. For $\gamma = 0.5$ and $\gamma = 1$ the acceptance probability at $p = 0.05$ is increased to 0.063 and 0.149, resp. Each symbol represents the acceptance frequency on the basis of 1000 simulation runs.

The total standard deviaton S is negatively biased for σ and its standard deviation is smaller than that of the unbiased estimator $\hat{\sigma}$ of σ according to (28): $E(S) < \sigma$, $V(S) < V(\hat{\sigma})$. Therefore, the use of S instead of $\hat{\sigma}$ in the test statistic (22) increases the acceptance probability even more.

When no knowledge about γ exists one should start with a sampling plan with $n_{s2} = 1$ because its OC does not depend on γ. Unfortunately, application of this sampling plan does not allow to estimate σ_w^2 and γ. Therefore, it is recommended to repeat all or some of the n_{s1} measurements in order to estimate σ_w^2 and γ and to adjust the sampling plan accordingly. If the estimate for γ turns out to be 0.5 a plan out of the series given above could be applied.

References

1. Graf, U., Henning, H.-J., Stange, K., Wilrich, P.-Th. (1987). Formeln und Tabellen der angewandten mathematischen Statistik. Springer, Berlin
2. International Standard ISO/DIS 11608-1.3 (1998). Pen-injectors for medical use – Part 1: Requirements and test methods. International Organization for Standardization, Geneva
3. International Standard ISO 3951 (1989). Sampling procedures and charts for inspection by variables for percent nonconforming. International Organization for Standardization, Geneva
4. Liebermann, G. J., Resnikoff, G. J. (1955). Sampling plans for inspection by variables. Journal of the American Statistical Association **50,** 457-516
5. Military Standard MIL-STD-414 (1957). Sampling procedures and tables for inspection by variables for percent defective. United States Government Printing Office, Washington, D.C.
6. Satterthwaite, F.E. (1946). An approximate distribution of estimates of variance components. Biometrics Bulletin **2,** 110-114
7. Wilrich, P.-Th. (1970). Nomogramme zur Ermittlung von Stichprobenplänen für messende Prüfung bei einer einseitig vorgeschriebenen Tolerenzgrenze. Teil 1: Pläne bei bekannter Varianz der Fertigung – Teil 2: Pläne bei unbekannter Varianz der Fertigung. Qualität und Zuverlässigkeit **15,** 61-65, 181-187
8. Wilrich, P.-Th. (2000). Single sampling plans for inspection by variables in the presence of measurement error. Allgemeines Statistisches Archiv **84,** 239-250

Joint Determination of Economic Design of $\bar{X}$ - Control Charts, Economic Production Quantity, and Production Run Length for a Deteriorating Production System

K.K. Lam and M.A. Rahim*

Faculty of Administration, University of New Brunswick, Fredericton, New Brunswick, Canada E3B 5A3

1 Introduction

In this study, we present an integrated economic model for the joint optimization of production quantity, quality control, and maintenance policy. The economic Production Quantity (EPQ) model is a very useful model in production planning and traditionally, it is based on the assumption that the production process is perfect and produces items of perfect quality. However, product quality may not always perfect. It is usually dependent on the operating state of a production process. Once the process shifts to an out-of-control state, an increased number of defective items are produced until some inspection and maintenance activities are performed to bring the process back to an in-control state. The effect of production-lot-size on the quality of the product may also be significant. If the production process shifts to an out-of-control state at the beginning of the production cycle, the entire lot will contain more defective items. Hence, it is wiser to reduce the production cycle to decrease the percentage of defective items and thus, improve output quality. On the other hand, reduced production cycle may result in an increase in costs due to frequent setups. A balance must be maintained so that the total cost is minimized. The production of quality goods may also depends upon the operating condition of the machine tools. It is assumed that the cost of maintaining the equipment increases with age, therefore, an economic age replacement strategy is also needed to minimize the total cost of the system.

This study considers a deteriorating production process that possesses an increasing failure rate. The operating state of the process is classified as either in-control or out-of-control. An $\bar{X}$ - control chart is used to monitor the output quality of the product. Some maintenance/inspection policies are introduced to monitor

* Contract/grant sponsor: NSERC (Natural Science and Engineering Research Council) of Canada

and restore the process. After each inspection, a decision is made whether or not to restore the process at that time. It is assumed that maintenance and inspection are done simultaneously either at a regular (uniform) intervals of time or at non-regular (non-uniform) intervals of time. The production cycle begins when a new component is installed and ends with a repair after the detection of a failure or after a specified number of inspection intervals, *m*, whichever occurs first. An expected average cost model is developed. The objective is to jointly determine the economic production quantity, the production cycle, monitoring and inspection policy, and design parameters of $\overline{X}$- control charts by minimizing the expected average cost. Numerical examples are provided to evaluate the performance of the cost model. Sensitivity analyses are conducted to study the effects of various model parameters.

2 Literature Review

The issues of quality control, maintenance, and economic production quantity in a deteriorating process were studied separately a decade ago. The classical economic manufacturing quantity (EMQ) model has been studied by many researchers. However, the effect of an imperfect process on the EMQ has not been studied extensively. The effect of an imperfect production process was initially studied by Rosenblatt and Lee [27]. The elapsed time to the process shift was assumed to be exponentially distributed, and the optimal production cycle was found to be shorter than that of the classical EMQ model. Porteus [21] found the same result by analyzing the relationship between quality and lot size. Lee and Rosenblatt [12-13] considered both an optimal production cycle and maintenance schedule. They used an equally spaced maintenance and inspection policy, resulting in cost effectiveness while continuing to produce large lots.

Duncan [6] was the first to introduce the economic design of the $\overline{x}$ - control chart for controlling the process mean. He assumed that the occurrence time of the assignable cause follows an exponential distribution (i.e. a Markovian shock) resulting in an uniform inspection interval. In a Markovian shock assumption, the rate of process deterioration remains constant throughout the production cycle. Whereas in a non-Markovian shock assumption, the rate of deterioration varies as the system ages if there is no maintenance inspection intervention. Lorenzen and Vance [16] have developed a unified approach for control chart design so that the model can accommodate the situation in which production continues/ceases during search/repair. Some authors have analyzed the economic design of control charts when the "time to shift" or in-control time follows a Weibull distribution (i.e. non-Markovian shock). These include Banerjee and Rahim [1], Rahim [22-23], Rahim and Banerjee [24], Rahim and Ben-Daya [25], and Zhang and Berardi [32].

An optimal maintenance policy restores the performance of equipment to its original level to enhance product quality. Tseng [29] considered a preventive maintenance policy for a deteriorating production system with an increasing failure distribution. Tseng *et al.* [30] proposed two maintenance strategies: one is a consideration of a constant hazard criterion and the other considers a cumulative hazard rate criterion. Makis [17] provided a model for the optimal replacement of a tool subject to random failure. Kim et al. [8-10] have extended the EMQ model for a failure-prone machine with a general lifetime distribution.

The relationship between the optimal inspection and the production cycle has also been studied extensively by various researchers. Makis and Fung [18] integrated the EMQ model by considering generally distributed machine failure time, replacement time, and periodic inspections. In the above section, it has been assumed that the inspection/maintenance activities are error free. Ohta and Ogawa [20], Lee and Park [14], and Liou *et al.* [15] have integrated Type I and Type II inspection errors into the EMQ model. Tagaras [28] considered an economic model for quality control with a preventive maintenance policy under the Markovian deterioration assumption. Rahim [22] developed a model for incorporating an inspection schedule and control chart design into an economic production quantity of an imperfect production process under a non-Markovian deterioration assumption with an increasing failure rate. These results indicate that the use of non-uniform inspection intervals is more economical than the use of uniform inspection intervals. Later, Rahim and Ben-Daya [25] generalized the Rahim [22] model by determining that stoppage of the machine during the in-control phase of a false alarm. Ben-Daya and Rahim [2] have also introduced maintenance policy into a quality control model under a non-Markovian shock assumption. They found that preventive maintenance reduces the probability of moving to an out-of-control state by an amount proportional to the preventive maintenance level. Also, the difference between uniform and non-uniform inspection schemes with and without preventive maintenance policies was studied. However, maintenance cost was assumed to be a constant. Maintenance and inspections were assumed to be performed simultaneously. Lee and Rahim [11] integrated both the age-dependent maintenance cost and salvage value of manufacturing equipment with the economic design of $\bar{x}$- control charts for additional economic benefits derived from joint preventive maintenance and quality control. Cassady *et al.* [3] integrated both preventive maintenance and control charts for a process that shifts to an out-of-control condition. They found that using a combined policy is superior to using only a control chart or a preventive maintenance policy. Collani and Kiesmüller [5] provided an integrated approach for simultaneously determining a monitoring and maintenance policy that minimizes the expected cost per unit time for a deteriorating production process. Recently, Wang and Sheu [31] studied an optimal production-inspection-maintenance policy with inspections errors.

In this paper, we present an integrated economic model for the joint optimization of production quantity, quality control, and maintenance policy. We provide a sensitivity analysis of the model to study the effects of various parameters on the decision variables and cost function. A general distribution of the in-control period is assumed and the salvage value of the equipment is introduced. The model allows for age-dependent replacement before failure. The paper is organized as follows. In section 3, the notation of the model is presented. The derivation of an optimal production cycle is developed in section 4. Numerical examples are discussed in section 5, and a sensitivity analysis of the model is illustrated in section 6. Conclusions are drawn in the last section.

3 Notation

The following notation is used to develop an economic model for production run length related to $\bar{x}$ - control chart design, maintenance, and replacement policy.

Decision variables:

m	number of inspection intervals
n	sample size
h_j	length of the *j*-th sampling (inspection) interval
k	control limit coefficient for the $\bar{x}$ - control chart

Other notation:

Z_1	the expected search time associated with a false alarm
Z_2	the expected time to repair the process if a failure is detected (or the expected preventive maintenance time at the end of the *m*-th sample if no true failure is detected)
a	the fixed sampling cost
b	the cost per unit sampled
Y	the cost per false alarm
W	the cost to locate and repair the assignable cause
D_0	the quality cost per hour while producing in an in-control state
D_l	the quality cost per hour while producing in an out-of-control state
α	P (exceeding control limits I process in control)
β	P (not exceeding control limits I process out of control)
ω_j	$= \sum_{i=1}^{j} h_i, \quad j = 1, 2..., m; \quad \varpi_0 = 0$

p_j	conditional probability that the process shifts to the out-of-control state during the time interval (ω_{j-1}, ω_j) given that it was in the in-control state at time ω_{j-1}
$F(\omega_j)$	the cumulative distribution function
$\Delta F(\omega_j)$	$= F(\omega_j) - F(\omega_{j-1}); \quad j = 1, 2, \ldots, m$
$\bar{F}(\omega_j)$	$= 1 - F(\omega_j)$
$S(t)$	the salvage value for working equipment of age t
$E(T)$	the expected total length of the production cycle including repair time
$E(QC)$	the total expected quality control cost per production cycle including the repair cost
K	replacement cost
$C(u)$	cost function representing the maintenance cost of an item of age u
$E(MR)$	the expected total maintenance and replacement cost
μ_0	standard process mean
σ	process standard deviation
δ	the shift parameter in the process mean

The following notation is used for the EPQ model:

D	the demand rate
P	the production rate
C_h	the holding cost per item per unit time
C_o	the setup cost for each production cycle
T_P	the length of the production run including the m-th inspection, during which inventory is built up
I_j	expected inventory level at the end of the j-th interval
T_I	the total length for the inventory cycle
$E(HC)$	expected inventory holding cost per production cycle
$E(QC)$	expected quality control cost per cycle
$E(T_P)$	expected production run time in a cycle
ETC	the expected total cost per unit time for integrated production, inventory, quality control cycle, and maintenance cost

4 Integrated Model Development

4.1 Economic Design of $\bar{x}$ - Control Charts

The design of an $\bar{x}$ - control chart requires the determination of various design parameters. These include the size of the sample (n) drawn in each sampling or inspection interval (h), and the upper and lower control limits coefficient (k). Process monitoring by sampling inspection is necessary to determine the state of the process in order that the penalties associated with the probabilities of Type I and Type II errors can be minimized. The following costs are considered for determining the decision variables: sampling cost, search cost, and cost of operating both in control and out of control.

Duncan [6] assumed that output quality is measurable on a continuous scale and is normally distributed. When the process is in control, the initial mean is μ_0; however, due to the occurrence of an assignable cause, the initial mean may be shifted from μ_0 to $\mu_0 + \delta\sigma$ or $\mu_0 - \delta\sigma$ (out-of-control state), where δ is the shift parameter, and σ is the standard deviation.

The control limit of the $\bar{x}$-control charts is set at $\mu_0 \pm k$ times the standard deviation of the sample means, where k is known as the control limit coefficient, such that,

$$UCL = \mu_0 + \frac{k\sigma}{\sqrt{n}}, \qquad LCL = \mu_0 - \frac{k\sigma}{\sqrt{n}}. \tag{1}$$

Traditionally, the symbol α is used to represent the probability of a Type I error and β is used to represent the probability of a Type II error. The expressions for α and β are as follow:

$$\alpha = 2F(-k), \qquad \text{and} \qquad \beta = F(k - \delta\sqrt{n}) - F(-k - \delta\sqrt{n}).$$

The expected cost per hour due to production of nonconforming items when the process is in control, D_0, should be taken into consideration in terms of cost. Let D_1 be the expected cost per hour due to the production of nonconforming items when the process is out of control.

In the economic design of control charts, the objective is to determine the optimal design parameter values of n, h, and k so as to minimize the expected cost per unit time. Rose [26] introduced a renewal reward process to express the expected cost per unit of time. Equation (2) illustrates this condition:

$$\text{E(Cost per unit time)} = \frac{\text{E(Cost per cycle)}}{\text{E(Length of cycle)}}. \tag{2}$$

The optimal design parameter values may be obtained by using any pattern search technique for minimizing the objective function given in equation (2).

4.2 Weibull Shock Model and Frequency of Sampling

We assumed a general distribution of the in-control period. However, because of its versatility and numerous applications in quality and reliability analysis, a Weibull distribution is used for the specific purpose of numerical illustration. In Duncan's approach, a fixed sampling interval and a constant failure rate over each interval were assumed. Duncan's model was extended by Banerjee and Rahim [1], by assuming a process with a Weibull distribution time shift and an increasing hazard rate. They found that by increasing the sampling frequency according to the age of the system results in a lower operational cost per hour. The density function for the in-control period is expressed as:

$$f(t) = \lambda v t^{v-1} \exp(-\lambda t^{v}), \quad t > 0, \quad v \geq 1, \tag{3}$$

where $\lambda > 0$ is the scale parameter and v is the shape parameter (when $v = 1$, the distribution becomes exponential).

Uniform sampling intervals for Markovian shock models provide a constant integrated hazard rate over each interval. This principle may be applied to non-Markovian shock models by choosing the length of sampling intervals such that the integrated hazard rate over an interval is constant.

Banerjee and Rahim [1] have proposed that the lengths of the sampling intervals $h_j, j = 2,\ldots,m$ are related to h_1, where the time that the process remains in the in-control state follows a Weibull distribution. Thus,

$$h_j = \left[j^{1/v} - (j-1)^{1/v} \right] h_1, \quad j = 2,\ldots,m. \tag{4}$$

The production process is inspected at time $\omega_1 = h$, $\omega_2 = h_1 + h_2, \ldots$ to determine its state. If the inspection identifies that the state of the process is out of control, production ceases until the accumulated on-hand inventory is depleted to zero. Otherwise, production continues until a predetermined level of inventory is achieved. Production ceases temporarily during the in-control phase for a constant amount of time,Z_1, whenever there is a false alarm. However, It is assumed that a production cycle ceases permanently either with a true alarm or at ω_m, whichever occurs first. According to Rahim's [22] model, it is assumed that m is infinity. For instance, a production cycle ceases when the true alarm is triggered. However, in the present model, m could be a decision variable along with n, k and h_j, $j = 1, 2,\ldots,m$.

4.3 Derivation of E(T), E(QC) and Inspection Intervals

This section presents the expressions for expected cycle length *E(T)* and the expected quality control cost per cycle *E(QC)*. The expected cycle length consists of the following periods: the expected time for inspection intervals when the process is in control, the expected time elapsed for searching the false alarm, the expected time to detect the presence of an assignable cause, and the repair of the machine. Similarly, the expected quality control cost consists of the following categories: (1) the expected cost to operate while in control with no alarm, (2) the expected cost of a false alarm, (3) the expected cost to operate while out of control with no alarm, (4) the repair cost, (5) the cost of sampling.

The *E(QC)* is derived from combining these cost components together (Rahim [23]):

$$\begin{aligned} E(QC) &= (D_0 - D_1)\int_0^{\omega_m} tf(t)dt + (D_1 - D_0)\sum_{j=1}^{m} \omega_j \Delta F(\omega_j) \\ &+ D_0 \sum_{j=1}^{m} h_j \bar{F}(\omega_{j-1}) + \alpha Y \sum_{j=1}^{m-1} \bar{F}(\omega_j) \\ &+ D_1 \beta \left(\sum_{j=1}^{m-1} \Delta F(\omega_j) \sum_{i=j+1}^{m} h_i \beta^{i-j-1} \right) + W \\ &+ (a+bn)\left[1 + \sum_{j=1}^{m-2} \bar{F}(\omega_j) + \beta \sum_{j=1}^{m-2} \Delta F(\omega_j) \right. \\ &\left. \left((1-\beta) \sum_{i=1}^{m-j-1} i\beta^{i-1} + (m-j-1)\beta^{m-j-1} \right) \right]. \end{aligned} \tag{5}$$

The expected quality control cycle length, $E(T)$, consists of the following components: (1) the expected time for inspection intervals when the process is in control with no false alarm, (2) the expected time elapsed for search during a false alarm, (3) the expected time to detect the presence of an assignable cause, (4) the expected repair time.

By combining these periods together (Rahim [23]), the expected cycle length is obtained:

$$\begin{aligned} E(T) &= \sum_{j=1}^{m} h_j \bar{F}(\omega_{j-1}) + Z_1 \alpha \sum_{j=1}^{m-1} \bar{F}(\omega_j) \\ &+ \beta \sum_{j=1}^{m-1} \Delta F(\omega_j) \sum_{i=j+1}^{m} h_i \beta^{i-j-1} + Z_2 , \end{aligned} \tag{6}$$

where

$$\omega_j = \sum_{i=1}^{j} h_i , \quad j = 0, 1, \ldots, \omega_0 = 0 . \tag{7}$$

4.4 Inventory Holding Cost

The total expected inventory holding cost *E(HC)* is expressed as

$$E(HC) = C_h \int_0^{T_I} I(t)dt = C_h E(A) . \tag{8}$$

The expression of *E*(*A*) can be found in Rahim and Ben-Daya [25].

$$E(A) = \sum_{j=1}^{m} U_j \bar{F}(\omega_{j-1}) + (1-\beta) \sum_{j=1}^{m-1} B_j \Delta F(\omega_j) + \beta \sum_{j=1}^{m-1} \Delta F(\omega_j)$$
$$\times \left[\sum_{i=j+1}^{m} \beta^{i-j-1} U_i + \beta^{m-j-1} B_m \right] + \alpha \sum_{j=1}^{m-1} G_j \bar{F}(\omega_j) + B_m \bar{F}(\omega_{m-1}) . \tag{9}$$

The quality and inventory control cycle is depicted in Figure 1.

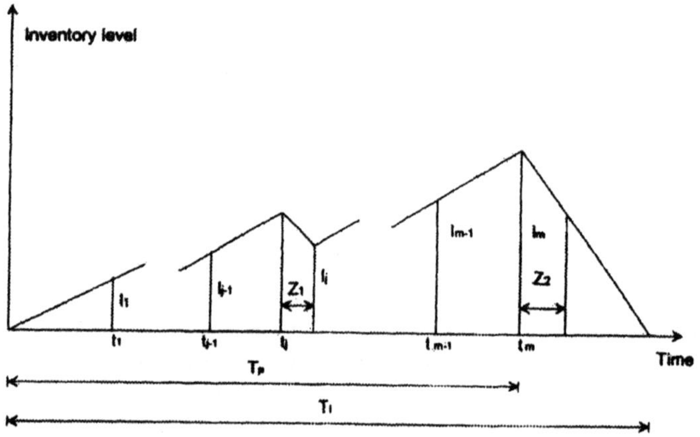

Figure 1. Quality and Inventory Control Cycle.

4.5 Maintenance Model

The most widely recognized preventive maintenance policy for a single piece of equipment is an age-replacement preventive maintenance policy. There is a relationship between the age-replacement model and the quality control problem. Since the cost of maintaining equipment increases over time, preventive maintenance cost should include an optimal maintenance policy. The optimal replacement time is derived by comparing the replacement cost with the total maintenance cost over a finite horizon. According to the mathematical expression of Nahmias [19], the total replacement cost per cycle is denoted as *K*. In addition, it is assumed that the maintenance cost *C*(*u*) has a cost function with age *u*. The equation is given by

$$\begin{aligned} &\text{Total replacement cost per cycle } = K, \\ &\text{Total maintenance cost per cycle } = \int_0^{\omega_m} c(u)du. \end{aligned} \tag{10}$$

Furthermore, if the machine has a salvage value provided it does not fail, then the cost function will be reduced. The expected total maintenance and replacement cost function is expressed as:

$$\begin{aligned} E(MR) &= KF(\omega_m) + \left(\int_0^{\omega_m} c(u) - S(\omega_m) \right) \bar{F}(\omega_m) \\ &= KF(\omega_m) + \bar{F}(\omega_m) \int_0^{\omega_m} c(u)\bar{F}(\omega_m) - S(\omega_m)\bar{F}(\omega_m) . \end{aligned} \tag{11}$$

4.6 An Integrative Model

The expected total cost for integrated production, inventory, the maintenance model and the quality control cycle per unit time can be described as the setup cost, the expected inventory cost, the expected quality control cost and the expected maintenance cost. The expected total cost per unit time is given by

$$ETC = \frac{C_0 + E(HC) + E(QC) + E(MR)}{E(T_1)}. \tag{12}$$

The model is as follows:

$$ETC = \frac{C_0 + \frac{1}{2}\frac{P}{D}(P-D).E(T_p^2).C_h + E(QC) + E(MR)}{E(T_1)}. \tag{13}$$

The expression $E(T_1)$ is given by

$$E(T_1) = \frac{P}{D}\left[E(T_p) - (m-1)\alpha Z_1 \right], \tag{14}$$

where

$$T_P = T - Z_1$$

$$E(T_P) = E(T) - Z_1 .$$

The derivations of $E(T_p^2)$ and $E(T_p)$ are shown in Appendix B in Rahim and Ben-Daya [25].

The objective function given by formula (13) include the decision variables *m, n, k* and h_1. Furthermore, the cost function can be minimized by using the pattern search technique of Hooke and Jeeves [7]. The search is terminated when the step size is reduced to a pre-specified value or when the number of iterations equals a predetermined value, whichever occurs first (see Rahim [23]). Nevertheless, due to the characteristics of the cost function, some modifications to the standard method have to be made to limit the sample size to integer value and to account for the constraint that there will be no shortages, that is $(P-D)\omega_m > 0$. The optimal value of $m \geq 2$ is determined by the inequalities:

$$ETC(m-1) \geq ETC(m) \leq ETC(m+1) . \qquad (15)$$

The objective of this paper is to determine the optimal decision variables, including the sample size *n*, the control limit coefficient *k*, the number of inspection intervals *m*, the optimal production run time ω_m^*, and the length of the *j*th sampling interval $h_j (j = 1, 2, \ldots, m)$. The results will be determined by minimizing the expected cost per hour *ETC(m)* while combining the models from Lee and Rahim [11] and Rahim and Ben-Daya [25]. Once ω_m is determined, the nearly optimal production quantity Q^* can easily be found.

5 Numerical Illustration

5.1 Numerical Examples

To illustrate the application of the present model, an industrial example from Lee and Rahim [11] will be used for determining the economic production quantity, the design parameters of an $\bar{x}$ - control chart, and the maintenance policy under non-Markovian shock models. Suppose the process failure mechanism is followed by a Weibull distribution with an increasing hazard rate $r(t) = \lambda \nu t^{\nu-1}$. Let the scale parameter be $\lambda = 0.05$ and the shape parameter be $\nu = 2$. The manufacturer has used an $\bar{x}$ - control chart to monitor the process. The average time required to search for a false assignable cause, Z_1, is 0.25 hours. However, the average time required to repair the process if a failure is detected, Z_2, is 1 hour. It is assumed that the average cost of finding an assignable cause, or *W*, is \$1100.00. Whereas, the cost of searching for an assignable cause during a false alarm is estimated to be $Y = \$500.00$. The magnitude of the shift in the process mean is found to be $\delta = 0.5$. The expected cost of operating the process in an in-control state is $D_0 = \$50.00$, and the expected cost of operating the process in an out-of-control state is $D_1 = \$950.00$. The fixed cost of the sampling is assumed to be $a = \$2.00$, and the variable cost of sampling is $b = \$0.50$. The salvage value of the equipment is time dependent, and it is closely approximated by $S(\omega_m) = 1100.0 \exp(-\omega_m)$. Furthermore, the cost of maintaining the machine increases exponentially with the length of the production cycle; thus, the machine's maintenance function is $C(t) = 500 \exp(0.3365t)$. The constant demand rate is $D = 1400$ units per hour, and the constant production rate is $P = 1500$ units per hour. A fixed setup cost, $C_0 = \$20.00$, and a unit inventory holding cost, $C_h = \$0.10$, are assumed.

The optimal solution results in a production run length of $\omega_m = 4.40$ time units, where $m = 23$, and $EPQ = Q^* = 6525_m$. The optimal design parameters of the control charts are as follows: $n = 57$, $h_1 = 0.65$ hours, and $k = 2.64$. Total inventory cycle = 4.66 time units, $E(HC) = \$22.17$, $E(MR) = \$64.13$, the expected quality cost per hour, $E(QC) = 357.32$, and the expected minimum total cost of this design, $E(TC) = \$443.61$.

5.2 Uniform Vs. Non-Uniform Sampling Scheme

Table 1 is prepared to demonstrate a comparison of the two inspection schemes: uniform inspection scheme denoted by scheme A and a non-uniform inspection scheme denoted by scheme B. The table shows that a non-uniform inspection scheme provided a better result than a uniform inspection scheme.

Rahim [22] assumed that the time for searching for an assignable cause when it is false is negligible. In reality, the assumption of a zero search time may not be realistic. If the process goes out of control, it may be necessary to shut down the system for a considerable amount of time in order to perform the work of identifying an assignable cause. Rahim and Ben-Daya [25] have modified Rahim's [22] model allocating a fixed amount of time to search for an assignable cause (Z_1). Thus, if $Z_1 > 0$, for example, let $Z_1 = 0.75$. The resulting plan is as follows.

Table 1. The Effect of Searching Time $Z_1 > 0$

Scheme	Z_l	m	n	k	h	α	$1-\beta$	Q^*	$E(T_P)$	$E(QC)$	$E(HC)$	$E(TC)$
A	0.75	10	52	2.46	0.69	0.0139	0.8740	6305	4.30	353.15	86.46	439.61
B		14	55	2.70	1.91	0.0069	0.8424	6311	4.27	348.14	86.32	434.46

Note: Inspection Scheme: A – Uniform; B – Non-uniform

However, age dependent maintenance cost was not considered in the above table. Although machinery is subject to failure, it can be restored to a "good as new" condition by performing maintenance. Sometimes the change in the age of the equipment will affect the time to shift distribution and, consequently, the design of the control chart. Thus, the present model allows the integration of the optimal production quantity, the economic design of the control chart, a preventive maintenance policy, and, simultaneously, determines the optimal production run length for a deteriorating system. The effect of maintenance cost is depicted in Table 2.

Table 2. The Effect of Preventive Maintenance Cost under a Uniform Inspection Scheme

Z_1	Scheme	m	n	k	h	α	$1-\beta$	Q^*	$E(T_P)$	$E(QC)$	$E(HC)$	$E(MR)$	$E(TC)$
0	No PM	10	52	2.46	0.69	0.0139	0.8740	6305	4.30	353.15	86.46	-	439.61
0	PM	35	52	2.50	0.63	0.0124	0.8654	6561	4.37	357.26	23.19	60.09	440.54
0.25	PM	23	57	2.64	0.65	0.0086	0.8676	6525	4.40	357.32	22.17	64.22	443.61

PM = Preventive Maintenance

6 Sensitivity Analysis

In this section, a sensitivity analysis of the model is conducted to study the effect of the different cost parameters, search time parameters, and failure distribution parameters. The industrial example of Lee and Rahim [11] is used as the basis for analysis.

A factorial arrangement is made to perform the analysis. The effects of changes of each parameter are studied. To do so, we considered three levels (values) for each parameter as shown in Table 3.

Table 3. Simple Parameter Design Experiment

Parameter	δ	λ	V	Z_1	Z_2	A	B	D_0	D_1	W	Y	C_0	C_h
Level 1	0.50	0.05	2.00	0.25	1	2	0.5	50	950	100	500	20	0.10
Level 2	0.75	0.10	2.25	0.50	3	10	1	100	1500	2000	1000	50	0.25
Level 3	1.00	0.25	2.50	1.50	5	100	5	300	2000	3100	1500	100	0.50

The effects of changes in each parameter's values are shown in Table 4.

Effects of the magnitude of the shift parameter, δ, for the process mean

The relative sensitivity of the magnitude of the shift parameter for the process mean, δ, is studied and the results are presented in cases 1-3. The size of the shifts (δ) is a factor which influences the proportion of output produced while the system is running out of control. The smaller the shift, the harder it is to detect. Thus, the smallest value of δ (e.g. 0.50) in Table 3 coincides with the highest percentage of the occurrence of Type I error, the greatest number of samples, and the longest sampling interval. As the value of the shift parameter for the process mean increases, α decreases. Thus, a higher expected quality cost and higher production due to a longer production cycle results. On the other hand, as the size of the shift increases, less sampling is required, which increases the control limit, k, in the control chart; this results in a production quantity and reductions in expected quality cost, maintenance cost, and expected total cost.

Effects of the scale parameter, λ, of the failure distribution

The effects of the changes in the scale parameter, λ, are analyzed and the results are depicted in cases 1, 4 and 5. An increase in the scale parameter will drive the failure rate of the machine higher; thus, the relative Type II error is increased, as well as the sampling intervals. In addition, an increase in both expected quality and expected total cost, with a lower production quantity, results. Moreover, a higher risk of machine breakdown may occur and increase maintenance costs.

Effects of the shape parameter, v, of the failure distribution

In equation (3), the failure mechanism exhibits a decreasing function when the shape parameter lies between $0 < \nu 1$. On the other hand, for a shape parameter > 1, the failure mechanism increases over time. Thus, it is important to analyze the sensitivity of the shape parameter in relation to the proposed model. A comparison of cases 1, 6 and 7 provides the results of different shape parameters for sensitivity. As the shape parameter increases, the optimal design parameters (n, h, and k) decrease, and the inspection schedule decreases as well. The likelihood of an increase in Type I error is evident through increasing α. Although the production cycle length deteriorates, as does production quantity, these factors do not reduce the expected quality and/or expected total cost. The increase in the failure mechanism also causes the expected maintenance cost to sky-rocket.

Effects of search time, Z_1

As stated earlier, searching time should be considered whenever the process is found to be out of control. When the production process fails, the machine must be shut down in order to search for the assignable cause, even though occasionally the alarm turns out to be false. Cases 1, 8 and 9 illustrate an analysis of the effect of changes in search time, Z_1. The result shows that Z_1 has an inverse effect on the number of inspection intervals, *m*. In addition, the control chart design parameters *n*, *h*, and *k* are increased with the search time inherent to the false

Table 4. Sensitivity Analysis of Main Factors

Case	Vp^a	Level	m	n	k	h_1	α	$1-\beta$	Q^*	$E(Tp)$	$E(QC)$	$E(HC)$	$E(MR)$	$E(TC)$
1	Basic model*	1	23	57	2.64	0.65	0.0086	0.8676	6525	4.30	357.32	22.17	64.13	443.61
2	δ	2	32	30	2.90	0.48	0.0037	0.8863	6361	4.27	332.08	22.06	60.97	415.11
3	"	3	41	19	3.08	0.38	0.0020	0.8988	6274	4.20	318.60	22.00	59.31	399.91
4	λ	2	16	58	2.66	0.64	0.0079	0.8754	4790	3.22	427.13	15.10	109.45	551.68
5	"	3	9	58	2.67	0.77	0.0078	0.8677	3390	2.28	537.90	9.60	201.24	748.74
6	v	2	17	56	2.61	0.64	0.0091	0.8712	5612	3.78	389.31	17.68	83.06	490.05
7	"	3	14	56	2.61	0.63	0.0092	0.8719	4999	3.36	416.08	14.86	101.38	532.32
8	Z_1	2	22	59	2.71	0.66	0.0066	0.8700	6512	4.41	357.74	21.58	66.24	445.56
9	"	3	20	67	2.91	0.71	0.0036	0.8806	6509	4.44	360.14	20.83	69.15	450.12
10	Z_2	2	23	57	2.64	0.65	0.0083	0.8720	6525	4.40	260.70	16.17	166.74	443.61
11	"	3	23	57	2.64	0.65	0.0083	0.8720	6525	4.40	205.20	12.73	225.68	443.61
12	a	2	19	62	2.61	0.79	0.0093	0.9039	6609	4.46	367.37	22.56	64.18	454.11
13	"	3	10	80	2.53	1.59	0.0117	0.9728	7170	4.81	435.47	24.75	65.38	525.60
14	b	2	17	47	2.37	0.86	0.0177	0.8545	6730	4.56	385.55	22.16	67.80	475.51
15	"	3	8	25	1.59	1.73	0.1119	0.8187	7592	5.26	475.13	21.04	80.01	576.18
16	D_0	2	22	57	2.65	0.68	0.0081	0.8704	6559	4.42	394.44	22.31	69.24	485.99
17	"	3	17	59	2.69	0.88	0.0072	0.8760	6763	4.54	542.64	23.24	86.95	652.82
18	D_1	2	31	58	2.66	0.47	0.0078	0.8740	6333	4.28	391.23	21.33	73.42	485.98
19	"	3	37	58	2.67	0.39	0.0075	0.8714	6247	4.23	415.85	20.88	79.82	516.55
20	W	2	18	58	2.68	0.84	0.0073	0.8701	6727	4.52	525.57	23.08	85.31	633.96
21	"	3	10	58	2.69	1.58	0.0074	0.8640	7466	4.99	730.72	25.86	97.47	854.05
22	Y	2	22	63	2.80	0.69	0.0051	0.8786	6578	4.41	361.93	22.73	62.66	447.32
23	"	3	22	67	2.91	0.71	0.0036	0.8807	6600	4.42	364.51	22.97	62.13	449.61
24	C_0	2	23	57	2.64	0.65	0.0083	0.8717	6529	4.40	357.36	22.19	70.50	450.04
25	"	3	22	57	2.63	0.67	0.0085	0.8735	6546	4.41	357.60	22.24	80.91	460.75
26	C_h	2	24	53	2.55	0.61	0.0107	0.8613	6482	4.38	356.52	54.19	71.06	481.78
27	"	3	24	45	2.31	0.57	0.0211	0.8528	6386	4.38	357.37	107.13	78.72	543.21

Vp^a = Parameter whose value changes from that in case 1

Basic model* = All parameters are at level 1

alarm. Since a greater amount of search time has been allocated to seek for the assignable cause, the possibility of Type I error, α, may be reduced, and, thereby, the system becomes reliable. Furthermore, the power of the test, $1-\beta$, is increased, and the production cycle length is also accelerated. Nevertheless, the quality control, the maintenance, and the total cost of the system have the same relationship with search time whenever a false alarm occurs. The proposition indicates that maintenance work and searching work is performed simultaneously, thereby increasing the maintenance cost.

Effects of search time, Z_2

The results (cases 1, 10 and 11) indicate that Z_2 has no effects on the inspection schedule of optimal design parameters, and a minimal effect on expected cycle length. Since increasing the repair time of the system will increase the expected total cycle length, there are no influences on sampling, and control limits of the inspection intervals. However, prolonging Z_2 will reduce the expected quality cost of the system and cause a significant increase in maintenance cost since more maintenance work is assigned during the search and repair time interval. This scenario is not applicable to expected total cost because the same amount will be produced by using the same equipment and the same amount of raw materials.

Effects of fixed sampling cost, a

Cases 1, 12 and 13 provide a varied fixed sampling cost and the relationship between fixed sampling cost and other cost functions. The inspection schedule has an indirect impact on fixed sampling cost whenever it increases. Since the sampling cost has a positive relationship with production overhead cost, an increase in a indicates more samples will be required for testing or sampling. However, an increased sample size imply more chances to detect a false alarm. Therefore, Type I error, α, is increased. The frequency of sampling will increase since fewer inspections are undertaken as a is raised, which in turn, lengthens the expected cycle length of the process. In addition, because the fixed sampling cost is one of the components of expected quality cost, an increase in a will increase the expected quality cost. The results in Table 4 also illustrate that the increase in the fixed sampling cost is similar proportionally to the increase in the expected quality cost. Thus, an increase in a will unavoidably increase the expected total cost. This situation may not be relevant to the maintenance cost.

Effects of variable cost of sampling, b

Cases 1, 14 and 15 illustrate the sensitivity results that occur from varying the cost of sampling, b. More frequent sampling is done initially with respect to low variable sampling cost; thus, more sampling will identify the nonconforming items immediately and reduce the occurrence of false alarms. However, an increase in the variable cost of sampling will force a reduction of sample size for testing and create a demand for more sampling intervals instead. In addition, the power of the test of the process decreases, thus increasing the likelihood of Type I error.

Consequently, there are more frequent shutdowns during the process, and the production cycle is extended. Since b is also a component of expected quality cost, an increase in b, obviously, will increase both the expected quality cost and the overall cost of the system extensively. In addition, the expected maintenance cost increases significantly due to more frequent shutdowns during the process when searching or repairing the assignable cause.

Effects of cost of operating in an in-control state, D_0

The analysis of the expected cost of producing nonconforming items when the process is in control, D_0, is shown in cases 1, 16 and 17. The inspection interval is inversely related to D_0. Because more effort has been spent on maintaining the in-control state, less inspection time is required. However, the optimal design parameters (n, h, and k) are less responsive to the effect of D_0 cost, even a rapid change in D_0. This scenario suggests that the economic model is robust. It is capable of coping with temporary changes in the process mean when the system is still in control, even though there is an increased cost of D_0. In addition, the expected cycle run is increased when the cost of operating in an in-control state is raised. The reason for this increase is that the system becomes more secure and reliable; thus, the economic production quantity, Q^* is increased as well. However, this relationship will cause increases in the expected quality cost, expected maintenance cost, and the expected total cost. It is obvious that the increase in expected quality cost is the proportional increase in D_0.

Effects of cost of operating in an out-of-control state, D_1

A comparison of 1, 18 and 19 shows that D_1 increases; more frequent inspections are undertaken to limit the duration of out-of-control operation, thereby, to ensure that the system is reliable. There is less influence the on sample size and the control limits due to the lower likelihood of a Type I error. On the other hand, this is not the case for the sampling interval; the frequency of sampling is reduced for the purpose of cost saving. A decrease in the cost of operating in an out-of-control state indicates a higher possibility that the production system will shift to an out-of-control state. This proposition can be proven by increasing α. The increases in the production cycle length is because of the increased need to shut down to search for an assignable cause. The situation reverses as D_1 increases. More will be spent on quality control and the improvement of the equipment's reliability which increases the overall cost of the system.

Effects of cost to locate and repair the assignable cause, W

Cases 1, 20 and 21 demonstrate a test of the sensitivity of cost to identify and repair the assignable cause, W. As W increases, the inspection schedule becomes shorter due to the additional costs allocated to repair the production system. The decreasing Type I error (false alarm), α, demonstrates that these repairs reduce the failure rate. However, the sampling interval may be increased to monitor the state of the system. The system becomes more reliable when more time and money are allocated to identify and to repair the assignable cause; thus, the production cycle length increases relative to the increase in cost, W. However, the increase in W also causes both the expected quality and total cost increase. When W increased from \$1100 to \$2000 (an 81% increase), it resulted in a 47% increase in expected quality cost and a 43% increase in total cost. Thus, W is a very sensitive cost function in the proposed model.

Effects of cost of false alarm, Y

When comparing Cases 1, 22 and 23, one can see that, as the cost of false alarms increases, it may be necessary to increase the sample size. The control limit k also moves in the same direction to tighten the quality control. In addition, as Type I error, α, becomes lower, the likelihood of a false alarm occurring and the high cost associated with false alarms is greatly reduced. The economic production quantity increases due to a more secure system, and, therefore, fewer defective items are produced. The increase in expected quality cost also implies that more resources have been allocated to the quality control department for testing and monitoring the process to ensure that the occurrence of false alarms will be reduced. Consequently, a positive relationship between $E(QC)$ and $E(TC)$ is created. In addition, the reduction of machine failure rate or lower α may reduce the maintenance cost.

Effects of the setup cost for the production cycle, C_0

The result of a sensitivity analysis of the setup cost for the production cycle C_0 is presented in Cases 1, 24 and 25. Despite an increase in setup cost, the control chart parameters (n, h, and k), inspection schedule, and probability of false alarms have proven to be less elastic in response to the changes in C_0 cost. When the setup cost is higher, the machinery may need more maintenance work to perform before a new production run begins. Thus, the maintenance work obviously prolongs the production cycle and also increases the maintenance cost. The expected quality cost seems less responsive to the changes in setup cost for the production cycle.

Effects of the inventory holding cost, C_h

Cases 1, 26 and 27 depict various cost components of the inventory holding cost, C_h. The inspection schedule increases relative to the increased cost of holding inventory. However, the optimal design parameters (n, h, and k) decrease. This scenario implies that less frequent sampling and smaller sample size increase the occurrence of false alarms. Thus, the number of false alarms increases and may lead to increased maintenance cost for restorative work of the production system. In addition, although the production cycle run is lengthened, it does not increase the optimal production quantity, which, in turn, drives the expected total cost higher as a result of the higher cost for holding inventory. C_h, however, is less correlated to expected quality cost because inventory holding cost is not a cost component in expected quality costs.

7 Conclusion

This paper presents an integrated economic model for jointly determining a monitoring and maintenance policy which minimizes cost per unit time for a deteriorating process.

An extensive sensitivity analysis of the model is performed to study the effects of various parameters of the model. This may serve the interests of industries where the manufacturers are concerned about which parameters are more sensitive or not. It can also be seen that the consequences of misspecification of the cost parameters and time parameters could give misleading values of the decision variables. A comparison of two inspection/maintenance schemes is presented to show the relative advantages of non-uniform inspection/maintenance schemes over a uniform inspection/maintenance scheme.

One of the most common criticisms of economic control chart models is that there are too many parameters to be estimated, and adequate estimation of these terms is often difficult. Because this model has additional parameters for maintenance and production decisions, the problem is even more severe, making the challenge substantial. Generally, a large number of different input parameters makes the determining economic design more cumbersome. However, a simplified approach was developed by Collani [4]. This approach could be used to show the direction for further research on the relevant interrelationship between input parameters and design parameters.

Acknowledgement

The financial support for this paper by the Natural Sciences and Engineering Research Council (NSERC) of Canada and the financial assistance of the Faculty of Administration at the University of New Brunswick is greatly appreciated. The editorial assistance of Janet Gamache, Christopher Hegler, and Kok-Wai Wong is greatly appreciated. The authors wish to express their sincere appreciation to the anonymous referee and to the editor for their value comments and suggestions.

References

1. Banerjee, P.K., Rahim, M.A. (1988) Economic design of $\bar{x}$ - control charts under Weibull shock models. Technometrics **30**(4) 407-414
2. Ben-Daya, M., Rahim, M.A. (2000) Effect of maintenance on the economic design of $\bar{x}$ -control charts. European Journal of Operational Research **120**(1) 131-143

3. Cassady, C.R., Bowden, R.O., Liew, L., Pohl, E.A. (2000) Combining preventive maintenance and statistical process control: a preliminary investigation. IIE Transaction **32**(6) 471-478
4. Collani, E.V. (1997) Determination of the economic design of control charts simplified. In: Optimization in Quality Control, edited by Al-Sultan, K.S., Rahim, M.A. Kluwer Academic Publishers, Norwell Massachusetts 89-143
5. Collani, E.V., Kiesmüller (2001) Improving and maintaining process profitability, integrated models. In: Production Planning, Inventory, Quality, and Maintenance, edited by Rahim, M.A., Ben-Daya, M. Kluwer Academic Publishers, Norewell Massachusetts 245-275
6. Duncan, A.J. (1956) The economic design of $\overline{X}$ - charts used to maintain current control of a process. Journal of the American Statistical Association **51** 228-242
7. Hooke, R., Jeeves, T.A. (1961) Direct search solution of numerical and statistical problems. Journal of the Association for Computing Machinery **8** 212-229
8. Kim, C.H., Hong, Y. (1997) An extended EMQ model for a failure prone machine with general lifetime distribution. International Journal of Production Economics **49**(3) 215-223
9. Kim, C.H., Hong, Y. (1999) An optimal production run length in deteriorating production processes. International Journal of Production Economics **58**(2) 183-189
10. Kim, C.H., Hong, Y., Chang, S.Y. (2000) Optimal production run length and inspection schedules in a deteriorating production processes. IIE Transactions **33**(5) 421-426
11. Lee, B.H., Rahim, M.A. (2000) An integrated economic design model for quality control, replacement, and maintenance. Quality Engineering **13**(4) 581-593
12. Lee, H.L., Rosenblatt, M.J. (1987) Simultaneous determination of production cycle and inspection schedules in a production system. Management Science **33**(9) 1125-1136
13. Lee, H.L., Rosenblatt, M.J. (1989) A production and maintenance planning model with restoration cost dependent on detection delay. IIE Transactions **21**(4) 368-375
14. Lee, J-S., Park, K.S. (1991) Joint determination of production cycle and inspection intervals in a deteriorating production system. Journal of Operational Research Society **42**(9) 775-783
15. Liou, M-J., Tseng, S-T., Lin, T-M. (1994) The effects of inspection errors to the imperfect EMQ model. IIE Transactions **26**(2) 42-51
16. Lorenzen, T.J., Vance, L.C. (1986) The economic design of control charts: a unified approach. Technometrics **28**(1) 3-10
17. Makis, V. (1995) Optimal replacement of a tool subject to random failure. International Journal of Production Economics **41**(1-3) 249-256
18. Makis, V., Fung, J. (1998) An EMQ model with inspections and random machine failures. Journal of Operational Research Society **49**(1) 66-76
19. Nahmias, S. (1997) Production and Operation Analysis, 3rd Edition. Irwin, Boston, Massachusetts
20. Ohta, H., Ogawa, S. (1991) Joint determination of production cycle and inspection schedules under inspection errors. In: Pridhan, M., Brien, C.O. (Eds.) Production Research: Approaching the 21st Century, Taylor and Francis, 179-188

21. Porteus, E.L. (1986) Optimal lot sizing, process quality improvement and setup cost reduction. Operations Research **34**(1) 137-144
22. Rahim, M.A. (1994) Joint determination of production quantity, inspection schedule, and control chart design. IIE Transactions **26**(6) 2-11
23. Rahim, M.A. (1998) Economic design of $\bar{x}$ - control charts under truncated Weibull shock models. Economic Quality Control **13**(2) 79-105
24. Rahim, M.A., Banerjee, P.K. (1993) A generalized model for the economic design of $\bar{x}$ - control charts for production systems with increasing failure rate and early replacement. Naval Research Logistics **40**(6) 787-809
25. Rahim, M.A., Ben-Daya, M. (1998) A generalized economic model for joint determination of production run, inspection schedule and control chart design. International Journal of Production Research **36**(1) 277-289
26. Rose, S.M. (1970) Applied Probability Models with Optimization Application, Holden-Day San Francisco
27. Rosenblatt, M.J., Lee, H.L. (1986) Economic production cycles with imperfect production processes. IIE Transactions **18**(1) 48-55
28. Tagaras, G. (1988) An integrated cost model for the joint optimization of process control and maintenance. Journal of Operational Research Society **39**(8) 757-766
29. Tseng, S-T. (1996) Optimal preventive maintenance policy for deteriorating production systems. IIE Transactions **28**(8) 687-694
30. Tseng, S-T., Yeh, R.H., Ho, W-T. (1998) Imperfect maintenance policies for deteriorating production systems. International Journal of Production Economics **55**(2) 191-201
31. Wang, C.H., Sheu, S.H. (2003) Determining the optimal production-maintenance policy with inspection errors: using a Markov chain. Computers & Operations Research **30** 1-17
32. Zhang, G., Berardi, V. (1997) Economic statistical design of $\bar{x}$ - control charts for systems with Weibull in-control times. Computers and Industrial Engineering **32**(3) 575-686

The Robustness and Performance of CUSUM Control Charts Based on the Double-Exponential and Normal Distributions

Zachary G. Stoumbos[1] and Marion R. Reynolds, Jr.[2]

[1] Rutgers, The State University of New Jersey, Piscataway, NJ 08854-8054, USA
[2] Virginia Polytechnic Institute & State University, Blacksburg, VA 24061-0439, USA

Abstract. We investigate the effects of non-normality on the statistical performance of cumulative sum (CUSUM) control charts for monitoring the process mean μ. Our objective is to maintain tight control of the process so that detecting both small and large shifts in μ is important. We consider the standard CUSUM chart, which is based on the normal distribution, and a second CUSUM chart, which is based on the heavy-tailed double-exponential distribution. We also investigate the issue of whether it is preferable from the perspective of statistical performance to base these control charts on samples of $n > 1$ observations or on individual observations ($n = 1$). We assume that the sampling rate in terms of the number of observations per unit time is fixed, so using $n = 1$ means that samples can be taken more frequently than when $n > 1$. We show that using $n > 1$ in either chart does not give a significant improvement in in-control robustness, compared to using $n = 1$. We also show that, unlike the double-exponential CUSUM chart, for $n = 1$, the standard CUSUM chart can be designed to be highly robust to non-normality and very effective at detecting shifts of all sizes, even for highly skewed and extremely heavy-tailed process distributions.

1 Introduction

Control charts are graphical tools widely used to monitor processes to detect special causes that produce changes in the process output. Their original applications were primarily in manufacturing, but now their practical applications extend far beyond manufacturing into engineering, environmental science, genetics, epidemiology, medicine, finance, and even law enforcement and athletics (see Gibbons (1994), Ryan (2000), Montgomery (2001), and Stoumbos and Reynolds (2003)). The most commonly used types of control charts are the Shewhart charts, proposed by Shewhart (1931), the cumulative sum (CUSUM) charts, initially investigated by Page (1954), and the exponentially weighted moving average (EWMA) charts, originating in the work of Roberts (1959). Lai (1995, 2001), Stoumbos et al. (2000), and Stoumbos, Reynolds, and Woodall

(2003) give reviews of developments in the theory and application of control charting methods.

In statistical process control (SPC) applications in which a continuous process variable, say X, is being measured, it is usually assumed that X is normally distributed. The normality assumption is used for designing a control chart and for evaluating its statistical performance. The statistics plotted on control charts are often based on samples (subgroups) taken from the process, and these samples are usually of a fixed size, say n, with a fixed sampling interval, say d, between successive sampling points. For example, samples of $n = 4$ or 5 observations might be taken using a convenient sampling interval of, say, $d = 1$ hour. The Shewhart $\overline{X}$ chart, on which sample means are plotted, is the traditional control chart used to monitor the process mean μ. In some applications, only one observation is available at each sampling point ($n = 1$), in which case the $\overline{X}$ chart reduces to the individuals X chart.

The Shewhart $\overline{X}$ chart is simple to understand and used very extensively for the on-line monitoring of μ, but has the disadvantage that it is not effective for the quick detection of small changes in μ. This disadvantage arises because the sample mean plotted at each sampling point is a function of data taken only at that sampling point. The ability to detect small changes in μ can be significantly improved by using a control chart on which a statistic is plotted that incorporates information across past samples, in addition to the information in the current sample. In particular, on the EWMA chart for monitoring μ, a weighted average of current and past sample means is plotted, and in the CUSUM chart for monitoring μ, information is accumulated over current and past sample means. Runs rules, which are based on patterns of points in an $\overline{X}$ chart, can help improve somewhat the $\overline{X}$ chart's ability to detect small changes in μ (see Champ and Woodall (1987)), but use of these rules is not the best method for detecting small changes in μ. The EWMA and CUSUM charts are much more effective for detecting small and moderate changes in μ (see Stoumbos et al. (2000), Stoumbos, Reynolds, and Woodall (2003), and references therein).

Control charts are usually designed under the assumption that the process variable X is normally distributed. However, in many applications, the actual distribution of X may not be close to normal, and this may have a significant effect on the properties of the control chart. Stoumbos and Reynolds (2000) and Reynolds and Stoumbos (2003) studied the robustness of various Shewhart and EWMA control charts for monitoring μ and/or σ when the process data are not normally distributed. They showed that the performance of the Shewhart charts is very sensitive to the assumption of normality and, for practical implementation, recommended a pair of EWMA charts with excellent performance that is robust to the normality assumption.

One major objective of this paper is to investigate the effects of non-normality on the statistical performance of CUSUM control charts for monitoring μ. We consider two types of CUSUM charts: The standard CUSUM chart, which is based on the normal distribution, and a second CUSUM chart, which is based on the double-exponential distribution; i.e., a heavy-tailed distribution. The latter CUSUM will be referred to as the double-exponential CUSUM chart. For the numerical results presented in this paper, heavy-tailed symmetric as well as skewed non-normal distributions are considered. Furthermore, in this investigation, we assume that the objective of monitoring is to maintain tight control of the process so that detecting both small and large process changes is important. Thus, Shewhart control charts are not considered. For in-depth robustness studies of Shewhart charts, the reader is referred to Stoumbos and Reynolds (2000), Stoumbos and Sullivan (2002), and Reynolds and Stoumbos (2003).

Control chart robustness and detection performance are closely related to the sample size n and sampling interval d being used. Although there are traditional choices for n and d based on heuristics, a frequent question asked by practitioners is, "How often should we sample from our process and what sample size should we use?" There is, of course, no simple universal answer to this question because there are numerous process factors that affect the answer. In many cases, there will be a choice between taking process observations in groups of $n > 1$ and taking observations individually ($n = 1$). For example, if it is desirable to sample at the rate of one observation per hour, then one option is to take one observation every hour, while another option could be to take 4 observations every 4 hours. In addition to statistical considerations, other considerations may suggest that one option is preferred over the other. For example, in some cases it may be more administratively convenient to take observations in groups, and in other cases, it may be preferable to spread out the sampling effort as evenly as possible.

A second major objective of this paper is to investigate the issue of whether it is preferable from a statistical viewpoint to base a standard or double-exponential CUSUM chart on samples of $n > 1$ observations or to base the CUSUM charts on individual observations. It is assumed that the sampling rate in terms of the number of observations per unit time must be held constant. This means that the ratio n / d is constant, so when $n = 1$, points will be plotted more frequently than when $n > 1$. The basic issue that is considered here is the efficient allocation of fixed sampling resources over time.

Our findings lead to several conclusions that have important implications for the practical implementation of CUSUM charts in SPC applications. In particular, for monitoring μ:

- The standard CUSUM chart greatly outperforms the double-exponential CUSUM chart, when the process data are normally distributed.

- The standard CUSUM chart can be designed to be considerably more robust than the double-exponential CUSUM chart to the assumption of normality, when the process data actually follow a heavy-tailed and, especially, a skewed distribution.
- Using $n > 1$ does not improve the shift-detection ability of the standard CUSUM chart.
- Using $n > 1$ does not significantly improve the robustness of the standard CUSUM chart to the assumed normal distribution of the process data.

Some of these conclusions are initially rather surprising and seem to contradict some of the conventional wisdom about control charts.

There are few papers in the SPC literature dealing with the choice of n and d, and these papers are limited in scope. Hawkins (1992) considered the choice of n in a standard CUSUM chart used to detect shifts in μ, and recommended using $n = 1$. For the same problem, Prabhu, Runger, and Montgomery (1997) stated that a value of $n > 1$ is often the best choice. Hawkins and Olwell (1998) and Reynolds and Stoumbos (2000) considered the choice of n when a CUSUM chart is being used to monitor a process proportion p. Stoumbos and Reynolds (1996) investigated the choice of n and d in a standard CUSUM chart for detecting shifts in μ.

The remainder of this paper is organized as follows. First, the process model is described and definitions are provided for the standard and double-exponential CUSUM charts to be considered. Then, various statistical measures of control chart performance and methods for their evaluation are discussed. Next, robustness and detection performance comparisons are conducted among the standard and double-exponential CUSUM charts, for different values of n and d and for normal as well various non-normal process data. Finally, the conclusions from the paper are summarized, and a discussion of some extensions is presented.

2 The Process Model and CUSUM Control Charts

2.1 The Process Model

Suppose that the process variable X being measured has a continuous distribution with a mean μ and a stable standard deviation σ. Assume that if a special cause occurs, then any resulting change in the process is reflected by a change in μ from its in-control value μ_0. The ability of a control chart to detect a special cause is usually evaluated under the assumption that the effect of the special cause is to produce a sustained shift in a process parameter. This sustained shift (or simply, shift) is assumed to persist until the chart detects it and action is taken to eliminate the special cause. In this paper, we will focus on the detection of sustained shifts in μ from μ_0. The detections of transient shifts or drifts in μ were recently investigated by Reynolds and Stoumbos (2001a, 2003) and Stoumbos, Reynolds, and Woodall (2003) in the context of the problem of monitoring μ either individually or together with σ.

In practice, σ (and sometimes μ_0) would need to be estimated from process data during a preliminary phase in which data are collected for this purpose. The resulting estimates of the process parameters will be meaningful only if the process is actually in control during the time period when these data are collected, so it is also necessary to use these data to try to determine whether the process is in control. Here, we assume that this initial phase has been completed and the in-control parameter values have been estimated accurately enough that they can be treated as known quantities for future monitoring. We also assume here that the process will be monitored over a relatively long period of time, and that it is desirable to maintain tight control of the process, so that it is important to detect small as well as large changes in μ.

When samples of $n \geq 1$ observations are taken using a sampling interval d, we define the unit of time to be d/n. Then, without loss of generality, the number of observations per unit time will always be $n/d = 1.0$. For example, if the sample size is $n = 4$ observations and the sampling interval is $d = 4$ hours, then the time unit is taken to be one hour, so that the sampling rate per unit time is $n/d = 4/4 = 1.0$ observation per hour. Let $\mathbf{X}_k = (X_{k1}, X_{k2}, \ldots, X_{kn})$ denote the sample taken at the kth sampling point, and $\bar{X}_k$ denote the sample mean of $\mathbf{X}_k$. We assume that all observations within and between samples are independent. We comment on the situation in which observations are not independent in the last section of this paper.

For most of the numerical results presented in this paper, we assume that the time between the individual observations in a sample is negligible relative to d, so this time is taken to be 0. The numerical results given here are also based on the assumption that the time required to test or measure an item is negligible relative to d, so this time is also taken to be 0. These are standard assumptions used in the SPC literature for the evaluation and comparison of the performance of control charts.

2.2 The Standard CUSUM Control Chart

We first consider the standard CUSUM chart for monitoring μ, which is based on the normal distribution. This chart is constructed using two separate one-sided CUSUM statistics. The upper one-sided CUSUM statistic for detecting increases in μ can be expressed as

$$C_k^{STD+} = \max\{0, C_{k-1}^{STD+}\} + (\bar{X}_k - \mu_0 - \gamma\sigma), \quad k = 1, 2, \ldots,$$

where $\gamma\sigma$ is referred to as the *reference value* (γ is explained below), and the starting value is usually taken to be $C_0^{STD+} = 0$. The standard CUSUM chart was originally developed by Page (1954), and many papers on CUSUM charts have since been published. The recent monograph by Hawkins and Olwell (1998) provides an extensive list of references to the literature on CUSUM charts (see also Lai (1995, 2001), Stoumbos et al. (2000), and the references therein).

The reference value in the standard upper CUSUM chart for μ is usually chosen so that $\gamma = |\mu_1 - \mu_0|/(2\sigma)$, where μ_1 is a value of μ that should be detected quickly. This expression for the reference value comes from the representation of the CUSUM chart as a sequence of sequential probability ratio tests for testing H_0: $\mu = \mu_0$ versus H_1: $\mu = \mu_1$ (see Page (1954)). In practice, γ can be used as a tuning parameter: Small values of γ make the chart sensitive to small shifts in μ, and large values of γ make the chart sensitive to large shifts in μ. A signal is generated at sampling point k if C_k^{STD+} exceeds an upper control limit constructed at

$$h_{STD}\ \sigma/\sqrt{n},$$

where the chart parameter h_{STD} is usually determined to attain specified in-control properties.

The lower one-sided CUSUM statistic for detecting decreases in μ can be expressed as

$$C_k^{STD-} = \min\{0, C_{k-1}^{STD-}\} + \left(\bar{X}_k - \mu_0 + \gamma\sigma\right), \quad k = 1, 2, \ldots,$$

where $\gamma = |\mu_1 - \mu_0|/(2\sigma)$, and the starting value is usually taken to be $C_0^{STD-} = 0$. A signal is generated at sampling point k if C_k^{STD-} falls below a lower control limit constructed at

$$-h_{STD}\ \sigma/\sqrt{n}.$$

The standard two-sided CUSUM chart that is based on using concurrently the control statistics C_k^{STD+} and C_k^{STD-} will be referred to as the CUSUM$_{STD}$ chart. Note that, in general, it is not necessary that this two-sided CUSUM$_{STD}$ chart be symmetric with the same value of γ and the same value of h_{STD} in the two individual one-sided CUSUM charts, but we focus on the symmetric case here. The CUSUM$_{STD}$ chart reduces to the Shewhart $\bar{X}$ chart, when $h_{STD} = 0$ and $\gamma = h_{STD} / \sqrt{n}$. For $n = 1$, the CUSUM$_{STD}$ chart reduces to the standard individuals CUSUM chart.

2.3 The Double-Exponential CUSUM Control Chart

A second CUSUM chart for monitoring the process mean μ can be developed using the heavy-tailed, symmetric double-exponential distribution. The probability density function of the double-exponential distribution can be written as

$$f_{DE}(x) = \frac{1}{\sqrt{2}\sigma} e^{-\sqrt{2}|x-\mu|/\sigma}, \quad -\infty < x < \infty,$$

in terms of its mean μ and standard deviation σ. In this paper, the double-exponential distribution with mean μ and variance σ^2 will be denoted by $DE(\mu, \sigma^2)$.

Like the standard CUSUM chart, the double-exponential CUSUM chart is constructed using two separate one-sided CUSUM statistics. By expressing the double-exponential CUSUM chart as a sequence of sequential probability ratio tests for testing H_0: $\mu = \mu_0$ versus H_1: $\mu = \mu_1$, the upper one-sided CUSUM statistic for detecting increases in μ can be written as

$$C_k^{DE+} = \max\{0, C_{k-1}^{DE+}\} + \sum_{i=1}^{n} \left(|X_{ki} - \mu_0| - |X_{ki} - \mu_0 - 2\gamma\sigma| \right)/n, \quad k = 1, 2, \ldots,$$

where $\gamma = |\mu_1 - \mu_0|/(2\sigma)$, and the starting value would usually be taken to be $C_0^{DE+} = 0$. As in the standard CUSUM chart, in the double-exponential CUSUM chart, μ_1 would usually be chosen to be a value of μ that should be detected quickly, and in practice, γ could be used as a tuning parameter that adjusts the chart to be more sensitive to small or to large shifts. A signal is generated at sampling point k if C_k^{DE+} exceeds an upper control limit constructed at

$$h_{DE}\ \sigma/\sqrt{n},$$

where the chart parameter h_{DE} is usually determined to attain specified in-control properties.

The lower one-sided CUSUM statistic for detecting decreases in μ can be expressed as

$$C_k^{DE-} = \min\{0, C_{k-1}^{DE-}\} + \sum_{i=1}^{n} \left(|X_{ki} - \mu_0| - |X_{ki} - \mu_0 + 2\gamma\sigma| \right)/n, \quad k = 1, 2, \ldots,$$

where the starting value would usually be taken to be $C_0^{DE-} = 0$. A signal is generated at sampling point k if C_k^{DE-} falls below a lower control limit constructed at

$-h_{DE}\sigma/\sqrt{n}$.

The two-sided double-exponential CUSUM chart that is based on using concurrently the control statistics C_k^{DE+} and C_k^{DE-} will be referred to as the CUSUM$_{DE}$ chart. Note that, in general, it is not necessary that this two-sided CUSUM$_{DE}$ chart be symmetric with the same value of γ and the same value of h_{DE} in the two individual one-sided CUSUM charts, but we consider the symmetric case here. For $n = 1$, the CUSUM$_{DE}$ chart reduces to the individuals double-exponential CUSUM chart.

3 Measures of Performance

3.1 The ATS, ANOS, and ANSS

When there is a sustained shift in a process parameter, an obvious measure of the effectiveness of a control chart in detecting this shift is the expected time required for detection. A fair comparison among different control charts can be made when all of the charts have the same rate of false alarms and the same sampling rate when the process is in statistical control. Control charts with different sample sizes and different sampling intervals will be compared here. Thus, it is necessary to separately consider the time, the number of observations, and the number of samples required by a control chart to produce a signal.

We define the *average time to signal* (ATS) to be the expected length of time from the start of process monitoring until a signal is generated. Similarly, we define the *average number of observations to signal* (ANOS) and the *average number of samples to signal* (ANSS) to be, respectively, the expected number of observations and the expected number of samples from the start of process monitoring until a signal is given. For control charts that can be modeled as a renewal process, the in-control value of 1/ATS is the long-run-average *false-alarm rate* per *unit time*, because a signal is a false alarm when the process is in control (see Stoumbos and Reynolds (1996, 1997, 2001)). Similarly, the in-control value of 1/ANSS is the long-run-average *false-alarm rate* per *sample*. The *average sampling rate*, in terms of the number of observations per unit time, is the ratio ANOS/ATS.

For example, if a standard Shewhart $\bar{X}$ chart with the traditional "three-sigma" control limits is based on samples of $n = 4$ observations taken every $d = 4$ hours, then the in-control ANSS is 370.4 samples, the in-control ANOS is (4)(370.4) = 1481.6 observations, and the in-control ATS is 1481.6 hours. In this case, the false-alarm rate is 1/1481.6 = 0.000675 false alarms per hour, or

1/370.4 = 0.00270 false alarms per sample. The in-control average sampling rate is ANOS/ATS = 1481.6/1481.6 = 1.0 observation per hour.

For the control chart comparisons presented in this paper, the in-control average sampling rate is set at 1.0 observation per unit time, and the in-control ATS is set very close to 1481.6 time units, corresponding to a false-alarm rate of 0.000675 false alarms per unit time. That is, the in-control ANSS is very close to 370.4 samples and the in-control ANOS is very close to 1481.6 observations.

3.2 The Steady-State ATS

The ATS provides a measure of the time required to detect a parameter shift when the shift is present at the time that monitoring starts. However, a parameter shift may occur at some random time in the future, after process monitoring has started. In this case, the appropriate measure of detection time is the expected length of time from the random point in time that the shift occurs until the time that the control chart generates a signal.

For CUSUM and EWMA charts, the expected length of time from the parameter shift to the signal will depend on the value of the control statistic at the time that the shift occurs. When the parameter shift occurs a relatively long period of time after monitoring has started, the effect of the value of the control statistic at the time of the shift can be modeled by using the *steady-state ATS* (SSATS). The SSATS is computed assuming that the control statistic has reached its steady-state or stationary distribution by the random time point that the process shift occurs.

The SSATS also allows for the possibility that the process shift can occur in a time interval between samples. In particular, it is assumed that when a parameter shift occurs within a particular sampling interval, the position of the shift within this interval is uniformly distributed over the interval. Using the SSATS as a measure of detection time is especially important when comparing control charts with different sample sizes and sampling intervals. For example, suppose that a special cause produces a very large shift in μ and this shift occurs between sampling times. If one control chart takes one observation every hour ($n = 1$ and $d = 1$), then the maximum time from this shift to the next sampling point is only one hour. However, for a chart taking samples of $n = 4$ every $d = 4$ hours, the maximum time from this shift to the next sampling point is 4 hours. Thus, the latter control chart with the long sampling interval would be at a disadvantage for detecting this shift. The SSATS accounts for the possibility of a parameter shift occurring between samples and accounts for the fact that different control charts being compared have different values of d. Additional discussion on the SSATS can be found in Reynolds (1995), Stoumbos and Reynolds (1996, 2001), and Stoumbos, Mittenthal, and Runger (2001).

4 Methods for Evaluating the Measures of Performance

For a single Shewhart control chart such as the $\overline{X}$ chart, it is typically possible to express performance measures such as the ATS and SSATS simply in terms of probabilities involving the distribution of the control statistic. For a more complex control chart, such as a CUSUM chart or an EWMA chart, the integral equation method or the Markov chain method can frequently be used. These are well known methods in the SPC literature that are based on modeling the control statistic as a Markov process. The Markov chain method offers flexibility for evaluating certain performance measures that are very difficult or even impossible to compute with the integral equation method. When applicable, however, the integral equation method is usually preferred to the Markov chain method because it provides higher accuracy for the same computational effort (see, for example, Reynolds (1995), Stoumbos and Reynolds (1996, 2001), and Stoumbos, Mittenthal, and Runger (2001)).

When a pair of control statistics, such as the statistics C_k^{STD+} and C_k^{STD-} used in the CUSUM$_{STD}$ chart or the statistics C_k^{DE+} and C_k^{DE-} used in the CUSUM$_{DE}$ chart, is applied together, it can be very difficult to use the integral equation or even the Markov chain method. Thus, we used simulation for the CUSUM$_{STD}$ and CUSUM$_{DE}$ charts that are applied with non-normal process data. In general, for all CUSUM charts considered here, either the Markov chain method or simulation (using 100,000 runs) was appropriately used to evaluate their ANOS, ANSS, ATS, and SSATS, as to ensure the high accuracy of the numerical results. Since the standard deviation of the time to signal is approximately equal to the ATS (see Reynolds, Amin, and Arnold (1990), Reynolds (1995), and Stoumbos and Reynolds (1996)), the standard deviation of the simulated results is approximately equal to $\text{ATS}/\sqrt{100{,}000}$. The same result is also true for the SSATS, unless the parameter change is very large and the SSATS is very small. Hence, the simulation results presented in this paper are very accurate for meaningful comparisons.

The evaluation of the SSATS depends on the assumption made regarding the behavior of the control statistics before the process goes out of control. In particular, the distribution of each control statistic over the states of the Markov process at the sample immediately before the parameter shift must be known. For the Markov chain method, we assumed that the distribution of each control statistic immediately before the shift in μ is the stationary distribution conditional on no false alarms (see, for example, Stoumbos and Reynolds (2000, 2001) and Reynolds and Stoumbos (2001a, b)). For the simulations, 100 initial in-control observations were generated before the parameter shift was introduced. If a false alarm occurred in the initial observations, this sequence of observations was discarded and a new one was generated. After a sequence of 100 in-control observations was obtained, a shift point (the point of the beginning of the shift in μ) was chosen randomly from within the next sampling interval.

5 Robustness and Performance Comparisons

5.1 The Parameter Values for the CUSUM$_{STD}$ and CUSUM$_{DE}$ Charts

In this section, we evaluate the robustness and compare the statistical performance of the standard CUSUM$_{STD}$ chart and the double-exponential CUSUM$_{DE}$ chart for the problem of monitoring the process mean μ. Our objective is to recommend a robust CUSUM chart design that is very effective at detecting small as well as large sustained shifts in μ, for implementation in practical applications where the observations sampled from the process may or may not be normally distributed.

For simplicity in explaining the comparisons made here, it will be convenient to take the sampling interval to be one hour and measure time in hours. It will also be convenient to standardize μ and express a shift in μ in terms of $(\mu-\mu_0)/\sigma$. That is, all charts considered here have been constructed to have an in-control average sampling rate of 1.0 observation per hour and an in-control ATS very close to 1481.6 hours, as discussed in Section 3.1. When all control charts are matched to have the same in-control average sampling rate and false-alarm rate, their performance can be evaluated by computing the SSATS of the charts for various shift sizes, to determine which chart is the most efficient at detecting shifts in μ. The chart with the smallest SSATS for a given shift in μ is the most efficient chart for that shift. The ATS and SSATS values presented here also apply to other cases, if the unit of time is taken to be d/n.

In evaluating the CUSUM$_{STD}$ and CUSUM$_{DE}$ charts, we consider all combinations of the reference values γ= 0.15, 0.25, and 0.50 with the sample sizes $n = 1$ and 4. When the CUSUM$_{STD}$ chart is applied with normal process data, the reference values γ= 0.15, 0.25, and 0.50 respectively offer the best performance for detecting shifts of sizes $|(\mu-\mu_0)/\sigma| \approx 0.30$, 0.50, and 1.00, but also give good performance for a wide range of other shift sizes. This is also the case with the reference values of the CUSUM$_{DE}$ chart, although for double-exponential process data. Assuming that there is a negligible cost difference between taking samples of size $n = 1$ every hour and samples of size $n > 1$ every $d = n$ hours, we intend to show that there is no advantage to using $n > 1$. Thus, it seems to be sufficient to consider only $n = 4$ for comparison with $n = 1$. The values of the parameters h_{STD} and h_{DE} respectively used to construct the control limits of the CUSUM$_{STD}$ and CUSUM$_{DE}$ charts are provided throughout Tables 1-4.

Table 1. In-Control ATS Values with Chart Parameter Values for Various $CUSUM_{STD}$ and $CUSUM_{DE}$ Control Charts and for Various Distributions of Process Data

Control Chart	Sample Size	Distribution of Process Data							Control Chart Parameters
		$N(\mu_0,\sigma^2)$	$DE(\mu_0,\sigma^2)$	$t(10)$	$t(4)$	$Gam(4,1)$	$Gam(2,1)$	$Gam(1,1)$	
$CUSUM_{STD}$ $\gamma = 0.15$	$n = 1$	1481.6	1469.7	1476.9	1393.9	1448.2	1421.4	1353.7	$h_{STD} = 15.290$
	$n = 4$	1481.6	1428.8	1458.4	1366.5	1445.9	1421.5	1353.7	$h_{STD} = 7.065$
$CUSUM_{DE}$ $\gamma = 0.15$	$n = 1$	1481.6	10031.1	1964.7	4344.4	296.6	172.3	101.4	$h_{DE} = 5.013$
	$n = 4$	1481.6	10758.6	1956.6	4343.2	296.8	172.4	101.4	$h_{DE} = 2.309$
$CUSUM_{STD}$ $\gamma = 0.25$	$n = 1$	1481.6	1288.4	1406.2	1095.5	1294.5	1162.5	971.3	$h_{STD} = 10.703$
	$n = 4$	1481.6	1255.6	1393.6	1090.1	1328.5	1211.8	1029.8	$h_{STD} = 4.775$
$CUSUM_{DE}$ $\gamma = 0.25$	$n = 1$	1481.6	14370.4	2164.3	6443.6	414.9	229.5	120.1	$h_{DE} = 5.697$
	$n = 4$	1481.6	13754.2	2160.9	6251.3	415.4	229.2	119.9	$h_{DE} = 2.536$
$CUSUM_{STD}$ $\gamma = 0.50$	$n = 1$	1481.6	715.9	1081.3	527.6	764.2	543.6	375.5	$h_{STD} = 6.148$
	$n = 4$	1481.6	791.3	1134.4	613.0	921.0	699.5	510.7	$h_{STD} = 2.517$
$CUSUM_{DE}$ $\gamma = 0.50$	$n = 1$	1481.7	10249.8	2418.2	9108.5	1226.3	1205.8	2804.4	$h_{DE} = 5.940$
	$n = 4$	1481.4	9517.8	2351.4	8280.0	1186.3	1131.7	2347.4	$h_{DE} = 2.426$

Table 2. In-Control ATS and Out-of-Control SSATS Values with Chart Parameter Values for Matched $CUSUM_{STD}$ and $CUSUM_{DE}$ Control Charts and for $N(\mu,\sigma^2)$ Process Data

$N(\mu,\sigma^2)$ Process Data	$CUSUM_{STD}$ $\gamma = 0.15$		$CUSUM_{DE}$ $\gamma = 0.15$		$CUSUM_{STD}$ $\gamma = 0.25$		$CUSUM_{DE}$ $\gamma = 0.25$		$CUSUM_{STD}$ $\gamma = 0.50$		$CUSUM_{DE}$ $\gamma = 0.50$	
$\lvert(\mu-\mu_0)/\sigma\rvert$	$n=1$	$n=4$	$n=1$	$n=4$	$n=1$	$n=4$	$n=1$	$n=4$	$n=1$	$n=4$	$n=1$	$n=4$
0.00	1481.6	1481.6	1481.6	1481.6	1481.6	1481.6	1481.6	1481.6	1481.6	1481.6	1481.7	1481.4
0.25	104.9	104.3	136.8	136.3	133.4	133.6	163.8	164.3	267.3	272.6	286.3	292.0
0.50	36.7	35.9	49.8	48.9	35.5	34.7	46.1	45.6	51.2	51.1	58.8	59.1
1.00	15.5	14.8	23.0	22.1	12.8	12.2	18.1	17.4	11.4	10.6	14.1	13.4
1.50	9.7	9.2	16.8	15.9	7.7	7.2	12.5	11.8	5.9	5.2	8.1	7.4
2.00	7.0	6.6	14.7	13.7	5.5	5.2	10.7	10.0	3.9	3.2	6.2	5.9
3.00	4.6	4.6	13.7	12.5	3.4	2.6	9.8	9.1	2.3	2.0	5.2	5.5
5.00	2.7	2.0	13.7	12.4	2.0	2.0	9.8	9.1	1.3	2.0	5.1	5.5
8.00	1.6	2.0	13.7	12.4	1.3	2.0	9.8	9.1	0.6	2.0	5.1	5.5
10.00	1.4	2.0	13.7	12.4	1.0	2.0	9.8	9.1	0.5	2.0	5.1	5.5
12.00	1.2	2.0	13.7	12.4	0.6	2.0	9.8	9.1	0.5	2.0	5.1	5.5
h_{STD}	15.290	7.065	—	—	10.703	4.775	—	—	6.148	2.517	—	—
h_{DE}	—	—	5.013	2.309	—	—	5.697	2.536	—	—	5.940	2.426

Table 3. In-Control ATS and Out-of-Control SSATS Values with Chart Parameter Values for Matched $CUSUM_{STD}$ and $CUSUM_{DE}$ Control Charts and for $DE(\mu, \sigma^2)$ Process Data

$DE(\mu,\sigma^2)$ Process Data	$CUSUM_{STD}$ $\gamma = 0.15$		$CUSUM_{DE}$ $\gamma = 0.15$		$CUSUM_{STD}$ $\gamma = 0.25$		$CUSUM_{DE}$ $\gamma = 0.25$		$CUSUM_{STD}$ $\gamma = 0.50$		$CUSUM_{DE}$ $\gamma = 0.50$	
$\lvert(\mu-\mu_0)/\sigma\rvert$	$n=1$	$n=4$	$n=1$	$n=4$	$n=1$	$n=4$	$n=1$	$n=4$	$n=1$	$n=4$	$n=1$	$n=4$
0.00	1481.6	1481.6	1481.7	1481.6	1481.3	1481.7	1481.8	1481.3	1481.6	1481.1	1481.7	1481.7
0.25	107.0	106.3	68.4	67.9	146.1	144.2	97.6	97.5	408.2	385.9	289.9	306.0
0.50	37.2	36.5	26.3	25.7	37.2	36.4	24.2	23.5	68.8	64.9	43.6	44.0
1.00	15.6	15.0	14.9	14.1	13.2	12.6	11.3	10.4	13.2	12.1	8.9	8.2
1.50	9.8	9.4	12.2	11.3	7.9	7.4	8.9	7.8	6.8	6.0	5.7	4.6
2.00	7.1	6.7	11.3	10.2	5.7	5.4	8.1	6.8	4.5	3.7	4.8	2.3
3.00	4.6	4.7	10.6	9.4	3.6	2.8	7.6	6.1	2.7	2.1	4.3	2.3
5.00	2.7	2.0	10.5	9.2	2.1	2.0	7.4	5.8	1.5	2.0	4.2	2.0
8.00	1.6	2.0	10.5	9.2	1.3	2.0	7.4	5.8	0.7	2.0	4.2	2.0
10.00	1.4	2.0	10.5	9.2	1.1	2.0	7.4	5.8	0.5	2.0	4.2	2.0
12.00	1.2	2.0	10.5	9.2	0.6	2.0	7.4	5.8	0.5	2.0	4.2	2.0
h_{STD}	15.318	7.128	—	—	10.991	4.947	—	—	7.021	2.899	—	—
h_{DE}	—	—	3.603	1.612	—	—	4.095	1.757	—	—	4.572	1.822

Table 4. In-Control ATS and Out-of-Control SSATS Values with Chart Parameter Values for Matched $CUSUM_{STD}$ and $CUSUM_{DE}$ Control Charts and for $t(4)$ Process Data

$t(4)$ Process Data	$CUSUM_{STD}$ $\gamma = 0.15$		$CUSUM_{DE}$ $\gamma = 0.15$		$CUSUM_{STD}$ $\gamma = 0.25$		$CUSUM_{DE}$ $\gamma = 0.25$		$CUSUM_{STD}$ $\gamma = 0.50$		$CUSUM_{DE}$ $\gamma = 0.50$	
$\lvert(\mu-\mu_0)/\sigma\rvert$	$n=1$	$n=4$	$n=1$	$n=4$	$n=1$	$n=4$	$n=1$	$n=4$	$n=1$	$n=4$	$n=1$	$n=4$
0.00	1481.7	1481.5	1481.7	1481.5	1481.8	1481.4	1481.4	1481.5	1481.7	1481.6	1481.6	1481.5
0.25	109.6	108.4	89.9	89.2	160.0	157.3	112.4	112.4	595.9	540.1	252.2	266.1
0.50	38.0	37.2	32.6	31.9	39.1	38.2	29.5	28.8	87.2	80.7	42.3	43.0
1.00	15.9	15.2	16.4	15.8	13.9	13.2	12.4	11.4	15.2	13.6	9.4	8.6
1.50	10.0	9.5	13.2	12.8	8.3	7.8	9.6	8.1	7.7	6.8	5.7	4.6
2.00	7.2	6.8	12.2	11.9	5.9	5.6	8.7	6.9	5.2	4.4	4.8	3.2
3.00	4.7	4.8	11.6	11.5	3.8	3.0	8.3	6.1	3.1	2.1	4.3	2.3
5.00	2.7	2.0	11.4	11.4	2.2	2.0	8.2	5.9	1.6	2.0	4.2	2.0
8.00	1.6	2.0	11.4	11.4	1.4	2.0	8.1	5.8	1.1	2.0	4.2	2.0
10.00	1.4	2.0	11.4	11.4	1.2	2.0	8.1	5.8	0.5	2.0	4.2	2.0
12.00	1.2	2.0	11.4	11.4	0.7	2.0	8.1	5.8	0.5	2.0	4.2	2.0
h_{STD}	15.500	7.204	—	—	11.432	5.151	—	—	7.963	3.281	—	—
h_{DE}	—	—	4.107	1.858	—	—	4.519	1.964	—	—	4.634	1.844

5.2 The Robustness of the CUSUM$_{STD}$ and CUSUM$_{DE}$ Charts

To investigate the robustness of the CUSUM$_{STD}$ and CUSUM$_{DE}$ control charts to the normality assumption, both heavy-tailed symmetric and skewed distributions will be used. In particular, the double-exponential and *t* distributions will be used for the heavy-tailed symmetric case, and the gamma and exponential distributions will be used for the skewed case.

The double-exponential distribution with mean μ and variance σ^2, $DE(\mu,\sigma^2)$, is symmetric about its mean μ, and the *t* distribution is symmetric about its mean 0. Both of these distributions have more probability in the tails than the standard normal distribution. The probability density function of the $DE(\mu,\sigma^2)$ distribution is given in Section 2.3. The probability density function of the *t* distribution with *k* degrees of freedom ($k = 1, 2, \ldots$) is

$$f_t(x) = \frac{\Gamma[(k+1)/2]\cdot[1+(x^2/k)]^{-(k+1)/2}}{\sqrt{k\pi}\,\Gamma(k/2)}, \qquad -\infty < x < \infty,$$

where the gamma function is defined by

$$\Gamma(\alpha) = \int_0^{\infty} x^{\alpha-1} e^{-x} dx, \quad \alpha > 0 .$$

Here, the *t* distribution with *k* degrees of freedom will be denoted by $t(k)$. The variance of this distribution is $k/(k-2)$, for $k = 3, 4, \ldots$, and as *k* increases, $t(k)$ converges in distribution to the standard normal distribution.

The probability density function of the gamma distribution with parameters α and λ is

$$f_\Gamma(x) = \begin{cases} \dfrac{\lambda^\alpha}{\Gamma(\alpha)} x^{\alpha-1} e^{-\lambda x}, & x > 0 \\ 0, & x \le 0. \end{cases}$$

The mean and variance of this gamma distribution, which will be denoted here by $Gam(\alpha, \lambda)$, are α/λ and α/λ^2, respectively. As α increases, the shape of the gamma distribution increasingly resembles that of a normal distribution. For $\alpha = 1$, the $Gam(\alpha,\lambda)$ distribution reduces to the exponential distribution with mean $1/\lambda$ and variance $1/\lambda^2$. In what follows, we will take $\lambda = 1$ without any loss of generality, and the normal distribution with mean μ and variance σ^2 will be denoted by $N(\mu,\sigma^2)$.

Columns 3-9 of Table 1 give in-control ATS values for various CUSUM$_{STD}$ and CUSUM$_{DE}$ control charts and for process observations that follow the

$N(\mu_0,\sigma^2)$, $DE(\mu_0,\sigma^2)$, $t(10)$, $t(4)$, $Gam(4,1)$, $Gam(2,1)$, and $Gam(1,1)$ (exponential with mean 1.00) distributions. To study the robustness of these charts to the normality assumption, the same control limits that were determined under normality to give an in-control average sampling rate of 1.0 observation per hour and an in-control ATS of 1481.6 hours have also been used in evaluating all in-control ATS values for the $DE(\mu_0,\sigma^2)$, $t(10)$, $t(4)$, $Gam(4,1)$, $Gam(2,1)$, and $Gam(1,1)$ distributions. The values of the chart parameters h_{STD} and h_{DE} used in these normality-based control limits are given in Column 10 of Table 1.

Control chart robustness is considered to be closely related to the sample size n being used, so we first consider the issue of using $n = 1$ versus $n = 4$. From Table 1, it can be seen that the in-control ATS performance of the CUSUM$_{STD}$ chart with $n = 1$ and $\gamma = 0.15$ is slightly more robust than the in-control ATS performance of the CUSUM$_{STD}$ chart with $n = 4$ and $\gamma = 0.15$, but the difference is extremely small. The difference is also extremely small when the CUSUM$_{STD}$ chart with $n = 1$ and $\gamma = 0.25$ is compared to the CUSUM$_{STD}$ chart with $n = 4$ and $\gamma = 0.25$, especially for the heavy-tailed distributions. Regardless of the sample size n being considered, the in-control ATS performance of the CUSUM$_{STD}$ charts with $\gamma = 0.15$ and $\gamma = 0.25$ is much more robust than that of the CUSUM$_{STD}$ charts with $\gamma = 0.50$, with the most robust of these control charts being the CUSUM$_{STD}$ chart with $\gamma = 0.15$.

As γ decreases, the in-control ATS values of the CUSUM$_{STD}$ chart corresponding to each non-normal distribution increase towards the normal-theory design value of 1481.6 hours, and for $\gamma = 0.15$, are all well within 9% of 1481.6 hours for both $n = 1$ and $n = 4$, even for the most skewed and heavy-tailed distributions. In most applications, the performance of the CUSUM$_{STD}$ chart with $\gamma = 0.15$ would be regarded as very close to the anticipated in-control ATS value under normality. In fact, in many cases, the in-control ATS values for the CUSUM$_{STD}$ chart with $\gamma = 0.25$ would also be regarded as sufficiently close to the nominal design value of 1481.6 hours, especially compared to the in-control ATS performance of the traditional Shewhart X and $\overline{X}$ charts. Designing the X chart or $\overline{X}$ chart under normality, when the process data actually follow a heavy-tailed or skewed distribution, results in considerably more false alarms than anticipated (see Stoumbos and Reynolds (2000) and Reynolds and Stoumbos (2003)). An excessive number of false alarms can lead to unneeded process adjustment, loss of confidence in the control scheme, and eventually low productivity and increased cost.

Thus, using a smaller value of γ in the CUSUM$_{STD}$ chart gives excellent robustness in in-control ATS performance, but using $n > 1$ in this chart does not give any significant increase in robustness, compared to using $n = 1$. The implication of this is that the robustness of the CUSUM$_{STD}$ chart should come primarily from the reference value γ, not from the sample size n.

From Table 1, it can also be seen that, using $n > 1$ in the $CUSUM_{DE}$ chart does not contribute significantly to its in-control ATS robustness, compared to using $n = 1$. For all values of γ and n, the respective in-control ATS values of the $CUSUM_{DE}$ chart corresponding to the heavy-tailed $DE(\mu_0,\sigma^2)$, $t(10)$, and $t(4)$ distributions are much larger than the nominal value of 1481.6 hours attained under normality. That is, designing the $CUSUM_{DE}$ chart under normality, when the process observations actually follow a heavy-tailed distribution, can result in a much lower false-alarm rate than anticipated, which would hurt the shift-detection ability of this control chart.

On the other hand, for all combinations of $\gamma = 0.15$ and $\gamma = 0.25$ with $n = 1$ and $n = 4$, the in-control ATS values of the $CUSUM_{DE}$ chart corresponding to the skewed $Gam(4,1)$, $Gam(2,1)$, and $Gam(1,1)$ distributions are much smaller than the design value of 1481.6 hours. This means that, in these cases, false alarms will occur much more frequently than anticipated. For example, when the process observations follow the $Gam(2,1)$ distribution, in the long run, with $\gamma = 0.15$ or $\gamma = 0.25$ and $n = 1$ or $n = 4$, there will be anywhere between six and nine times as many false alarms as the in-control ATS of 1481.6 hours would indicate.

Although still smaller than 1481.6 hours, for $\gamma = 0.50$ and either value of n, the in-control ATS values for the $CUSUM_{DE}$ charts corresponding to the $Gam(4,1)$ and $Gam(2,1)$ distributions are much larger than those for the $CUSUM_{DE}$ charts with $\gamma = 0.15$ and $\gamma = 0.25$. The opposite is true about the $CUSUM_{DE}$ chart with $\gamma = 0.50$, when this chart is applied to process data that follow the exponential $Gam(1,1)$ distribution. In this case, the two in-control ATS values for $n = 1$ and $n = 4$ are much larger than the nominal value of 1481.6 hours.

Thus, unlike the $CUSUM_{STD}$ chart, a larger value of $\gamma = 0.50$ is needed for the $CUSUM_{DE}$ chart to attain an in-control ATS performance that is reasonably close to that anticipated under normality, when the process data actually come from a skewed distribution. As discussed above, however, when the process data actually follow a heavy-tailed distribution or a skewed exponential distribution, using $\gamma = 0.50$ in the $CUSUM_{DE}$ chart can result in a much lower false-alarm rate than anticipated, which would compromise the shift-detection performance of this control chart.

The overall conclusion that we obtain from Table 1 is that, unlike the $CUSUM_{DE}$ chart, the $CUSUM_{STD}$ chart can be designed to be highly robust to non-normality in the process data. A reference value γ between 0.10 and 0.30 with a sample size $n = 1$ ensures an in-control ATS performance comparable to that with normal data, even for highly skewed and extremely heavy-tailed process distributions. Rather surprisingly, using $n > 1$ in the $CUSUM_{STD}$ chart or, for that matter, in the $CUSUM_{DE}$ chart does not provide a significant improvement in the in-control ATS robustness of these charts, compared to using $n = 1$.

5.3 Performance Comparisons for Process Shifts

After studying the in-control ATS robustness of the various $CUSUM_{STD}$ and $CUSUM_{DE}$ control charts included in Table 1, we now investigate the out-of-control SSATS performance of these charts. Table 2 presents SSATS values for a wide range of sustained shifts in μ, for $CUSUM_{STD}$ and $CUSUM_{DE}$ charts that are matched to have an in-control average sampling rate of 1.0 observation per hour and an in-control ATS of 1481.6 hours, when the process data are normally distributed. The various values of the chart parameters h_{STD} and h_{DE} used in constructing the control limits of these charts are given in the last two rows of Table 2.

Tables 3 and 4 have the same structure as Table 2, except that the ATS and SSATS values and the values of the chart parameters h_{STD} and h_{DE} presented in Tables 3 and 4 pertain to process data from the $DE(\mu_0,\sigma^2)$ and $t(4)$ distributions, respectively. In the interest of brevity, we have not included results for other t or gamma distributions. We found that our conclusions for other process distributions are similar to those we obtain below from Tables 2- 4.

We first consider the issue of whether the $CUSUM_{STD}$ and $CUSUM_{DE}$ charts should use $n = 1$ or $n = 4$. The SSATS values in Tables 2-4 show that for small shifts in μ, there is almost no difference between using $n = 1$ and $n = 4$, for both types of CUSUM charts and all three distributions of the process observations. This is particularly true for values of γ between 0.10 and 0.30 recommended above for robust in-control performance. For intermediate shifts in μ, $n = 4$ is slightly better than $n = 1$ for both types of control charts and all three distributions of the process observations.

For the $CUSUM_{STD}$ chart and large shifts in μ, $n = 1$ is much better than $n = 4$ for all three process distributions considered in Tables 2-4, respectively. The poorer performance of $n = 4$ for large shifts in μ occurs, of course, because $n = 4$ implies $d = 4$, and therefore the lower bound on the SSATS is $d/2 = 2$ hours, while the lower bound on the SSATS for $n = 1$ and $d = 1$ is only $d/2 = 0.5$ hours.

On the other hand, for the $CUSUM_{DE}$ chart and large shifts in μ, $n = 4$ is better than $n = 1$, especially for the heavy-tailed $DE(\mu_0,\sigma^2)$ and $t(4)$ distributions, but the improvement is very marginal. Note that since the $CUSUM_{DE}$ chart is defined in terms of the heavy-tailed $DE(\mu_0,\sigma^2)$ distribution, it is relatively insensitive to large shifts in μ. A large sample size $n > 1$ and large reference value $\gamma > 0.50$ are required for somewhat improved SSATS performance for large shifts in μ. Such a control chart design, however, would greatly compromise the ability of the $CUSUM_{DE}$ chart to detect small shifts in μ.

From Tables 2-4, we see that, regardless of the sample size n, most SSATS values of the $CUSUM_{STD}$ charts corresponding to the $DE(\mu_0,\sigma^2)$ and $t(4)$

distributions exceed the respective SSATS values corresponding to the normal distribution, particularly for $\gamma = 0.50$ and small shifts in μ. The inflation of the SSATS values, which increases as γ increases, is extremely small for $\gamma = 0.15$ and remains quite small for $\gamma = 0.25$. For example, for $n = 1$, $\gamma = 0.15$, and shift sizes $|(\mu - \mu_0)/\sigma| = 1.00$, 2.00, and 5.00, the SSATS values respectively are 15.5, 7.0, and 2.7 for the normal distribution (see Column 2 of Table 2), as compared to 15.6, 7.1, and 2.7 for the $DE(\mu_0, \sigma^2)$ distribution (see Column 2 of Table 3), and 15.9, 7.2, and 2.7 for the $t(4)$ distribution (see Column 2 of Table 4). For $n = 1$, $\gamma = 0.25$, and shift sizes $|(\mu - \mu_0)/\sigma = 1.00$, 2.00, and 5.00, the SSATS values respectively are 12.8, 5.5, and 2.0 for the normal distribution (see Column 6 of Table 2), as compared to 13.2, 5.7, and 2.1 for the $DE(\mu_0, \sigma^2)$ distribution (see Column 6 of Table 3), and 13.9, 5.9, and 2.2 for the $t(4)$ distribution (see Column 6 of Table 4). For $n = 4$, the non-normal SSATS values are even closer to the corresponding non-normal SSATS values. In general, for values of γ between 0.10 and 0.30, the CUSUM$_{STD}$ chart maintains an SSATS performance very close to that under normality, for heavy-tailed as well as skewed process distributions, for a wide range of shift sizes, and for any sample size $n \geq 1$.

The opposite seems to be true about the CUSUM$_{DE}$ charts in Tables 2-4. From these tables, we see that, regardless of the sample size n, almost all SSATS values of the CUSUM$_{DE}$ charts corresponding to the $DE(\ _0, \sigma^2)$ and $t(4)$ distributions are considerably smaller than the respective SSATS values corresponding to the normal distribution, especially for $\gamma = 0.15$ and $\gamma = 0.25$. Furthermore, the SSATS values corresponding to the $DE(\mu_0, \sigma^2)$ distribution are much more similar to the SSATS values corresponding to the $t(4)$ distribution, as compared to the respective normal SSATS values. These conclusions are attributed to the fact that the CUSUM$_{DE}$ chart is based on the heavy-tailed $DE(\mu_0, \sigma^2)$ distribution, and the $t(4)$ distribution is another heavy-tailed distribution.

Finally, the SSATS values in Table 2 reveal that, as expected, the CUSUM$_{STD}$ chart uniformly, substantially outperforms the CUSUM$_{DE}$ chart in the detection of shifts in μ of all sizes when the process data are normally distributed. Furthermore, the SSATS values in Tables 3 and 4 show that for values of γ between 0.10 and 0.30 and $n = 1$, the CUSUM$_{STD}$ chart can be designed so that it outperforms the CUSUM$_{DE}$ chart, even when the process data come from a heavy-tailed distribution such as the $DE(\mu_0, \sigma^2)$ or $t(4)$ distribution. Spreading out the sampling will not hurt robustness as long as the reference value γ is adjusted properly. The sensitivity of the CUSUM$_{STD}$ chart with $n = 1$ to large shifts in μ permits the use of relatively small values of γ. This way, in addition to reliable robustness, excellent performance for both small and large shifts is achieved, even for highly skewed and extremely heavy-tailed process distributions.

6 Conclusions and Recommendations for Applications

We have investigated the interrelated issues of robustness, detection performance, and choice of sampling pattern for the standard $CUSUM_{STD}$ chart and the double-exponential $CUSUM_{DE}$ chart, in the context of the problem of monitoring the process mean μ. We considered these issues under the assumption that various choices of n and d are feasible, provided that the sampling rate per unit time, n/d, remains fixed.

For practical applications, the $CUSUM_{STD}$ chart offers much better overall performance than the $CUSUM_{DE}$ chart in detecting sustained shifts in μ. Unlike the $CUSUM_{DE}$ chart, the $CUSUM_{STD}$ chart can be designed to be robust to non-normality for both in-control and out-of-control process observations. In general, a reference value γ between 0.10 and 0.30 will give statistical performance comparable to that with normal data, even for extremely heavy-tailed or highly skewed distributions. Moreover, the out-of-control performance of the $CUSUM_{STD}$ chart is much less sensitive than that of the $CUSUM_{DE}$ chart to the choice of n, regardless of the shift size and the type of underlying process distribution. For the $CUSUM_{STD}$ chart, the effect of n is significant only for large shifts in μ, in which case, $n = 1$ is best. Interestingly, using $n > 1$ in either CUSUM chart does not give a notable improvement in in-control robustness, compared to using $n = 1$.

For the best predictable performance, sampling for $CUSUM_{STD}$ chart should be spread out as evenly as possible, with the sample size $n = 1$ being used whenever feasible. For example, if the current practice is to take samples of $n = 4$ every 4 hours and apply a Shewhart $\bar{X}$ chart, then it would be better to switch to samples of $n = 1$ every hour and apply a $CUSUM_{STD}$ chart. If, for some reason, it is not feasible to sample more often than every 2 hours, then the best option in this case is to take samples of $n = 2$ every 2 hours and use the $CUSUM_{STD}$ chart. It appears that a sample size $n > 1$ can be justified only in cases in which it is more economical or more administratively convenient to take observations in groups rather than individually.

The CUSUM charts considered in Section 5 had their control limits constructed so that they all have an in-control average sampling rate of 1.0 observation per time unit (say, per hour) and an in-control ATS of 1481.6 time units. The bottom two rows of Tables 2-4 give the chart parameters for the control limits corresponding to these values. If these average sampling rate and ATS values (and the values of γ and n) are acceptable for a certain practical application, then the control limits obtained from Tables 2-4 can be used for this application. Detailed examples on the design and implementation of the standard $CUSUM_{STD}$ chart and of other Shewhart and CUSUM control charts are presented in Stoumbos and Reynolds (1996), Reynolds and Stoumbos (2003), and Stoumbos, Reynolds, and Woodall (2003) for the problem of monitoring μ either individually or together with σ.

7 Additional Discussion

The results on CUSUM chart performance that have been presented here are based on the assumption of independent process data. In many processes, there may be some level of autocorrelation, and this autocorrelation could affect the performance of the charts used to monitor the process. A number of papers have been published in recent years on control charts for autocorrelated processes (see, for example, Lu and Reynolds (1999) and Stoumbos and Reynolds (2000) for some references to this work). Most of this work assumes a fixed sampling pattern, and the question is which control chart should be used. An important question that has not received much attention in the SPC literature is how to sample from such a process. For example, with current measurement technology, it is often possible to measure some process variables using extremely short sampling intervals. But a very short sampling interval could result in high autocorrelation, so increasing the sampling frequency may not necessarily result in significant increases in information. More research is needed on the issue of the choice of the sampling plan when there is high autocorrelation. This is a complex problem, because the solution depends on the assumptions made about the structure of the autocorrelation in the process, the objectives of process monitoring, and the effects of potential special causes.

Another assumption that has been made here in evaluating the CUSUM charts is that, prior to their implementation to monitor a process, there has been an initial phase of data collection when process parameters were estimated, so μ_0 and σ can be treated as known values. This is a standard assumption in evaluating the performance of control charts, but it is reasonable only if the size of the initial data set used to estimate process parameters is sufficiently large. When this data set is not large enough, the sampling variability of the estimates can notably affect the properties of control charts. Further research is needed on the issue of parameter estimation when the initial data set is small.

Acknowledgements

Zachary G. Stoumbos' work was funded in part by the Law School Admission Council (LSAC) and by a 2001 Rutgers Faculty of Management Research Fellowship. The opinions and conclusions contained in this publication are those of the authors and do not necessarily reflect the position or policy of LSAC.

References

1. Champ, C.W., and Woodall, W.H. (1987) Exact Results for Shewhart Control Charts With Supplementary Runs Rules. Technometrics **29**, 393-399
2. Gibbons, R.D. (1994) Statistical Methods for Groundwater Monitoring. Wiley, New York, NY
3. Hawkins, D.M. (1992) A Fast Accurate Approximation for Average Run Lengths of CUSUM Control Charts. Journal of Quality Technology **24**, 37-43
4. Hawkins, D.M., and Olwell, D.H. (1998) Cumulative Sum Control Charts and Charting for Quality Improvement. Springer-Verlag, New York, NY
5. Lai, T.L. (1995) Sequential Changepoint Detection in Quality Control and Dynamical Systems (with discussion). Journal of the Royal Statistical Society, Ser. B **57**, 613-658
6. Lai, T.L. (2001) Sequential Analysis: Some Classical Problems and New Challenges (with discussion). Statistica Sinica **11**, 303-408
7. Lu, C.W., and Reynolds, M.R., Jr. (1999) EWMA Control Charts for Monitoring the Mean of Autocorrelated Processes. Journal of Quality Technology **31**, 166-188
8. Montgomery, D.C. (2001) Introduction to Statistical Quality Control, 4th Edition. Wiley, New York, NY
9. Page, E.S. (1954) Continuous Inspection Schemes. Biometrika **41**, 100-114
10. Prabhu, S.S., Runger, G.C., and Montgomery, D.C. (1997) Selection of the Subgroup Size and Sampling Interval for a CUSUM Control Chart. IIE Transactions on Quality and Reliability Engineering **29**, 451-457
11. Reynolds, M.R., Jr. (1995) Evaluating Properties of Variable Sampling Interval Control Charts. Sequential Analysis **14**, 59-97
12. Reynolds, M.R., Jr., Amin, R.W., and Arnold, J.C. (1990) CUSUM Charts With Variable Sampling Intervals. Technometrics **32**, 371-384
13. Reynolds, M.R., Jr., and Stoumbos, Z.G. (2000) A General Approach to Modeling CUSUM Charts for a Proportion. IIE Transactions on Quality and Reliability Engineering **32**, 515-535
14. Reynolds, M.R., Jr., and Stoumbos, Z.G. (2001a) Individuals Control Schemes for Monitoring the Mean and Variance of Processes Subject to Drifts. Stochastic Analysis and Applications **19**, 863-892
15. Reynolds, M.R., Jr., and Stoumbos, Z.G. (2001b) Monitoring the Process Mean and Variance Using Individual Observations and Variable Sampling Intervals. Journal of Quality Technology **33**, 181-205
16. Reynolds, M.R., Jr., and Stoumbos, Z.G. (2003) Control Charts and the Efficient Allocation of Sampling Resources. Preprint
17. Roberts, S.W. (1959) Control Charts Based on Geometric Moving Averages. Technometrics **1**, 239-250
18. Ryan, T.P. (2000) Statistical Methods for Quality Improvement, 2nd Edition. Wiley, New York, NY
19. Shewhart, W.A. (1931) Economic Control of Quality of Manufactured Product. Van Nostrand, New York, NY

20. Stoumbos, Z.G., Mittenthal, J., and Runger, G.C. (2001) Steady-State-Optimal Adaptive Control Charts Based on Variable Sampling Intervals. Stochastic Analysis and Applications **19**, 1025-1057
21. Stoumbos, Z.G., and Reynolds, M.R., Jr. (1996) Control Charts Applying a General Sequential Test at Each Sampling Point. Sequential Analysis **15**, 159-183
22. Stoumbos, Z.G., and Reynolds, M.R., Jr. (1997) Control Charts Applying a Sequential Test at Fixed Sampling Intervals. Journal of Quality Technology **29**, 21-40
23. Stoumbos, Z.G., and Reynolds, M.R., Jr. (2000) Robustness to Non-normality and Autocorrelation of Individuals Control Charts. Journal of Statistical Computation and Simulation **66**, 145-187
24. Stoumbos, Z.G., and Reynolds, M.R., Jr. (2001) The SPRT Control Chart for the Process Mean with Samples Starting at Fixed Times. Nonlinear Analysis: Real World Applications **2**, 1-34
25. Stoumbos, Z.G., and Reynolds, M.R., Jr. (2003) Economic Statistical Design of Adaptive Individuals Control Schemes for Monitoring the Process Mean and Variance. To appear in Nonlinear Analysis: Real World Applications
26. Stoumbos, Z.G., Reynolds, M.R., Jr., Ryan, T.P., and Woodall, W.H. (2000) The State of Statistical Process Control as We Proceed into the 21st Century. Journal of the American Statistical Association **95**, 992-998
27. Stoumbos, Z.G., Reynolds, M.R., Jr., and Woodall, W.H. (2003) Control Chart Schemes for Monitoring the Mean and Variance of Processes Subject to Sustained Shifts and Drifts. To appear in: Handbook of Statistics: Statistics in Industry **22**, C.R. Rao and R. Khattree (Eds.). Elsevier Science, Amsterdam, Netherlands
28. Stoumbos, Z.G., and Sullivan, J.H. (2002) Robustness to Non-normality of the Multivariate EWMA Control Chart. Journal of Quality Technology **34**, 260-276

An Application of Confirmation Sample Control Chart in the High-Yield Processes and Its Economic Design

Hiroshi Ohta and Etsuko Kusukawa

Osaka Prefecture University, Sakai, Osaka 599-8531, Japan

Abstract. This paper presents a modified CS(Confirmation Sample) chart for detecting sensitively small or moderate changes in p for both of upward and downward shifts in high-yield processes. The simplified design of the modified CS chart is done by applying the design method of the CS chart for the Shewhart-type charts to the CCC(Cumulative Count of Conforming)-r chart dealing with the cumulative count of items inspected until observing r nonconforming items. As r gets larger, the modified CS chart, named the $CS_{CCC\text{-}r}$ chart, becomes more and more sensitive to small or moderate changes in p for both of upward and downward shifts, but the $CS_{CCC\text{-}r}$ chart needs more and more observations to obtain a plotting point on the $CS_{CCC\text{-}r}$ chart, so the cost is fairly high. In this paper, a simplified optimal economic design method for the $CS_{CCC\text{-}r}$ chart is also proposed. For illustrative purposes, some numerical results for the optimal design parameter values are provided.

1 Introduction

Shewhart-type charts are known well as the typical charting techniques to monitor process variations due to assignable cause(s) in SPC(Statistical Process Control). Traditionally, when variable quality characteristics are normally distributed, an $\overline{x}$ chart is used to monitor the process mean and an R chart is used to monitor the process variance. Joint $\overline{x}$ -R charts are used for monitoring simultaneous changes in the process mean and the process variance. These charts are effective to detect large changes from the in-control process, but do not respond quickly to small or moderate changes in the process output. Recently, Steiner (1999) has presented CS charts applied to Shewhart-type charts to detect sensitively small or moderate changes in quality characteristics for both of upward and downward shifts.

On the other hand, the Shewhart-type p and np charts are used to monitor and control the product attributes. However, in high-yield processes where p is substantially less than one per cent, such as the manufacturing process of today, few or even no items are nonconforming even for a fairly large sample. Therefore, all of the above control charts for monitoring and controlling the product attributes are inadequate for high-yield processes. Recently, Xie and Goh (1992, 1997a) have studied an alternative statistical process control method called the CCC chart based on a geometric distribution and have shown that this charting technique is very useful in the monitoring of high-yield manufacturing processes. They have determined that in order to monitor a process it is better to monitor the cumulative count of conforming items and plot this number than to monitor the occurrence of nonconformance. Xie *et al.* (1998) also have pointed out that the conventional CCC charting technique in which a point is plotted whenever a nonconforming item is observed, is not sensitive enough to detect small changes in the process fraction defectives p. Therefore, they have extended the CCC chart to limit the number of nonconforming items produced whenever a process adjustment is made by monitoring the cumulative count of items produced until a fixed number of nonconforming items. This extended chart is referred to as a CCC-r chart, where r is the number of nonconforming items observed before a point is plotted. They have determined that the CCC-r chart is more suitable for monitoring the process fraction defectives in high-yield processes than the CCC chart. However, even the CCC-r chart is not sensitive enough to detect small or moderate changes in the process fraction defectives. It is desired to detect sensitively these shifts in p in high-yield processes. From this viewpoint, this paper presents a modified CS chart, named the $CS_{CCC\text{-}r}$ chart, for detecting sensitively small or moderate changes in the process fraction defectives for both of upward and downward shifts in high-yield processes. The simplified design of the $CS_{CCC\text{-}r}$ chart is done by applying the design method of the CS chart for the Shewhart-type charts to the CCC-r chart. The basic idea underlining the CS chart is to allow for a confirmation observation if the initial observation does not lead to a clear conclusion. The $CS_{CCC\text{-}r}$ chart can be operated using only confirmation control limits which are used to determine if the initial observation provides a clear evidence that the process is still in-control. The design of confirmation control limits for the $CS_{CCC\text{-}r}$ chart is based on the probability limit method using the negative binomial distribution. In use of the $CS_{CCC\text{-}r}$ chart, whenever an initial observation of poor or questionable quality is observed, an independent confirmation observation is immediately taken to verify that the original observation was not a fluke. The $CS_{CCC\text{-}r}$ charting technique can be readily implemented with the help of on-line computers and the modern inspection equipment commonly available in automated manufacturing environments.

As r increases, the $CS_{CCC\text{-}r}$ chart is more and more sensitive to small or moderate changes in p for both of upward and downward shifts. However, because more and more observations are required to obtain a plotting point on the $CS_{CCC\text{-}r}$ chart, the cost is fairly high. For this trade-off problem, it is necessary to determine the optimal number, r, of nonconforming items observed before a point is plotted, the sampling (inspection) interval, h, and the upper and lower confirmation control limits for the $CS_{CCC\text{-}r}$ chart. In this paper, we also present a simplified optimal design method for the $CS_{CCC\text{-}r}$ chart by applying the economic design methods proposed by Collani (1989), Collani and Dräger (1995, 1998), Xie and Goh

(1997b) and Tang *et. al* (2000) to control charts that monitor the discrete quality characteristics. For illustrative purposes, some numerical results for the optimal parameter values are provided. Furthermore, the expected profits per unit time of a process cycle, obtained by using the $CS_{CCC\text{-}r}$ chart, are compared with those obtained by using the CCC-*r* chart.

2 Outline of the CCC-*r* chart

Denote x as the cumulative count of items inspected until observing $r\,(\geq 2)$ nonconforming items. This count follows the negative binomial distribution NB(r, p) with parameters (r, p) and the probability function is given as

$$P\{x=n\} = \binom{n-1}{r-1} p^r (1-p)^{n-r} \quad (n=r, r+1, \ldots\ldots), \tag{1}$$

where p is the fraction defectives of the process being monitored. By using the exact probability limits based on the negative binomial distribution given byequation (1), the lower control limit $LCL_{NB(r,p)}$ and the upper control limit $UCL_{NB(r,p)}$ of the CCC-*r* chart are given as the solutions of the following equations:

$$\sum_{i=r}^{LCL_{NB(r,p)}} \binom{i-1}{r-1} p^r (1-p)^{i-r} = \alpha/2, \tag{2}$$

$$\sum_{i=r}^{UCL_{NB(r,p)}} \binom{i-1}{r-1} p^r (1-p)^{i-r} = 1-\alpha/2, \tag{3}$$

where α is the Type I error. Let p_0 be the fraction defectives of the in-control process being monitored. The decision procedure using a CCC-*r* chart applied in high-yield processes is different from that using the usual p and np charts. According to the CCC-*r* chart, if p decreases such that $p<p_0$, then the quality characteristic x for plotting a point on the CCC-*r* chart, that is, the cumulative count of items inspected until observing r nonconforming items, is expected to be larger, while if p increases such that $p>p_0$, then it is expected to be smaller. Therefore, according to the CCC-*r* chart, if a plotted point is above the upper control limit, the process is likely to have improved, while if a plotted point is below the lower control limit, the process has probably deteriorated. Thus, the decision procedure using a CCC-*r* chart is reversed compared to that using the usual p and np charts.

3 A modified CS chart for applying in high-yield processes

Recently, Steiner has presented CS charts applied to Shewhart-type charts to detect sensitively small or moderate changes in quality characteristics for both of upward and downward shifts. In this section, we present a modified CS chart, named the $CS_{CCC\text{-}r}$ chart, for detecting more sensitively small or moderate changes in p for both of upward and downward shifts in high-yield processes, based on applying the design method for the CS chart to the CCC-r chart.

3.1 Decision procedure for the $CS_{CCC\text{-}r}$ chart

The $CS_{CCC\text{-}r}$ chart can be designed by applying the design method for the CS chart to the CCC-r chart. In the $CS_{CCC\text{-}r}$ chart, an initial observation, x_A, based on the cumulative count of items inspected until observing r nonconforming items is taken from the process. Denote the upper and lower confirmation control limits of the $CS_{CCC\text{-}r}$ chart as $UCSL_{CCC\text{-}r}$ and $LSCSL_{CCC\text{-}r}$, respectively. If

$$\mathrm{LCSL}_{CCC-r} < x_A < \mathrm{UCSL}_{CCC-r}, \tag{4}$$

conclude that the process is in-control. Otherwise, an independent confirmation observation, x_B, based on the cumulative count of items inspected until observing r nonconforming items is taken from the process. Then, if

$$x_A \,\&\, x_B \le \mathrm{LCSL}_{CCC-r} \text{ OR } x_A \,\&\, x_B \ge \mathrm{UCSL}_{CCC-r} \tag{5}$$

conclude that the process is out-of-control. On the other hand, if the condition (5) is not satisfied, conclude that the process is in-control. The decision procedure for the $CS_{CCC\text{-}r}$ chart is summarized in flowchart form in Figure 1. Let α be the Type I error of the $CS_{CCC\text{-}r}$ chart. Based on the decision procedure for the $CS_{CCC\text{-}r}$ chart given in Figure 1, the probability of the Type I error assuming fixed values for the upper and lower confirmation control limits, $UCSL_{CCC\text{-}r}$ and $LCSL_{CCC\text{-}r}$, can be given as follows:

$$a = P(x_A \,\&\, x_B \ge \mathrm{UCSL}_{CCC-r})2 + P(x_A \,\&\, x_B \le \mathrm{LCSL}_{CCC-r})^2\,. \tag{6}$$

Here, assume that the Type I error in each direction should be equally likely. Therefore, from equation (6), the upper and lower confirmation control limits of the CS_{CCC-r} chart can be given as the solution of the following equations:

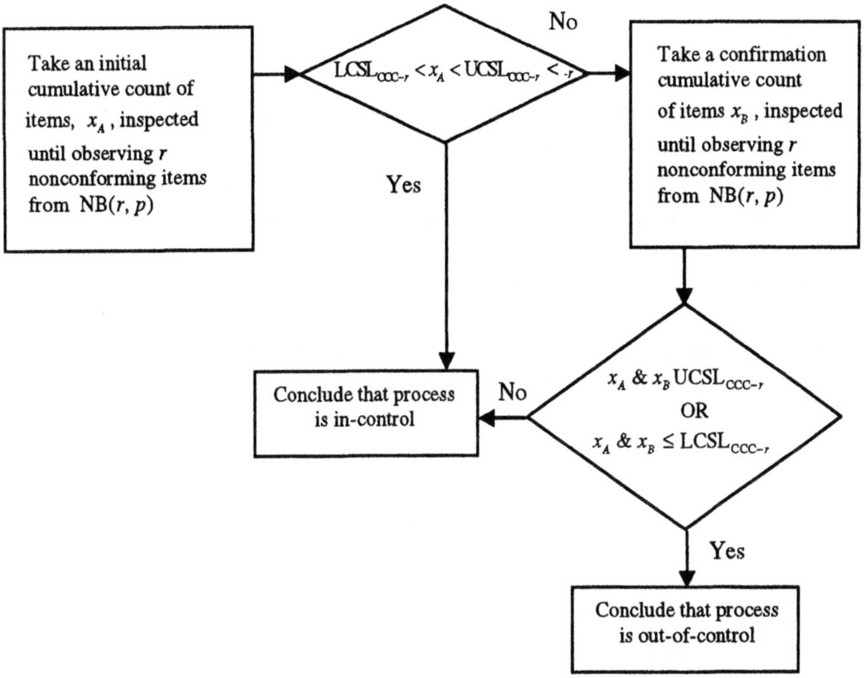

Figure 1 Decision procedure for the CS$_{CCC\text{-}r}$ chart

$$\sum_{i=r}^{LCSL_{CCC\text{-}r}} \binom{i-1}{r-1} p_0^{\,r}(1-p_0)^{i-r} = \sqrt{\alpha/2}\,, \tag{7}$$

$$\sum_{i=r}^{UCSL_{CCC\text{-}r}} \binom{i-1}{r-1} p_0^{\,r}(1-p_0)^{i-r} = 1-\sqrt{\alpha/2}\,. \tag{8}$$

3.2 Comparison with the CCC-*r* chart

For illustrative purposes, in this section we compare the performance of the proposed CS$_{CCC\text{-}r}$ chart with that of the CCC-*r* chart. We may express the null hypothesis H_0 (process is in-control) and the alternative hypothesis H_1 (process is out-of-control) in a formal manner as follows:

$$H_0 : p = p_0$$
$$H_1 : p = \kappa p_0$$

We compare the ANOS(Average Number of Observation to Signal) of the proposed $CS_{CCC\text{-}r}$ chart with those of the CCC-r chart. The ANOS is defined as the expected number of individual observations required for the chart to detect the first out-of-control observation. For the comparison of ANOSs of the CCC-r chart and those of the proposed $CS_{CCC\text{-}r}$ chart, let NB(r, p_0) with the following parameters : r(= 2, 5) and p_0 = 0.001 be the population distributions of the in-control process and we set the overall Type I error α to 0.0027, assuming that the probability of the Type I error in each direction is equally likely. The upper and lower ANOSs of the p_0 = 0.001 chart under each of the conditions of NB(r, p_1=$\kappa\, p_0$) for several values of κ can be given respectively as the solutions of the followings:

$$\text{lower ANOS} = \frac{1}{P(x \le \text{LCL}_{\text{NB}(r,\, p_0)})} = \frac{1}{\sum_{i=r}^{\text{LCL}_{\text{NB}(r,\, p_0)}} \binom{i-1}{r-1} p_1^r (1-p_1)^{i-r}}, \tag{9}$$

$$\text{upper ANOS} = \frac{1}{P(x \ge \text{UCL}_{\text{NB}(r,\, p_0)})} = \frac{1}{1-\left(\sum_{i=r}^{\text{UCL}_{\text{NB}(r,\, p_0)}} \binom{i-1}{r-1} p_1^r (1-p_1)^{i-r}\right)}. \tag{10}$$

On the other hand, for the $CS_{CCC\text{-}r}$ chart the upper and lower ANOSs can be given respectively as the solutions of the followings:

$$\text{lower ANOS} = \frac{1}{P(x_A \,\&\, x_B \le \text{LCSL}_{\text{CCC-}r})^2}$$

$$= \frac{1}{\left(\sum_{i=r}^{\text{LCSL}_{\text{CCC-}r}} \binom{i-1}{r-1} p_1^r (1-p_1)^{i-r}\right)^2}, \tag{11}$$

$$\text{upper ANOS} = \frac{1}{P(x_A \,\&\, x_B \ge \text{UCSL}_{\text{CCC-}r})^2}$$

$$= \frac{1}{1-\left(\sum_{i=r}^{\text{UCSL}_{\text{CCC-}r}} \binom{i-1}{r-1} p_1^r (1-p_1)^{i-r}\right)^2}. \tag{12}$$

Table 1 The upper and lower control limits and actual values of the Type I error (%) for the CCC-*r* chart

$p_0 = 0.001$	the CCC-*r* chart		
r	$\Pr(x \leq LCL_{NB(r,p_0)}) = 0.00135$	$\Pr(x \geq UCL_{NB(r,p_0)}) = 0.99865$	actual value of the Type I error (%)
	$LCL_{NB(r,p_0)}$	$UCL_{NB(r,p_0)}$	
2	53	8896	0.2683
5	793	14387	0.2697

where the Type I error α is specified as 0.27 (%) and x is distributed as NB(r, p_o).

Table 2 The upper and lower confirmation control limits and actual values of the Type I error (%) for the $CS_{CCC\text{-}r}$ chart

$P_0 = 0.001$	the $CS_{CCC\text{-}r}$ chart		
r	$\Pr(x_A \,\&\, x_B \leq LCSL_{CCC\text{-}r}) = 0.03674$	$\Pr(x_A \,\&\, x_B \geq UCSL_{CCC\text{-}r}) = 0.96326$	actual value of the Type I error (%)
	$LCSL_{CCC\text{-}r}$	$UCSL_{CCC\text{-}r}$	
2	299	5112	0.2693
5	1805	9641	0.2697

where the Type I error α is specified as 0.27 (%) and both of x_A and x_B are distributed as NB(r, p_o).

Table 1 shows the upper and lower control limits and actual values of the Type I error α of the CCC-*r* chart under each of the conditions of NB(r, p_0). Table 2 shows the upper and lower confirmation control limits and actual values of the Type I error α of the CCC-*r* chart, respectively. It is evident from Tables 1 and 2 that all upper confirmation control limits of the proposed $CS_{CCC\text{-}r}$ chart for detecting the process improvement can be set lower than upper control limits of the CCC-*r* chart. Further, it is also evident from Tables 1 and 2 that all lower confirmation control limits of the proposed $CS_{CCC\text{-}r}$ chart for detecting the process deterioration can be set higher than lower control limits of the CCC-*r* chart. Therefore, it is expected that the upper and lower ANOSs of the proposed $CS_{CCC\text{-}r}$ chart can be smaller than those of the CCC-*r* chart. Tables 3(a)~(b) show the ANOSs of the CCC-*r* chart and the proposed $CS_{CCC\text{-}r}$ chart under the conditions of

Table 3 The ANOSs of the CCC–*r* chart and the CS_{CCC-r} chart

(a) In the case that the population distribution NB(r, p_0) is NB(2, 0.001)

κ	actual values of ANOS			
	CCC-*r* chart		CS_{CCC-r} chart	
	lower ANOS	upper ANOS	lower ANOS	upper ANOS
0.50	2950	16	9804	**13**
0.60	2058	33	4926	**28**
0.70	1517	70	2762	**61**
0.80	1166	152	1684	**138**
0.90	923	334	1093	**317**
1.00	0.2683 *		0.2693 *	
1.10	623	1653	**529**	1754
1.20	525	3704	**388**	4202
1.30	449	8333	**293**	10204
1.40	388	18519	**226**	25000
1.50	339	40000	**178**	62500

* denotes the Type I error (%)

(b) In the case that the population distribution NB (r, p_0) is NB (5, 0.001)

κ	actual values of ANOS			
	CCC-*r* chart		CS $_{CCC-r}$ chart	
	lower ANOS	upper ANOS	lower ANOS	upper ANOS
0.50	17241	6	166667	**4**
0.60	7353	15	38462	**10**
0.70	3636	36	11111	**26**
0.80	1992	94	3891	**73**
0.90	1179	258	1600	**224**
1.00	0.2697 *		0.2697 *	
1.10	492	2198	**381**	2597
1.20	339	6711	**212**	9615
1.30	243	20833	**126**	37037
1.40	178	66667	**79**	142857
1.50	134	200000	**52**	500000

* denotes the Type I error (%)

NB(r, $p_1 = \kappa p_0$) for several values of κ. It is evident from Tables 3(a)~(b) that the upper ANOSs of the proposed $CS_{CCC\text{-}r}$ chart are smaller than those of the CCC-r chart for the downward shifts in p (cases of $\kappa < 1$). It is also evident from Tables 3(a)~(b) that the lower ANOSs of the proposed $CS_{CCC\text{-}r}$ chart are smaller than those of the CCC-r chart for the upward shifts in p (cases of $\kappa > 1$). In the cases that $\kappa < 1$, the upper ANOS indicates the expected number of individual observations required for the chart until judging correctly that p is getting smaller. Similarly, in the cases that $\kappa > 1$, the lower ANOS indicates the expected number of individual observations required for the chart until judging correctly that p is getting larger. So, it can be seen that the smaller the values of the upper and lower ANOSs are, the more superior the performance of a control chart is. Further, it is also evident from Tables 3(a)~(b) that the proposed $CS_{CCC\text{-}r}$ chart is harder to make a misjudgment that p is getting larger in the cases that $\kappa < 1$ and that p is getting smaller in the cases that $\kappa > 1$ than the CCC-r chart, because the lower ANOSs in the cases that $\kappa < 1$ and the upper ANOSs in the cases that $\kappa > 1$ of the proposed $CS_{CCC\text{-}r}$ chart are larger than those of the CCC-r chart. Therefore, it can be seen that it is more adequate to apply the proposed $CS_{CCC\text{-}r}$ chart in high-yield processes than the CCC-r chart. From the comparison of ANOSs of the proposed $CS_{CCC\text{-}r}$ chart and those of the CCC-r chart, our first purpose has been successfully achieved.

4 Economic design of the $CS_{CCC\text{-}r}$ chart

In what follows, as our second purpose we present a simplified optimal design method for the $CS_{CCC\text{-}r}$ chart by applying the economic design methods proposed by Collani (1989), Collani and Dräger (1995, 1998), Xie and Goh (1997b) and Tang *et. al* (2000) to control charts that monitor the discrete quality characteristics.

4.1 Notation

The notation used in this section is defined as follows:

Decision variables

h : length of the sampling (inspection) interval for the $CS_{CCC\text{-}r}$ chart

r : the number of nonconforming items observed before a point is plotted on the $CS_{CCC\text{-}r}$ chart

$UCSL_{CCC\text{-}r}$: upper confirmation control limit of the $CS_{CCC\text{-}r}$ chart

$LCSL_{CCC\text{-}r}$: lower confirmation control limit of the $CS_{CCC\text{-}r}$ chart

Other notation

p_{I}: process fraction defectives in State I (in-control state)

p_{II}: process fraction defectives in State II (out-of-control state which indicates the process deterioration)

p_{III}: process fraction defectives in State III (out-of-control state which indicates the process improvement)

x : cumulative count of items inspected until observing r nonconforming items

$g_i(i\in$ **I, II, III**): profit per item produced in State i, where $g_{\mathrm{III}} > g_{\mathrm{I}} > g_{\mathrm{II}}$

$E[x|i \rightarrow j]$: the conditional expected value of x when the process changes from State i to State j

$x_i(i\in$ **I, II, III**): cumulative count of items inspected until observing r nonconforming items in State i

$A_i(i\in$ **I, II, III**): total number of samplings while the process is operating in State i

$E[A_i|i\rightarrow j]$: the conditional expected value of A_i when the process changes from State i to State j

A_F: total number of false alarms caused by the Type I error

$E[A_F|i\rightarrow j]$: the conditional expected value of A_F when the process changes from State i from State j

γ: search time for an assignable cause

e : search cost per alarm

a : sampling inspection cost per plotting a point

α: the probability of the Type I error when the process is in State I

β_{II}: the probability of the Type II error when the process is in State II

β_{III}: the probability of the Type II error when the process is in State III

λ_{II}: parameter of the exponential distribution used when the process changes from State I to State II

λ_{III}: parameter of the exponential distribution used when the process changes from State I to State III

$f(t)$: probability density function of the process failure time

$F(t)$: cumulative distribution function of the process failure time

v : number of items produced per unit time of operation

4.2 Assumptions

The assumptions for the model dealt with in this section are as follows:

(1) State I denotes the in-control state, where the process fraction defectives p_{I} are known. State II denotes the out-of-control state, where the known process fraction defectives p_{II} is greater than the known process fraction defectives p_{I}. State III denotes the out-of-control state, where the known process fraction defectives p_{III} is less than the known process fraction defectives p_{I}.

(2) When an alarm is observed, the process is stopped immediately, and a search for the assignable cause is undertaken. If an assignable cause of the process deterioration, in the case that the process changes from State I to State II, can be detected in search time γ, the process is repaired immediately. After a repair and an adjustment for the process, the process is returned to State I. If an assignable cause of the process improvement, in the case that the process changes from State I to State III, can be detected in search time γ, after standardizing the cause, the $\mathrm{CS}_{\mathrm{CCC}\text{-}r}$ chart must be redesigned by setting the State III to a new State I. If the assignable cause is not detected in search time γ, then it is judged that the alarm was false. In this case, the process restarts in the current state.

(3) The detection ability of the process dealt with in this paper does not change and unknown assignable causes do not occur. The process proceeds until a true alarm informs us that process deterioration or improvement has been found on the chart.

(4) A process cycle is defined to be the time period from the beginning of production (or after renewal) in State I to the correct detection after a process deterioration or a process improvement occurs and the process changes from State I to State II or State III. Note that in the case that a process improvement occurs and the process changes from State I to State III, the renewal process is terminated.

(5) The durations of the in-control period from starting the process in State I until changing the process in State II or in State III, denoted $\tau_{\mathrm{I}\to\mathrm{II}}$ and $\tau_{\mathrm{I}\to\mathrm{III}}$, are exponentially distributed with parameter λ_{II} or respectively. The probability density function $f(t)$ and the cumulative distribution function $F(t)$ of the exponential distribution with parameter λ are given respectively as

$$f(t) = \lambda e^{-\lambda t},$$

$$F(t) = 1 - \lambda e^{-\lambda t}.$$

4.3 Economic model

In this model, the expected length of the process cycle, $E[T \mid \mathbf{I} \to \mathbf{II}]$, when the process changes from State I to State II is given as

$$E[T \mid \mathbf{I} \to \mathbf{II}] = \{E[A_{\mathrm{I}} \mid \mathbf{I} \to \mathbf{II}] + E[A_{\mathrm{II}}]\}h + \{E[A_F \mid \mathbf{I} \to \mathbf{II}] + 1\}\,\gamma, \tag{13}$$

where +1 in the last term of equation (13) is based on the case that a true alarm is observed and that the process cycle is terminated. The expected profit per process cycle, $E[G \mid \mathbf{I} \to \mathbf{II}]$, when the process changes from State I to State II is obtained as follows:

$$\begin{aligned} E[G \mid \mathbf{I} \to \mathbf{II}] = \{\, & g_{\mathrm{I}} \cdot E[A_{\mathrm{I}} \mid \mathbf{I} \to \mathbf{II}] + g_{\mathrm{II}} \cdot E[A_{\mathrm{II}}]\}\, hv \\ & - e \cdot \{E[A_F \mid \mathbf{I} \to \mathbf{II}] + 1\} - a \cdot E[x \mid \mathbf{I} \to \mathbf{II}]\{E[A_I \mid \mathbf{I} \to \mathbf{II}] + E[A_{\mathrm{II}}]\}, \end{aligned} \tag{14}$$

where $E[x \mid \mathbf{I} \to \mathbf{II}]$ can be written as follows:

$$E[x \mid \mathbf{I} \to \mathbf{II}] = \frac{E[A_{\mathrm{I}} \mid \mathbf{I} \to \mathbf{II}]E[x_{\mathrm{I}}] + E[A_{\mathrm{II}}]E[x_{\mathrm{II}}]}{E[A_{\mathrm{I}} \mid \mathbf{I} \to \mathbf{II}] + E[A_{\mathrm{II}}]}, \tag{15}$$

where, from equation (1), $E[x_{\mathrm{I}}]$ and $E[x_{\mathrm{II}}]$ are given respectively as

$$E[x_{\mathrm{I}}] = r(1 - p_{\mathrm{I}})/p_{\mathrm{I}} \quad E[x_{\mathrm{II}}] = r(1 - p_{\mathrm{II}})/p_{\mathrm{II}}\ .$$

Furthermore, $E[A_{\mathrm{I}} \mid \mathbf{I} \to \mathbf{II}]$, $E[A_F \mid \mathbf{I} \to \mathbf{II}]$ and $E[A_{\mathrm{II}}]$ are given respectively as

$$E[A_{\mathrm{I}} \mid \mathbf{I} \to \mathbf{II}] = \frac{1 - \int_0^h \lambda_{\mathrm{II}} e^{-\lambda_{\mathrm{II}} t} dt}{\int_0^h \lambda_{\mathrm{II}} e^{-\lambda_{\mathrm{II}} t} dt} = \frac{e^{-\lambda_{\mathrm{II}} h}}{1 - e^{-\lambda_{\mathrm{II}} h}} = \frac{1}{e^{\lambda_{\mathrm{II}} h} - 1}, \tag{16}$$

$$E[A_F \mid \mathbf{I} \to \mathbf{II}] = \alpha \cdot E[A_{\mathrm{I}} \mid \mathbf{I} \to \mathbf{II}] = \frac{\alpha}{e^{\lambda_{\mathrm{II}} h} - 1}, \tag{17}$$

$$E[A_{\mathrm{II}}] = \frac{1}{1 - \beta_{\mathrm{II}}}, \tag{18}$$

where α and β_{II} can be obtained as follows:

$$\alpha = \left(\sum_{k=r}^{\mathrm{LCSL}_{\mathrm{CCC}-r}} \binom{k-1}{r-1} p_{\mathrm{I}}^{r} (1-p_{\mathrm{I}})^{k-r} \right)^{2} + \left(1 - \sum_{k=r}^{\mathrm{UCSL}_{\mathrm{CCC}-r}} \binom{k-1}{r-1} p_{\mathrm{I}}^{r} (1-p_{\mathrm{I}})^{k-r} \right)^{2}, \tag{19}$$

$$\beta_{\mathrm{II}} = 1 - \left(\sum_{k=r}^{\mathrm{LCSL}_{\mathrm{CCC}-r}} \binom{k-1}{r-1} p_{\mathrm{II}}^{r} (1-p_{\mathrm{II}})^{k-r} \right)^{2} - \left(1 - \sum_{k=r}^{\mathrm{UCSL}_{\mathrm{CCC}-r}} \binom{k-1}{r-1} p_{\mathrm{II}}^{r} (1-p_{\mathrm{II}})^{k-r} \right)^{2}. \tag{20}$$

Therefore, from equations (13) and (14), the expected profit per unit time of the process cycle, $\Pi\,(h, r, \mathrm{UCSL}_{\mathrm{CCC}-r}, \mathrm{LCSL}_{\mathrm{CCC}-r} \mid \mathrm{I} \rightarrow \mathrm{II})$, when the process changes from State I to State II can be given as

$$\Pi(h, r, \mathbf{UCSL}_{\mathrm{CCC}-r}, \mathbf{LCSL}_{\mathrm{CCC}-r} \mid \mathbf{I} \rightarrow \mathbf{II}) = E[G \mid \mathbf{I} \rightarrow \mathbf{II}] / E[T \mid \mathbf{I} \rightarrow \mathbf{II}]. \tag{21}$$

As the same way, the expected length per process cycle, $E[\,T \mid \mathbf{I} \rightarrow \mathbf{III}\,]$, when the process changes from State I to State III is obtained as

$$E[T \mid \mathbf{I} \rightarrow \mathbf{III}] = \{E[A_{\mathrm{I}} \mid \mathbf{I} \rightarrow \mathbf{III}] + E[A_{\mathrm{III}}]\} h + \{E[A_F \mid \mathbf{I} \rightarrow \mathbf{III}] + 1\}\, \gamma\,. \tag{22}$$

Also, the expected profit per process cycle, $E[G \mid \mathbf{I} \rightarrow \mathbf{III}\,]$, when the process changes from State I to State III is obtained as follows:

$$\begin{aligned} E[G \mid \mathbf{I} \rightarrow \mathbf{III}] &= \{\, g_{\mathrm{I}} \cdot E[A_{\mathrm{I}} \mid \mathbf{I} \rightarrow \mathbf{III}] + g_{\mathrm{III}} \cdot E[A_{\mathrm{III}}]\}\, hv \\ &- e \cdot \{E[A_F \mid \mathbf{I} \rightarrow \mathbf{III}] + 1\} - a \cdot E[x \mid \mathbf{I} \rightarrow \mathbf{III}]\{E[A_{\mathrm{I}} \mid \mathbf{I} \rightarrow \mathbf{III}] + E[A_{\mathrm{III}}]\}, \end{aligned} \tag{23}$$

where $E[x \mid \mathbf{I} \rightarrow \mathbf{III}\,]$ can be written as follows:

$$E[x \mid \mathbf{I} \rightarrow \mathbf{III}] = \frac{E[A_{\mathrm{I}} \mid \mathbf{I} \rightarrow \mathbf{III}] E[x_{\mathrm{I}}] + E[A_{\mathrm{III}}] E[x_{\mathrm{III}}]}{E[A_{\mathrm{I}} \mid \mathbf{I} \rightarrow \mathbf{III}] + E[A_{\mathrm{III}}]}, \tag{24}$$

where, from equation (1), $E[\,x_{\mathrm{III}}\,]$ is given as

$$E[x_{\mathrm{III}}] = r(1 - p_{\mathrm{III}}) / p_{\mathrm{III}}\,.$$

Furthermore, $E[A_{\mathrm{I}} \mid \mathbf{I} \rightarrow \mathbf{III}]$, $E[A_F \mid \mathbf{I} \rightarrow \mathbf{III}]$ and $E[A_{\mathrm{III}}]$ are given respectively as

$$E[A_{\mathrm{I}} \mid \mathrm{I} \to \mathrm{III}] = \frac{1 - \int_0^h \lambda_{\mathrm{III}} e^{-\lambda_{\mathrm{III}} t} dt}{\int_0^h \lambda_{\mathrm{III}} e^{-\lambda_{\mathrm{III}} t} dt} = \frac{e^{-\lambda_{\mathrm{III}} h}}{1 - e^{-\lambda_{\mathrm{III}} h}} = \frac{1}{e^{\lambda_{\mathrm{III}} h} - 1}, \tag{25}$$

$$E[A_F \mid \mathrm{I} \to \mathrm{III}] = \alpha \cdot E[A_{\mathrm{I}} \mid \mathrm{I} \to \mathrm{III}] = \frac{\alpha}{e^{\lambda_{\mathrm{III}} h} - 1}, \tag{26}$$

$$E[A_{\mathrm{III}}] = \frac{1}{1 - \beta_{\mathrm{III}}}, \tag{27}$$

where β_{III} can be obtained as follows:

$$\begin{aligned} \beta_{\mathrm{III}} = 1 &- \left(\sum_{k=r}^{\mathrm{LCSL}_{\mathrm{CCC}-r}} \binom{k-1}{r-1} p_{\mathrm{III}}^{\,r} (1 - p_{\mathrm{III}})^{k-r} \right)^2 \\ &- \left(1 - \sum_{k=r}^{\mathrm{UCSL}_{\mathrm{CCC}-r}} \binom{k-1}{r-1} p_{\mathrm{III}}^{\,r} (1 - p_{\mathrm{III}})^{k-r} \right)^2. \end{aligned} \tag{28}$$

Therefore, from equations (22) and (23), the expected profit per unit time of the process cycle, $\Pi(h, r, \mathrm{UCSL}_{\mathrm{CCC}-r}, \mathrm{LCSL}_{\mathrm{CCC}-r} \mid \mathrm{I} \to \mathrm{III})$, when the process changes from State I to State III can be given as

$$\Pi(h, r, \mathrm{UCSL}_{\mathrm{CCC}-r}, \mathrm{LCSL}_{\mathrm{CCC}-r} \mid \mathrm{I} \to \mathrm{III}) = E[G \mid \mathrm{I} \to \mathrm{III}] / E[T \mid \mathrm{I} \to \mathrm{III}] \tag{29}$$

From the relation that $p_{\mathrm{II}} > p_{\mathrm{I}} > p_{\mathrm{III}}$, $E[x_{\mathrm{I}}]$, $E[x_{\mathrm{II}}]$ and $E[x_{\mathrm{III}}]$ have the following relation:

$$E[x_{\mathrm{II}}] \;<\; E[x_{\mathrm{I}}] \;<\; E[x_{\mathrm{III}}]. \tag{30}$$

Therefore, the following restriction is required for obtaining observation as the test statistic from the process:

$$E[x_{\mathrm{III}}] \;<\; hv. \tag{31}$$

The probabilities that the process will change from State I to State II or State III can be given respectively as

$$Q_1 = 1 - \exp(-\lambda_{\mathrm{II}}\, E[A_{\mathrm{I}} \mid \mathrm{I} \to \mathrm{II}]\, h),$$

$$Q_2 = 1 - \exp(-\lambda_{\text{III}} E[A_{\text{I}} \mid \text{I} \to \text{III}] h) .$$

Denote п $(h, r, \text{UCSL}_{\text{CCC-}r}, \text{LCSL}_{\text{CCC-}r})$ as the expected profit per unit time of a process cycle that takes account of both changes to State II that indicates the process deterioration or State III that indicates the process improvement. In this case, п $(h, r, \text{UCSL}_{\text{CCC-}r}, \text{LCSL}_{\text{CCC-}r})$ can be given as

$$\begin{aligned} &\Pi(h, r, \text{UCSL}_{\text{CCC-}r}, \text{LCSL}_{\text{CCC-}r}) \\ &= (Q_1/(Q_1+Q_2))\Pi(h, r, \text{UCSL}_{\text{CCC-}r}, \text{LCSL}_{\text{CCC-}r} \mid \text{I} \to \text{II}) \\ &\quad + (Q_2/(Q_1+Q_2))\Pi(h, r, \text{UCSL}_{\text{CCC-}r}, \text{LCSL}_{\text{CCC-}r} \mid \text{I} \to \text{III}) . \end{aligned} \quad (32)$$

4.4 Optimal design of the $\text{CS}_{\text{CCC-}r}$ chart

Using equation (32), $(h^*, r^*, \text{UCSL}_{\text{CCC-}r}^*, \text{LCSL}_{\text{CCC-}r}^*)$, which satisfy the following condition

$$\begin{aligned} &\Pi(h^*, r^*, \text{UCSL}_{\text{CCC-}r}^*, \text{LCSL}_{\text{CCC-}r}^*) \\ &\qquad \geq \Pi(h, r, \text{UCSL}_{\text{CCC-}r}, \text{LCSL}_{\text{CCC-}r}) \\ &\qquad\qquad \forall\, h, r, \text{UCSL}_{\text{CCC-}r}, \text{LCSL}_{\text{CCC-}r} \end{aligned} \quad (33)$$

are the optimal values designed for the $\text{CS}_{\text{CCC-}r}$ chart.

4.5 Numerical examples of the optimal design

In this section, some hypothetical examples of the optimal design for the $\text{CS}_{\text{CCC-}r}$ chart are shown. The parameters of the exponential distribution when the process changes from State I to State II or State III are assumed to be $\lambda_{\text{II}} = 0.0125$ and $\lambda_{\text{III}} = 0.00625$, respectively. The number of items produced per unit time of operation is assumed to be $v = 1500000$. The sampling inspection cost per plotting a point is assumed to be $a = 50$. The search cost per alarm is assumed to be $e = 100000$. The search time is assumed to be $\gamma = 0.1$. Tables 4 and 5 show the optimal designs for the $\text{CS}_{\text{CCC-}r}$ and the CCC-r charts for several choices of the profits per item produced in States I, II and III ($g_{\text{I}}, g_{\text{II}}, g_{\text{III}}$), and the process fraction defectives in States I, II, III ($p_{\text{I}}, p_{\text{II}}, p_{\text{III}}$). It can be seen from Tables 4 and 5 that all expected profits per unit time of the process cycle of the $\text{CS}_{\text{CCC-}r}$ chart are larger than those of the CCC-r chart. Therefore, it can be noted that the proposed optimal economic designing method is more superior.

Table 4 The optimal design for the CS_{CCC-r} chart

g_I	1000	p_I	0.001	p_I	0.001	p_I	0.001
g_{II}	900	p_{II}	0.002	p_{II}	0.004	p_{II}	0.008
g_{III}	1100	p_{III}	0.0008	p_{III}	0.0006	p_{III}	0.0002
h*		10.41		3.52		9.04	
r*		12		19		8	
$UCSL_{CCC-r}$ *		7411		11641		5520	
$LCSL_{CCC-r}$ *		27727		43987		27097	
$\Pi_{CS_{CCC-r}}$ †		1560.70		1570.69		1566.41	
g_I	1000	p_I	0.001	p_I	0.001	p_I	0.001
g_{II}	500	p_{II}	0.002	P_{II}	0.004	p_{II}	0.008
g_{III}	2000	p_{III}	0.0008	p_{III}	0.0006	p_{III}	0.0002
h*		16.56		5.07		9.04	
r*		24		19		8	
$USCL_{CCC-r}$ *		16980		15960		5520	
$LCSL_{CCC-r}$ *		54007		44797		27097	
$\Pi_{CS_{CCC-r}}$		2188.41		2227.90		2215.33	
g_I	1000	p_I	0.001	p_I	0.001	p_I	0.001
g_{II}	100	P_{II}	0.002	p_{II}	0.004	p_{II}	0.008
g_{III}	10000	p_{III}	0.0008	p_{III}	0.0006	p_{III}	0.0002
h*		7.69		7.05		6.63	
r*		17		0		8	
$UCSL_{CCC-r}$ *		1820		14340		4230	
$LCSL_{CCC-r}$ *		41797		51637		36427	
$\Pi_{CS_{CCC-r}}$		8187.54		8236.83		8237.45	

† : $\Pi_{CS_{CCC-r}}$ denotes $\Pi(h^*, r^*, UCSL_{CCC-r}^*, LCSL_{CCC-r}^*)$ and is shown in millions

5 Conclusions

In this paper, we presented a modified CS chart, named the $CS_{CCC\text{-}r}$ chart, to detect sensitively small or moderate changes in p for both of upward and downward shifts in high-yield processes by applying the design method for the

Table 5 The optimal design for the CCC-*r* chart

g_I	1000	p_I	0.001	p_I	0.001	p_I	0.001
g_{II}	900	p_{II}	0.002	p_{II}	0.004	p_{II}	0.008
g_{II}	1100	p_{III}	0.0008	p_{III}	0.0006	p_{III}	0.0002
h^*		6.09		8.39		6.22	
r^*		14		11		13	
$UCL_{CCC\text{-}r}{}^*$		7261		3151		6900	
$LCL_{CCC\text{-}r}{}^*$		41287		42667		28387	
$\Pi_{CCC\text{-}r}$ †		1560.35		1564.11		1495.29	
g_I	1000	p_I	0.001	p_I	0.001	p_I	0.001
g_{II}	500	p_{II}	0.002	p_{II}	0.004	p_{II}	0.008
g_{II}	2000	p_{III}	0.0008	p_{III}	0.0006	p_{III}	0.0002
h^*		5.55		8.30		11.67	
r^*		19		11		3	
$UCL_{CCC\text{-}r}{}^*$		12091		3421		1830	
$LCL_{CCC\text{-}r}{}^*$		40477		37897		58237	
$\Pi_{CCC\text{-}r}$		2064.20		2178.86		2141.85	
g_I	1000	p_I	0.001	p_I	0.001	p_I	0.001
g_{II}	100	p_{II}	0.002	p_{II}	0.004	p_{II}	0.008
g_{III}	10000	p_{III}	0.0008	p_{III}	0.0006	p_{III}	0.0002
h^*		3.16		11.86		35.57	
r^*		22		9		13	
$UCL_{CCC\text{-}r}{}^*$		9841		3268		22586	
$LCL_{CCC\text{-}r}{}^*$		62797		30337		101347	
$\Pi_{CCC\text{-}r}$		8185.40		7696.85		7394.37	

† : $\Pi_{CCC\text{-}r}$ denotes $\Pi(h^*, r^*, UCL_{CCC\text{-}r}{}^*, LCL_{CCC\text{-}r}{}^*)$ and is shown in millions

CS chart for the Shewhart-type charts to the CCC-*r* chart dealing with the cumulative count of items inspected until observing *r* nonconforming items. However, the $CS_{CCC\text{-}r}$ chart has a trade-off problem that as *r* gets larger, the

$CS_{CCC\text{-}r}$ chart becomes more and more sensitive to small or moderate changes in p for both of upward and downward shifts, but at the same time, the $CS_{CCC\text{-}r}$ chart needs more and more observations to obtain a plotting point on the chart, making it fairly expensive to use. So, a simplified optimal economic designing method of the $CS_{CCC\text{-}r}$ chart was developed for determining the optimal values of the number of nonconforming items observed before a point is plotted, r, the sampling (inspection) interval, h, and the upper and lower control limits.

This research has been supported by the Grant-in-Aid for Scientific Research No. 14580484 from the Japan Society for the Promotion of Science.

References

1. Collani, E.V. (1989) Economically Optimal c-Chart and np-Control Chart. Metrika **36**, 215 - 232.
2. Collani, E.V. and Dräger, K. (1995) A Simplified Economic Design of Control Charts for Monitoring the Nonconforming Probability. Economic Quality Control **10**(4), 231 - 292.
3. Collani, E.V. and Dräger, K. (1998) A Simplified Economic Design of Control Charts for Monitoring the Number of Nonconformities. Economic Quality Control **12**, 15 - 46.
4. Steiner, S.H. (1999) Confirming Sample Control Charts. International Journal of Production Research **37**(4), 737 – 748.
5. Tang, X.Y., Xie, M. and Goh, T.N. (2000) A Note on Economic-Statistical Design of Cumulative Count of Conforming Chart. Economic Quality Control **78**(1-2), 3 - 14.
6. Xie, M. and Goh, T.N. (1992) Some Procedure for Decision Making in Controlling High Yield Processes. Quality and Reliability Engineering- International **8**, 355 - 360.
7. Xie, M. and Goh, T.N. (1997a) The Use of Probability Limits for Process Control Based on Geometric Distribution. International Journal of Quality & Reliability Management **14**(1), 64 - 73.
8. Xie, M and Goh, T.N. (1997b) A Study of Economic Design of Control Charts for Cumulative Count of Conforming Items. Communications in Statistics : Simulation and Computation **26**(3), 1009 - 1027.
9. Xie, M., Goh, T.N. and Lu X.S. (1998) Computer - Aided Statistical Monitoring of Automated Manufacturing Process. Journal of Computer & Industrial Engineering **35**(1-2), 189 – 192.

Statistical Design of Attribute Charts for Monitoring and Continuous Improvement When Count Levels Are Low

Erwin M. Saniga[1], Darwin J. Davis[1] and James M. Lucas[2]

[1] University of Delaware
[2] J. M. Lucas and Associates

Abstract. Consider the situation where an attribute chart is being used to monitor a process for special causes of variability and attempts at continuous improvement are being made. In certain cases where the count level is low no lower control limit will exist for standard Shewhart control charts.

Optimal methods for shift detection when attribute data is being used are traditional CUSUM methods but CUSUM methods have not enjoyed wide popularity when compared to Shewhart methods because they are difficult to apply and understand.

In this paper we show how one can design special CUSUMs that will enable the Shewhart attribute chart user to monitor decreases in the process parameter when count levels are small. These CUSUMs, like the Shewhart chart, are easy to understand, easy to use, and, with the application of the algorithm given in this paper, easy to design.

A comparison with traditional CUSUMs is given and a variety of example designs developed using this algorithm are presented.

1 Introduction

We present an algorithm to find statistical designs of a special CUSUM control chart for count data that is used to monitor decreases in the process parameter when a low count process results in a lower Shewhart control chart limit of zero. This enables the Shewhart control chart user to detect decreases in the process parameter. This special CUSUM method consists of using one of two counting rules initially suggested by Page (1955). Here, we apply them by counting the number of number of samples with zero counts of defectives or defects per unit. The first method is to signal when more than k samples in a row contain zero counts. The second method is to signal when at least two in t samples contain zero counts; this method has been discussed by Lucas (1989) for detecting increases in the process parameter. We call these low side schemes the "*k* in a row" scheme and the "2 in *t* " scheme.

Since this special CUSUM is to be used as an add-on to a Shewhart chart we wish to emphasize that our objective is to provide the user with a method that has

similar properties to the Shewhart chart. The particular properties of this special CUSUM are that it is, like the Shewhart chart, flexible, easy to use, easy to understand and, in the case of the "*k* in a row" scheme, is optimal in terms of providing the smallest detection ARL for large shifts. In addition, the use of the algorithm contained in this paper makes design of this CUSUM a simple task.
As we noted above these special CUSUM low side schemes allow for flexibility in design. Thus a control chart designer can tailor the design to fit the conditions of their organization. Thus, this design has an advantage over other methods proposed to handle situations where tail probabilities are unequal (due to the skewness of the corresponding discrete distributions); these include methods given by Shore (2000), Ryan (1989), Schewertman and Ryan (1997), and Eisenhart, Hastay and Wallis (1947). Similarly, designs based upon modified centerline counts such as those given by Nelson (1997) and Acosta-Mejia (1999) are not competitive in terms of ARL performance or flexibility.

Reynolds and Stoumbos (1999, 2000, and 2001) have studied the performance of binomial and Bernoulli CUSUM methods for count data including two sided methods that could be used to signal increases or decreases in the proportion defective. The optimality of the Bernoulli method is illustrated and was also shown by Bourke (2001). Lucas (1985) also studied CUSUM methods for counts but for Poisson data. In practice, sampling costs generally include a fixed and variable component thereby rendering a statistically optimal procedure such as the Bernoulli CUSUM suboptimal in an economic sense. While a binomial CUSUM is optimal for a particular shift for which it was designed it may be dominated by other methods in terms of ARL for other shifts. We argue that there are cases where our method is preferred; e.g., when one wishes to take the advantage of the simplicity of Shewhart methods combined with simple counting rules such as the "*k* in a row" or" 2 in *t*" schemes. This simplicity coupled with the optimal ARL performance for large shifts of these two-sided procedures makes them prime candidates for actual applications when count data is used and average counts are low.

The *k* zeros in a row and two zeros in *t* samples are both CUSUM procedures. For both procedures, the CUSUM can be written as:

$$S_i = \text{Max}\ \{0, = k_r - y_i + S_{(i-1)}\}\text{, and } S_0 = 0. \tag{1}$$

where a signal is given if $S_i \geq h$.

For the '*k* in a row' procedure:

y_i = observed count in the ith sample

k_r (the reference value) = $1/k$ (2)

For the 2 within t procedure:

$$y_i = \begin{cases} 0 \text{ if the observed count is } > 0 \\ -1 \text{ if the observed count is } 0 \end{cases}$$

$$k_r = -1/t \tag{3}$$

and $h = 1.0 - \varepsilon$

where $\varepsilon = 1 - 1/2\,k$ for the k in a row scheme and $\varepsilon = 1/2\,t$ for the 2 in t scheme.

Reference values for CUSUMs based upon likelihood ratios are given by Bourke (2001) and Gan (1993) for the binomial distribution and by Lucas (1985, 1989) for the Poisson distribution. For a particular shift in the process parameter these have shown to be optimal in terms of providing the smallest out-of -control ARL for a particular in control ARL by Moustakides (1986).

It is straightforward then to show that the "*k* in a row" scheme is optimal for particular ranges of p because it is a likelihood ratio based scheme in this region. For example, with $n = 50$ and $p = 0.025$ the $k = 6$ in a row scheme is optimal for all $p < 0.000013$. The Shewhart chart is a CUSUM with $h = 0$ and $k = 3$ so it is optimal for six sigma shifts but also performs well for shifts larger than 2 sigma. The "2 in *t*" scheme is not optimal since it ignores information within samples. Collectively though, these schemes perform very well in terms of providing small out of control ARLs for particular in-control ARLs for a wide range of conditions considered by the authors.

Figure 1 demonstrates these points by comparing the ARL curve for two control policies. The first policy is to use a traditional likelihood ratio based CUSUM for both the high and low side. The other policy is to use a Shewhart control chart with UCL = 5(signal at a count of 6) and a $k = 6$ in a row scheme for the low side. In this example we use an $n = 50$ and the in control $p = 0.025$. In addition we consider two traditional CUSUMs; these are chosen so that their in-control ARL brackets the in-control ARL of the Shewhart/"*k* in a row" combination. Note from this figure that the traditional CUSUM method provides lower ARLs for detecting smaller shifts-this is the case for which it was designed. For larger shifts both the Shewhart chart on the high side and the $k = 6$ in a row scheme on the low side have smaller ARLs. We also note that the traditional CUSUM method could be designed to detect these large shifts. Here it would yield ARLs equal to the ARLs of the Shewhart chart and the '*k* in a row' scheme because these are, in fact, likelihood ratio CUSUMs in this region. But the CUSUM method would not have the advantage of the Shewhart chart coupled with the low side schemes since it is not as easy to use, understand or design.

We recommend that these special CUSUM low side add-on schemes be considered as an alternative to the traditional CUSUM methods as discussed by Reynolds and Stoumbos (199, 2000, 2001) because of the advantages of their flexibility, their ease of understanding and use, their large shift detection optimality, and their ease of design using the algorithm given in this paper. We also note that these low side schemes can be easily used in conjunction with any type of high side scheme, including schemes designed around statistical, economic or economic-statistical criteria such as discussed by Saniga, Davis and McWilliams (1995).

In the next section we develop an algorithm to design control charts using these low side schemes in conjunction with a Shewhart type scheme on the high side. The high side scheme is designed using an algorithm published by Saniga, Davis and McWilliams (1995). This algorithm allows consideration of economic, statistical or economic statistical factors to be taken into account in the high side design and thus is very flexible. On the low side the design criterion is statistical.

In the next section we discuss the algorithm and some numerical results. Some brief conclusions are drawn in section 3.

2 Algorithm Description and Some Numerical Results

A high side design must first be specified before using this algorithm. An economic, statistical or economic statistical high side design can be found using the algorithm of Saniga, Davis, and McWilliams (1995). The sample size and acceptance number are then used in the low side design.

2.1 User Specified ARL Bounds

The low side design algorithm involves a simple search over potential Shewhart," two in *t*", and "*k* in a row" low side designs to find those designs which meet user specified constraints on low side ARLs. Like Woodall (1985), we allow users to specify a range of indifference where a small process parameter shift should not be signaled quickly. (Note that the Saniga, Davis, and McWilliams (1995) algorithm also allows the user to specify a range of indifference or range where no signal is desired for small shifts on the high side as well.) In the current version of the programmed version of the algorithm we allow the user to specify up to 25 ARL bounds. Each bound is either designated as a lower bound or an upper bound. Lower bounds are used for the in control state and for small process shifts for which a signal is not desired, while upper bounds are for shifts where a signal is wanted

If we let p equal the probability of a zero count, the lower ARL formula for the 'two in *t*' design is given by (see Lucas (1989), which follows Page (1955) and Brook and Evans (1972)):

$$ARL_{low} = \frac{2-(1-p)^{t-1}}{p-p(1-p)^{t-1}} \tag{4}$$

The lower ARL formula for the 'k in a row' design can be derived using the formula for the ARL vector given by Brook and Evans (1972): $\boldsymbol{ARL} = \lceil \boldsymbol{I}-\boldsymbol{R} \rceil^{-1} \boldsymbol{1}$. The first entry of the resulting ARL vector gives the lower ARL for the 'k in a row' design (without a fast initial response feature) as:

$$ARL_{low} = \frac{\sum_{j=0}^{k=1} p^j}{1-(1-p)\cdot\sum_{j=0}^{k=1} p^j} \tag{5}$$

2.2 Computational Considerations

When the value of p (in 5) is small and/or the value of t (in 4) or k (in 5) is large, direct calculation of equations (4) and (5) can produce underflows. To prevent underflows, terms such as p^j are calculated recursively with the resulting value being checked at each stage of the recursion to see if it has fallen below a small threshold (e.g., 1×10^{-10}). If the result falls below the threshold, calculation of the term ceases and either the current value is retained or set to zero (whichever value would understate the true ARL). In this event, our output indicates that the true ARL value is greater than the displayed value. In our computational experience, this has only affected very large ARLs where calculation precision is not an issue.

The combined ARL is calculated using the reciprocal formula

$$\frac{1}{ARL_{combined}} \cong \frac{1}{ARL_{high}} + \frac{1}{ARL_{low}} \tag{6}$$

which is exact for the 'k in a row' design and approximate for the two in t case.

2.3 User Input

The use of the program, which is written in FORTRAN, requires specification of the following input parameters:

Program Input Parameters	***Example Program Input***
$N, C_{\text{accept high}}$	50 5
$t_{max}, t_{min}, k_{min}, k_{max}$	100 2 2 100
Num	2

For $i = t$, to Num 0.025 1000 0
P_i, $ARLB_i$, $Bound_i$ 0.005 30 1
End for

The symbols N and $C_{accept\ high}$ represent the sample size and acceptance number from the high side design where a high side signal results from a count of $C_{accept\ high} + 1$ or higher. The parameters t_{max} and t_{min} are the upper and lower search limits, respectively, for feasible t values for 'two in t' designs. Similarly, k_{min} and k_{max} are the lower and upper search limits for 'k in a row' designs. Num is the number of constraints the user will specify for the low side design. For each constraint, P_i is the proportion defective, $ARLB_i$ is the bound the user desires to have on the ARL for the associated proportion defective, and $Bound_i$ indicates whether the bound is a lower or upper bound (0 = lower, 1 = upper).

2.4 Algorithm Search Order

The search for low side designs that meet user specified bounds proceeds from the design that is quickest to signal to the design that is slowest to signal. While this is the search logic that we use, we do not wish to imply that the design that signals most quickly will necessarily be preferred in terms of its ARL characteristics. If a feasible Shewhart design exists, it will signal more quickly than either a 'two in t' or a 'k in a row' design.

Step 1: Find N and C for the Shewhart high side scheme using the algorithm of Saniga, Davis, and McWilliams (1995).

Step 2: Specify bounds on ARL for specific in and out-of-control values of the process parameter.

Step 3: Search for Shewhart low side designs that meet all ARL bounds. Given the sample size N, we must find a feasible $C_{accept\ low} \geq 1$ for a Shewhart design (a signal occurs if the count is $C_{accept\ low}$-1 or lower). We search for feasible Shewhart designs from $C_{accept\ low} = C_{accept\ high}$ down to $C_{accept\ low} = 1$.

Step 4: Search for 'two in t' low side designs that meet all ARL bounds. 'Two in t' low side designs signal less quickly than Shewhart designs but more quickly than 'k in a row' designs. We search from $t = t_{max}$ down to $t = t_{min}$ (where t_{max} and t_{min} are user specified values).

Step 5: Search for 'k in a row' designs that meet all ARL bounds. We search from $k = k_{min}$ up to $k = k_{max}$.

Step 6: List feasible low side designs.

2.5 An Illustration

We now illustrate this search process using the example input shown above.

For this problem with characteristics given in Table 1 no Shewhart or 'two in *t*' designs were feasible. Results for the three feasible '*k* in a row' designs ($k = 6, 7$ and 8) are shown in Table 1. Choosing between the three may involve investigating the ARL curves for each *k* in conjunction with a history of process performance.

Table 2 show a design from an example of Saniga, Davis and McWilliams (1995) where the high side design with $p_0 = 0.0136$, $p_H = 0.113$ and ARL bounds of 400 and 5 respectively is $n = 19$ and $C_{\text{accept high}} = 2$. Here, any k between $k = 19$ and $k = 47$ yields a feasible low side designs that meet our low side ARL constraints of 500 and 50 respectively on $p_0 = 0.0136$ and $p_L = 0.0001$. Low side and combined ARLs for three of these ($k = 19, 36, 47$) are given. Note the tradeoff between the ARLs as k increases: ARL_{OCL} increases as ARL_{ICL} increases.

Table 3 shows an example with $p_0 = 0.12$, $p_H = 0.18$, and ARL bounds of 500 and 10 respectively. This yields $n = 84$ and $C_{\text{accept high}} = 19$. In this situation the feasible low side designs comprise the Shewhart chart with $C_{\text{accept low}} = 1$ or 2, any 'two in *t*' design between 2 and 100, and the $k = 2$ or 3 in a row design. Table 3 shows three cases, one each for the Shewhart, the 'two in *t*' and the '*k* in a row' design. Note that in this case the Shewhart design with $C = 1$ dominates the" 2 in $t = 100$" design as well as the "$k = 3$ in a row" design because it has lower ARLs for all decreases in the process parameter and approximately the same in-control ARL. The user would have the flexibility to choose between the $C = 2$ and $C = 1$ Shewhart design as there are tradeoffs to be considered in each of these.
Table 4 contains results for an example where one Shewhart design, any 'two in *t*' in the range explored and one k in a row design is feasible. Here $p_0 = 0.10$, $p_H = 0.30$ and the ARLs are 500 and 1.1 which yields $n = 66$ and $C_{\text{accept high}} = 14$. Once again the tradeoff in control and out of control ARLs is evident as we move from the Shewhart chart to the 'two in *t*' scheme to the '*k* in a row' scheme. The flexibility in allowing the user their choice of which design to employ is apparent.

3 Conclusions

In this paper we have presented an algorithm that can be used to design low side control schemes for count data. This algorithm can be used in conjunction with any high side design method; in this paper we use the algorithm of Saniga, Davis and McWilliams (1995) for high side design These low side schemes are powerful, easy to understand, easy to implement, and are flexible enough to be tailored to the requirements of any real industrial setting. The algorithm we provide makes control chart design for this situation a simple task.

References

1. Acosta-Mejia, C.A. (1999) Improved p Charts to Monitor Process Quality. IIE Transactions 31, 509-516
2. Bourke, P.D. (2001) Sample Size and the Binomial CUSUM Control Chart: The Case of 100% Inspection. Metrika 53, 51-70
3. Brook, D., Evans, D.A.(1972) An Approach to the Probability Distribution of CUSUM Run Length. Biometrika 59, 539-549
4. Eisenhart, C., Hastay, M.W., Wallis, W.A. (Eds.) (1947) Techniques of Statistical Analysis. McGraw-Hill, New York
5. Gan, F.F. (1993) An Optimum Design of CUSUM Control Charts for Binomial Counts. Journal of Applied Statistics 11, 16-31
6. Lucas, J.M. (1989) Control Schemes for Low Level Counts. Journal of Quality Technology 21, 199-201
7. Lucas, J.M. (1985) Counted Data CUSUMS. Technometrics 27, 129-144
8. Moustakides, G.V. (1986) Optimal Stopping Times for Detecting Changes in Distributions. The Annals of Statistics 14, 1379-1387
9. Nelson, L.S. (1997) Supplementary Runs Tests for np Charts. Journal of Quality Technology 29, 225-277
10. Page, E.S. (1955) Control Charts with Warning Lines. Biometrika 42, 243-257
11. Reynolds, M.R., Jr., Stoumbos, Z.G. (1999) A CUSUM Chart for Monitoring a Proportion When Inspecting Continuously. Journal of Quality Technology 31, 87-105
12. Reynolds, M.R., Jr., Stoumbos, Z.G. (2000) A General Approach to Modeling CUSUM Charts for a Proportion. IIE Transactions 32, 515-535
13. Reynolds, M.R., Jr., Stoumbos, Z.G. (2001) Monitoring a Proportion Using CUSUM and SPRT Control Charts. Frontiers in Statistical Quality Control 6, 155-175
14. Ryan, T.P. (1989) Statistical Methods for Quality Improvement. Wiley, New York
15. Saniga, E.M., Davis, D.J., McWilliams, T.P. (1995) Economic, Statistical, and Economic Statistical Design of Attribute Charts. Journal of Quality Technology 27, 56-73
16. Schwertman, N.C., Ryan, T.P. (1997) Implementing Optimal Attribute Control Charts. Journal of Quality Technology 29, 99-104
17. Shore, H. (2000) General Control Charts for Attributes. IIE Transactions 32, 1149-1160
18. Woodall, W.H. (1985) The Statistical Design of Quality Control Charts. The Statistician 34, 155-160

Table 1
ARL Results
***np* Charts with $n = 50$, $c = 5$ and $p = 0.025$**

		k=6		k=7		k=8	
$p_{defective}$	Upper ARL	Lower ARL	Combined ARL	Lower ARL	Combined ARL	Lower ARL	Combined ARL
0.00500	**	15.8	15.8	21.6	21.6	29.0	29.0
0.01000	91,769.8	49.1	49.1	82.8	82.7	138.5	138.3
0.01500	9,729.4	173.8	170.7	372.0	358.3	794.2	734.3
0.02000	2,091.1	672.7	509.0	1,850.0	981.6	5,082.7	1,481.6
0.02500	661.9	2,768.6	534.2	9,821.8	620.1	34,834.1	649.6
0.03000	267.6	11,892.2	261.8	54,539.7	266.3	**	267.4
0.03500	128.1	52,709.2	127.8	**	128.1	**	128.1
0.04000	69.4	**	69.4	**	69.4	**	69.4
0.04500	41.3	**	41.3	**	41.3	**	41.3
0.05000	26.5	**	26.5	**	26.5	**	26.5

** indicates ARL values > 100,000.0

Table 2
ARL Results
***np* Charts with $n = 19$, $c = 2$ and $p = 0.0136$**

		k=19		k=36		k=47	
$p_{defective}$	Upper ARL	Lower ARL	Combined ARL	Lower ARL	Combined ARL	Lower ARL	Combined ARL
0.00010	**	19.4	19.4	37.3	37.3	49.2	49.2
0.00250	68,058.9	31.6	31.6	97.8	97.6	179.8	179.3
0.00500	8,766.4	56.2	55.9	328.4	316.5	956.6	862.5
0.00750	2,676.6	106.1	102.1	1,285.4	868.4	6,228.3	1,872.1
0.01000	1,163.6	210.8	178.5	5,558.4	962.2	45,455.7	1,134.5
0.01360	483.0	607.8	269.1	51,012.5	478.5	**	482.7
0.02000	164.0	4,609.1	158.4	**	164.0	**	164.0
0.04000	26.0	**	26.0	**	26.0	**	26.0
0.06000	9.8	**	9.8	**	9.8	**	9.8
0.08000	5.2	**	5.2	**	5.2	**	5.2
0.11300	2.7	**	2.7	**	2.7	**	2.7

** indicates ARL values > 100,000.0

Table 3
ARL Results
***np* Charts with $n = 84$, $c = 19$ and $p = 0.12$**

$p_{defective}$	Upper ARL	Shewhart $C_{accept\ low} = 2$* Lower ARL	Shewhart $C_{accept\ low} = 2$* Combined ARL	Shewhart $C_{accept\ low} = 1$* Lower ARL	Shewhart $C_{accept\ low} = 1$* Combined ARL	t = 100 Lower ARL	t = 100 Combined ARL	k = 3 Lower ARL	k = 3 Combined ARL
0.02000	**	2.0	2.0	5.5	5.5	10.9	10.9	197.8	197.8
0.04000	**	6.9	6.9	30.9	30.9	62.9	62.9	30,335.2	30,335.1
0.06000	**	28.4	28.4	180.8	180.8	608.8	608.8	**	**
0.08000	**	132.6	132.5	1,101.1	1,091.4	13,900.8	12,502.2	**	24,254.9
0.10000	5,298.6	675.1	598.8	6,976.4	3,011.4	**	5,243.3	**	5,298.6
0.12000	520.6	3,699.4	456.4	46,074.0	514.7	**	520.6	**	520.6
0.14000	91.4	21,655.6	91.0	**	91.3	**	91.4	**	91.4
0.16000	24.6	**	24.6	**	24.6	**	24.6	**	24.6
0.18000	9.2	**	9.2	**	9.2	**	9.2	**	9.2

* signal at $C_{accept\ low} - 1$
** indicates ARL values > 100,000.0

Table 4
ARL Results
***np* Charts with $n = 66$, $c = 14$ and $p = 0.1$**

$p_{defective}$	Upper ARL	Shewhart $C_{accept\ low} = 1$* Lower ARL	Shewhart $C_{accept\ low} = 1$* Combined ARL	t = 100 Lower ARL	t = 100 Combined ARL	k = 2 Lower ARL	k = 2 Combined ARL
0.02000	**	3.8	3.8	7.6	7.6	18.2	18.2
0.04000	**	14.8	14.8	29.6	29.6	233.7	233.7
0.06000	**	59.4	59.4	132.3	132.2	3,584.1	3,499.8
0.08000	5,434.9	245.5	234.9	983.9	833.1	60,502.4	4,986.9
0.10000	529.2	1,047.1	351.5	12,648.9	507.9	**	528.9
0.15000	16.1	45,535.6	16.1	**	16.1	**	16.1
0.20000	3.0	**	3.0	**	3.0	**	3.0
0.25000	1.4	**	1.4	**	1.4	**	1.4
0.30000	1.1	**	1.1	**	1.1	**	1.1

* signal at $C_{accept\ low} - 1$
** indicates ARL values > 100,000.0

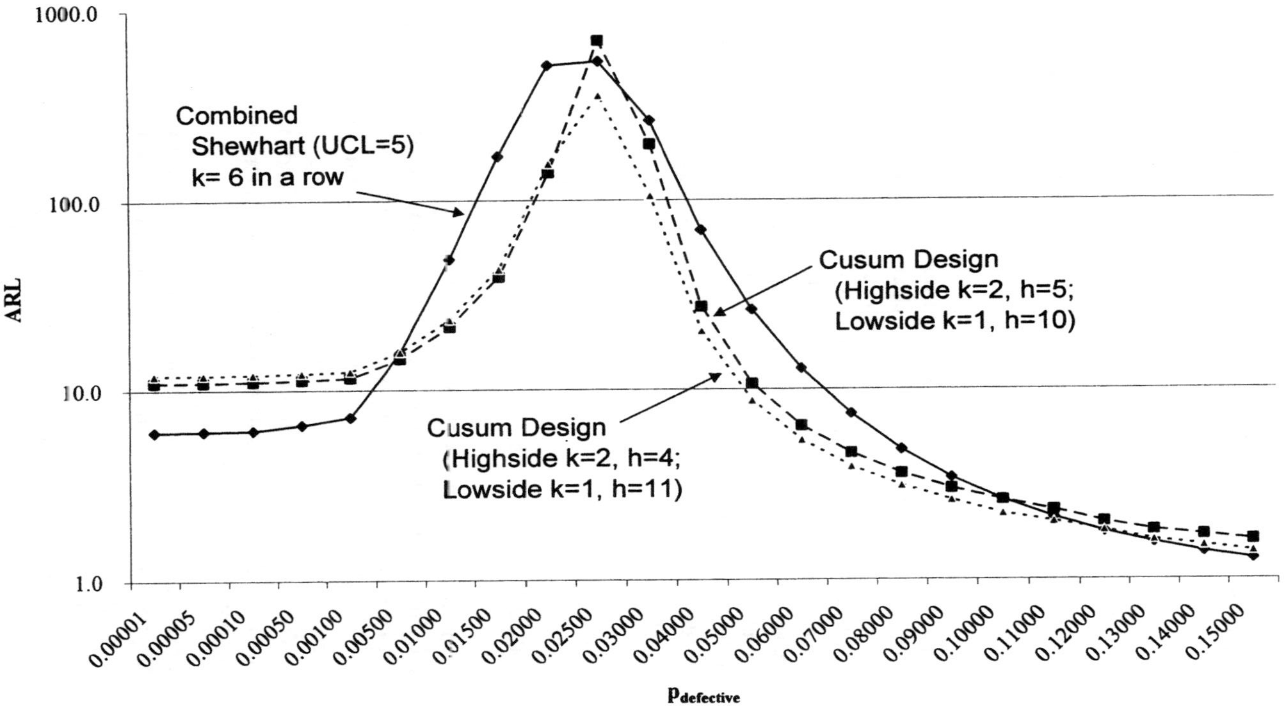

Figure 1: ARL Comparison of CUSUM Schemes

A Markov Approach to the Average Run Length of Cusum Charts for an AR(1) Process

Ken Nishina and Shin'ichiro Matsubara

Nagoya Institute of Technology, Gokiso-cho, Showa-ku, Nagoya 466-8555, Japan

Abstract. Some papers have presented the effects of autocorrelated observations to Cusum charts. In this paper, the performance of Cusum charts in a first-order autoregressive (AR(1)) model is examined in some more detail by representing the Cusum statistic by a two-dimensional Markov process. The distribution of Cusum statistic at any time and the Average Run Length (ARL) can be analytically derived. Thus a conditional out of control ARL for any change point can be obtained. The variation pattern of the out of control ARL can be presented until the ARL converges to the steady state ARL. It can be shown that the variation pattern in the case of the no autocorrelation is monotonously decreasing in the change point. In the presence of autocorrelation, the monotonious decrese still holds, however, only when the autocorrelation is not highly positive.

1 Introduction

It is well known that Cusum charts show high performance for detecting a relatively small shift of process mean, but they are not robust with respect to departures from independence. In many works, the performance of Cusum charts in the presence of autocorrelation has been shown. Especially the presence of positive autocorrelation has a significantly unfavorable effect on the ARL (for example, see Goldsmith and Whitfield (1961) and Johnson and Bagshaw (1974)). Lu and Reynolds (2001) have shown that autocorrelation also affects the determination of optimal parameters.

The statistical performance of Cusum charts such as ARL has been analytically derived by a Markov approach, which was firstly developed by Brook and Evans (1972). The ARL of Cusum charts in the presence of autocorrelation can also be done by the Markov approach (see Rowlands (1976), VanBrackle and Reynolds (1997) and Timmer, Pignatiello and Longnecker (1998)). Johnson and Bagshaw (1974) and Bagshaw and Johnson (1975) can analytically derive ARL and Run Length Distribution (RLD) using approximation of a Wiener process. But their approximation can be inadequate, especially for small values of the control limit (see Nishina (1982) and Rowlands and Wetherill (1991)).

The purpose of this study is to conduct more detailed examinations about the performance of Cusum charts in a first-order autoregressive (AR(1)) model by using a two-dimensional Markov process. The distribution of Cusum statistic at any time is derived, so a conditional out of control ARL can be obtained in the case of any change point.

2 Process model

Let the successive observations be denoted by X_t $(t = 1, 2, \ldots)$, then some different time series models have been considered in previous studies. As far as the first order autoregressive (AR(1)) model is concerned, the following models can be cited as typical ones;

$$X_t = \xi + r(\mu_{t-1} - \xi) + \eta_t + e_t \tag{1}$$

$$X_t = \mu_t + r(X_{t-1} - \mu_{t-1}) + e_t \tag{2}$$

where $-1 < r < 1$ and e_t is an inherent random process error. Harris and Ross (1991) and VanBrackle and Reynolds (1997) considered Equation (1), and Johnson and Bagshaw (1974) considered Equation (2). Equation (1) can be interpreted as a process model by which a process mean is autocorrelated. VanBrackle and Reynolds (1997) have derived the steady state ARL using a Markov approach. On the other hand, Equation (2) is a stationary AR(1) model, where the process mean is not autocorrelated but constant.

In this paper, Equation (1) is considered as a process model; however, since the out of control ARL is analytically calculated for any change point, the process model where the process mean makes an abrupt step shift is considered;

$$X_1 = \mu_0 + e_1 \tag{3}$$

$$X_t = \mu_0 + r(X_{t-1} - \mu_0) + e_t \quad (t = 2, \ldots, \tau) \tag{4}$$

$$X_{\tau+1} = \mu_0 + \delta + r(X_\tau - \mu_0) + e_{\tau+1} \tag{5}$$

$$X_t = \mu_0 + \delta + r(X_{t-1} - (\mu_0 + \delta)) + e_t \quad (t = \tau + 2, \ldots, T) \tag{6}$$

$$e_1 \sim N(0, \sigma^2) \tag{7}$$

$$e_t \sim N(0,\ (1 - r^2)\sigma^2) \quad (t = 2, \ldots, T) \tag{8}$$

where μ_0 is a known target value; δ is the unknown amount of shift; τ is called the change-point; and T, which is a random variable, stands for detection time of the mean shift. Then, without loss of generality we can assume that $\mu_0 = 0.0$, $\sigma^2 = 1.0$ and $\delta > 0$. As the process errors e_1 and e_t ($t = 2, \ldots, T$) are defined as shown in Equation (7) and Equation (8), respectively; we can set $Var(X_t)$ to be unity.

3 Cusum charts and Average Run Length (ARL)

The one-sided Cusum statistic, S_t, is given by

$$S_0 = 0,$$

$$S_t = \max\{S_{t-1} + (X_t - k),\ 0\}. \quad (t = 1, 2, \ldots), \tag{9}$$

An alarm is released by the Cusum chart, whenever

$$S_t \geq h,$$

where h and k are predetermined, positive design parameters, measured in units of σ.

The performance of Cusum charts is usually measured by the ARL, which is the expected number of observations required by the chart to release an alarm. When the process is in control, that is, in Equation (3), Equation (4), Equation (5) and Equation (6) $\delta = 0$, the ARL is the expectation value of T, which is called the in control ARL. When the process mean makes an abrupt step shift at a change point τ, the ARL is the expectation value of $T - \tau$. In the present study, the ARL is evaluated under the condition that no false alarm is released. The ARL is called the conditional out of control ARL (hereafter abbreviated as the out of control ARL). The conditional distribution of the Cusum statistic is asymptotically stationary as shown in Nishina (1986). However, the distribution is varied for the time t until it reaches the stationary distribution. Therefore, the out of control ARL is dependent on the change point τ.

4 Derivation of out of control ARL in AR(1) model

4.1 Markov representation of Cusum statistic

Nishina and Peng-Hsiung (1996) examined how much the distributions of the change point estimator by Cusum charts are affected by a sequence of autocorrelated observations. Following Nishina et al. (1996), the Cusum statistic S_t under the AR(1) model as shown in Equation (3), Equation (4), Equation (5) and Equation (6) can be represented by a two dimensional Markov process, because the AR(1) model can be regarded as a single Markov process. The two-dimensional Markov process means that the state at time t, say z, depends on the state at time t-1, say y, and the state at time t–2, say x. Therefore, the process can be explained as a transition from the state (x, y) to the state (y, z). From this explanation, it should be noted that S_t is dependent on both S_{t-2} and the Cusum statistic before the zero reset shown in Equation (9), say S^*_{t-1}. Therefore, Equation (9) is rewritten as follows:

$$S_{t-1}^* = S_{t-2} + (X_t - k)$$

$$S_t = \max\{S^*_{t-1},\ 0\} + (X_t - k)\,.$$

Let the probability density function of $\Pr\left\{S_t \le z,\ S^*_{t-1} \le y\ \ (y < h)\right\}$ be denoted by $\varphi(t,y,z)$, then

$$\varphi(t,y,z) = \int_{-\infty}^{h} \varphi(t-1,x,y) f^*(x,y,z;\delta_y,\delta_z)dx \quad (t = 3,4,...) \tag{10}$$

$$\varphi(1,0,z) = f(z-(\delta_Z - k)),\ \varphi(2,y,z) = f^*(0,y,z;\delta_y,\delta_z)$$

where

$$f^*(x,y,z;\delta_y,\delta_z) = f((z-(1+r)y + rx^+ - (\delta_z - k) + r(\delta_y - k))/\sqrt{(1-r^2)})$$

$$x^+ = \begin{cases} x, & 0 < x \le h \\ 0, & x \le 0 \end{cases},\quad y < h$$

and $f(x)$ is the density function of the standardized normal distribution; δ_z and δ_y stand for the process mean at time t and t–1, respectively.

As early described, we will derive the out of control ARL under the condition that no false alarm is released. For that preliminary, the probability for the Cusum statistic S_t not to reach h until the change point τ, say $\Phi(\tau)$, as follows;

$$\Phi(\tau) = \int_{-\infty}^{h}\int_{-\infty}^{h} \varphi(\tau, y, z)dydz \quad (\tau \geq 2,\ \delta_y = 0,\ \delta_z = 0)$$

$$\Phi(1) = \int_{-\infty}^{h} \varphi(1, 0, z)dz\ .\ (\delta_z = 0)$$

4.2 Derivation of out of control ARL considering the change point

As mentioned above, the out of control ARL is dependent on the change point τ. Therefore, the out of control ARL has to be derived for any change point in order to investigate effects of the autocorrelation in more detail.

At first, we derive the out of control ARL in the case of $\tau \geq 2$, which is a general formula for τ. Let the out of control ARL be denoted by $L(x, y)$ $(x < h, y < h)$ on the condition that $S_{\tau+1} = x$ and $S_{\tau+2}{}^{*=} y$. Then $L(x, y)$ is solution of the following integral equation;

$$L(x, y) = 1 + \int_{-\infty}^{h} L(y, z) f^{*}(x, y, z; \delta, \delta)dz\ .$$

The above integral equation can be solved by replacing it with a system of linear equations using Legendre-Gauss quadrature (Goel and Wu (1971)). In calculating it, $-\infty$ is replaced with -8, and the numbers of Gaussian points of the intervals $(-8, 0)$ and $(0, h)$ are 18 and 15, respectively. In addition to the solution, let us consider the probabilities $\Pr[S_{\tau+1} \geq h]$ and $\Pr[_{\tau+2} \geq h]$, say $p(\tau+1)$ and $p(\tau+2)$. They can be obtained as follows;

$$p(\tau+1) = \int_{h}^{\infty}\int_{-\infty}^{h}\int_{-\infty}^{h} \varphi(\tau, x, y) f^{*}(x, y, z; 0, \delta)dxdydz$$

$$= \int_{h}^{\infty}\int_{-\infty}^{h} \varphi(\tau+1, y, z)dydz$$

$$p(\tau+2) = \int_{h}^{\infty}\int_{-\infty}^{h}\int_{-\infty}^{h} \varphi(\tau+1, x, y) f^{*}(x, y, z; \rightleftharpoons \delta, \delta)dxdydz$$

$$= \int_{h}^{\infty}\int_{-\infty}^{h} \varphi(\tau+2, y, z)dydz$$

Consequently, the out of control ARL can be obtained as

$$\text{ARL} = \frac{p(\tau+1) + 2p(\tau+2) + \int_{-\infty}^{h}\int_{-\infty}^{h} (L(x, y) + 2)\varphi(\tau+2, x, y)dxdy}{\Phi(\tau)} \qquad (11)$$

The out of control ARL in the case that $\tau = 0$ can be obtained by setting $p(\tau + 1)$ and $p(\tau + 2)$ in Equation (11) as follows:

$$p(\tau+1) = \int_h^{\infty} f(z-(\delta-k))dz = \int_h^{\infty} \varphi(\tau+1,0,z)dz$$

and

$$p(\tau+2) = \int_h^{\infty}\int_{-\infty}^{h} f^*(0,y,z;\delta,\delta)dydz = \int_h^{\infty}\int_{-\infty}^{h} \varphi(\tau+2,y,z)dydz\,,$$

respectively; and setting $\Phi(\tau) = 1$. Similarly, where $\tau = 1$,

$$p(\tau+1) = \int_h^{\infty}\int_{-\infty}^{h} f^*(0,y,z;0,\delta)dydz = \int_h^{\infty}\int_{-\infty}^{h} \varphi(\tau+1,y,z)dydz$$

$$p(\tau+2) = \int_h^{\infty}\int_{-\infty}^{h}\int_{-\infty}^{h} \varphi(\tau+1,x,y) f^*(x,y,z;\delta,\delta)dxdydz$$

$$= \int_h^{\infty}\int_{-\infty}^{h} \varphi(\tau+2,y,z)dydz$$

are set in Equation (11). The above integrals are calculated by Legendre-Gauss quadrature, of which the numbers of Gaussian points are similar to the system of linear equations.

5 Numerical results and their consideration

Fig. 1 shows the variation pattern of the out of control ARL for the change point in the case of no autocorrelation. In Fig. 1, the Cusum charts with parameters $(h, k) = (5.0, 0.5)$ is considered, which is a standard scheme specified in ISO/TR 7871; and in the process model $\delta = 1.0$. Fig. 1 indicates that the out of control ARL monotonously decreases in the change point, and then it converges to the steady state ARL. Let the conditional p.d.f. of the Cusum statistic at time t in Equation (10) in the case of no autocorrelation be denoted by $\phi(t, x)$ $(0 < x < h)$, then $\phi(t, x)$ satisfies the following equation for time t (see Nishina (1986));

$$\int_v^h \phi(t,x)dx \le \int_v^h \phi(t+1,x)dx \tag{12}$$

$$\phi(t,x) \to \phi^*(x) \text{ as } t \to \infty \tag{13}$$

In addition, let the out of control ARL with the initial state v be denoted by ARL(v), then

$$\mathrm{ARL}(v_1) < \mathrm{ARL}(v_2) \quad (0 \le v_2 < v_1 < h)\,. \tag{14}$$

Equation (12), (13) and (14) explain the behavior indicated by Fig.1.

Next, in the presence of positive autocorrelation the variation patterns of out of control ARL for the change point are considered when $\delta = 1.0$. In Fig. 2 they are shown in the cases of $r = 0.4$ and 0.8. Fig. 2 indicates that the monotone does not hold in the case of the strongly positive autocorrelation, $r = 0.8$; on the other hand, when the positive autocorrelation is not so strong, $r = 0.4$, the monotone holds. Fig. 3 shows the variation patterns of out of control ARL for the change point in the negative autocorrelation, $r = -0.4$ and -0.8. As in the case of the no autocorrelation and the weakly positive autocorrelation, they indicate a monotonous decrease in the change point. Fig. 2 and Fig. 3 show the variation pattern of out of control ARL in the case of no autocorrelation, $r = 0.0$; also. Fig. 2 shows that the out of control ARL in the positive autocorrelation is larger than that in no autocorrelation for any change point. On the other hand, the out of control ARL for negative autocorrelation is smaller than that for no autocorrelation for small change point; however, as the change point becomes larger, the former is a little larger than the latter. But the differences are negligible.

6 Conclusion

In the previous studies the effects of the autocorrelation on the steady state ARL and the ARL with the initial state 0 has been presented. In this paper, the variation patterns of the out of control ARL until the steady state ARL are presented by using a two-dimensional Markov approach. These variation patterns in the case of no autocorrelation, negative autocorrelation and weak positive autocorrelation show monotonous decreases in the change point; however, for highly positive autocorrelation, the monotone does not necessarily hold.

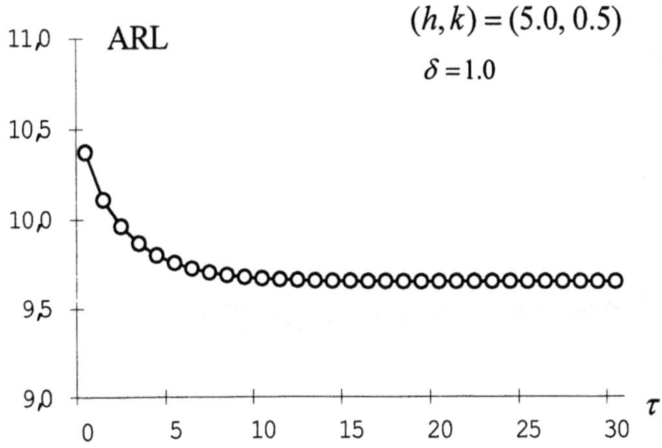

Fig. 1 Out of control ARL for the change point in the no autocorrelation

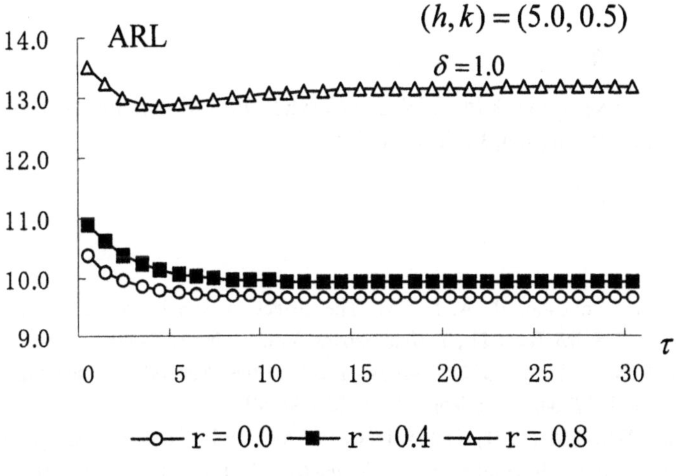

Fig. 2 Out of control ARL for the change point in the presence of positive autocorrelation

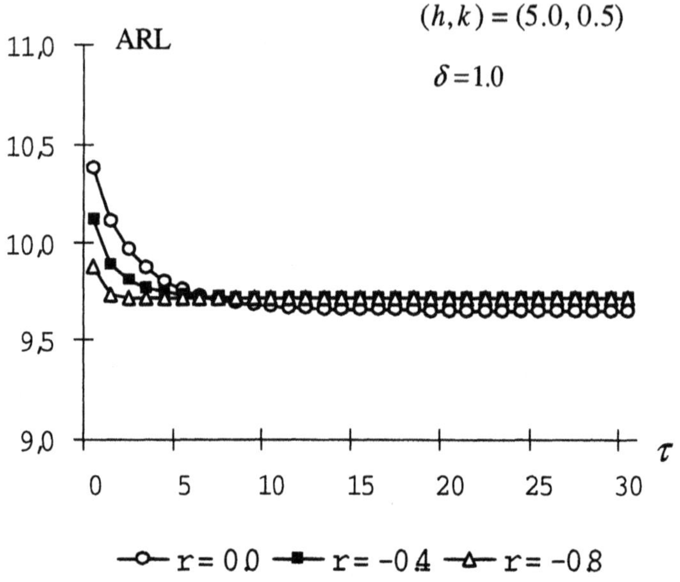

Fig. 3 Out of control ARL for the change point in the presence of negative autocorrelation

Acknowledgement

The authors would like to thank Prof. W. H. Woodall, Prof. S. Knoth and Prof. R. S. Spark for their providing helpful information.

References

1. Bagshaw, M. and Johnson, R. A. (1975): The Effect of Serial Correlation on the Performance of CUSUM Tests II, *Technometrics*, Vol. 17, No.1, 73-80.
2. Brook, D. and Evans, D.A. (1972): "An Approach to the Probability Distribution of Cusum Run Length," *Biometrika*, Vol. 59, No. 3, 539-549.
3. Goel, A. L. and Wu, S.M. (1971): Determination of ARL and a Contour Nomogram for Cusum Charts to Control Normal Mean, *Technometrics*, Vol. 13, No. 2, 221-230.
4. Goldsmith, P. L. and Whitfield, H. (1961): Average Run Length in Cumulative Chart Quality Control Schemes, *Technometrics*, Vol. 3, No. 1, 11-20.
5. Harris, T. J. and Ross, W.H. (1991): Statistical Process Control Procedures for Correlated Observations, *The Canadian Journal of Chemical Engineering*, Vol. 69, 48-57.

6. Johnson, R. A. and Bagshaw, M. (1974): The Effect of Serial Correlation on the Performance of CUSUM Tests, *Technometrics*, Vol. 16, No.1, 103-112.
7. Lu, C. W. and Reynolds, M. R. (2001): Cusum Charts for Monitoring an Autocorrelated Process. *Journal of Quality Technology*, Vol. 33, No. 3, 316-334.
8. Nishina, K. (1982): Design of CUSUM Charts in the Presence of Data Correlation, *Hinshitu*, Vol.12, No.3, 44-53 (in Japanese).
9. Nishina, K. (1986): "Estimation of the Change-point from Cumulative Sum Tests," *Reports of Statistical Application Research, JUSE.*, Vol. 33, No.4, 1-14.
10. Nishina, K. and Wang, P. H. (1996): "Performance of Cusum Charts from the Viewpoint of Change Point Estimation in the Presence of Autocorrelation," *Quality and Reliability Engineering International*, Vol. 12, 3-8.
11. Rowlands, R.J. (1976): Formulae for Performance Characteristica of CuSum Schemes When the Observations are autocorrelated, *Ph. D. thesis, University of Wales, Cardiff.*
12. Rowlands, R.J. and Wetherill, G.B. (1991): Quality Control, In: Ghosh, B.K. and Sen, P.K. (Eds.) *Handbook of Sequential Analysis*, New York etc.: Marcel Dekker.
13. Timmer, D. H., Pignatiello, J. and Longnecker, M. (1998): The Development and Evaluation of CUSUM-based Control Charts for an AR(1) process, *IIE Transactions*, Vol. 30, 525-534.
14. VanBrackle, L. N. and Reynolds, M. R. (1997): EWMA and Cusum Control Charts in the Presence of Correlation, *Communication in Statistics – Simulation and Computation*, Vol. 26, No. 3, 979-1008.

Effective Monitoring of Processes with Parts Per Million Defective. A Hard Problem!

Stefan H. Steiner and R. Jock MacKay

Dept. of Statistics and Actuarial Sciences, University of Waterloo,
Waterloo N2L 3G1, Canada

Abstract. Many processes have defective rates measured in parts per million (PPM). When the process yields such a high level of quality, traditional methods of process improvement, such as designed experiments, and control charts on the number of defectives or the time between defectives, are no longer effective. However, it may still be desirable to monitor and/or improve such processes. In this article, we take a critical look at attempts to apply control charts in this situation. As an alternative, we suggest that since defectives are so rare, we should carefully study any that are observed. By comparing the characteristics of the defectives to good units, both in terms of their physical dimensions and properties, and the process records from their production, we may be able to identify the key differences. Using this type of retrospective study, the goal is to identify an explanatory continuous variable or variables that can be monitored instead of the number of defectives or the time between defectives.

1 Introduction

Currently, many processes produce defectives at a rate less than 100 parts per million (PPM). We shall refer to such a process as a PPM process. For any process monitoring scheme, quickly detecting a deterioration in the defective rate is desirable. Since a PPM process produces so few defectives, looking at the number of defectives in a sample or the time between defectives does not provide much information on per unit basis. This information problem is accentuated if we are not employing 100% inspection. In this article we will address the question: "Can statistical methods be used to monitor changes in the rate of defectives in a PPM process?"

To answer this question we should first clearly define the problem. We assume that each part can be judged defective or not defective. This is a discrete environment where all non defective units are considered equivalent. The framework implies that either there is no known underlying continuous measurement or that it is not (cheaply) observable. For example, the environment does not include the situation where a part is called defective because a continuous measurement is outside of specification. If this were the case, the problem is greatly simplified by monitoring the underlying continuous measurement. We have also excluded the possibility of using compressed limits where it is possible to classify units based on artificial specification limits that are narrower than the actual specifications (Geyer et al. 1996). In this way, the problem of PPM defectives can be transformed into one with a much larger "defective" rate that is more amenable to standard monitoring methods.

One suggested monitoring tool for this discrete environment is a control chart of the number of defectives or the time between defectives (Montgomery, 1991, Nelson, 1994, McCool and Joyner-Motley, 1998). Unfortunately, as will be shown, these previously proposed control charts do not work well for monitoring a PPM process. In this article, we suggest that a better approach is to focus on the defectives produced and compare them in as many ways as possible with good units. The key is to find out how the defectives are different from good units in terms of some other characteristic (or combination of characteristics). Identifying such differences can provide a continuous measurement that can be monitored effectively using $\bar{X}$ and R control charts.

This article is organized in the following manner. In the next section previously suggested control chart based approaches are criticized. We demonstrate the weaknesses of using a *p*-chart to monitor the defective rate directly or Shewhart and sequential charts for the time between defectives. Next a possible remedy is discussed. The approach is based on using a retrospective analysis to identify explanatory variates whose values are related to the defective rate. An example from the automotive industry is given.

2 Critique of Previously Proposed Approaches

One approach to monitor the process performance of a PPM process is to use control charts on either the number of defectives or the time between defectives. Control charting has a long and successful history. The idea is that by quickly determining when the process performance deteriorates, the cause of the deterioration can be identified and eliminated. Control charts work well when the process output can be measured on some continuous scale or when the defective rate is not close to the extremes 0 or 1.

To implement any control chart, we must first observe the process in an in-control state for long enough to allow us to estimate the in-control process performance fairly accurately. However, with PPM processes there is not enough information in the occasional defective item either to set up a control chart or to use it effectively.

2.1 Control Chart to Monitor the Proportion Defective

In the case of good/defective process output, a p chart is recommended (Montgomery, 1991). Subgroups of size n are taken periodically from the process and the proportion of defectives is recorded. To set up a *p*-chart, the standard rule is to set the control limits at $p_0 \pm 3\sqrt{\left[p_0(1-p_0)\right]/n}$, where p_0, the in-control proportion defective is estimated as $\overline{p}$, the average defective rate in numerous subgroups from an in-control process.

These control limits, based on a normal approximation, are not totally satisfactory if np_0 is small as is likely the case in a PPM environment. The limits may be improved either by using an *arcsin* transformation or better yet, by using probability limits derived from the binomial distribution (Ryan, 1989).

However, even with probability limits, there are two major difficulties in using and setting up a p chart here. First, to obtain any reasonable power to detect changes in the PPM process, large subgroup sizes are needed. Montgomery (1991) recommended subgroup sizes large enough so that the probability of finding at least one defective in the subgroup is at least 95%. For a defective rate of p_0 this requirement translates into a subgroup size larger than $3/p_0$. When p_0 is small this rule of thumb results in enormous sample sizes. For example, assuming the process produces 50 PPM defectives when in control, then the $3/p_0$ rule implies subgroups of a least 60,000 units. A second problem is that to set up the chart, we need to estimate p_0 accurately. Assuming we follow the standard recommendation to collect initially at least 20 subgroups in order to set up the chart, in the above example, we must inspect 1.2 million units (from an in-control process) before we can begin to monitor. Similarly, when the defective rate is 5 PPM, the minimum sample size is 600,000 units and 18 million units are needed just to set up the control chart! In most applications, we suspect that these numbers are so large as to make the procedure inoperable.

2.2 Control Chart to Monitor the Time Between Defectives

A clever idea to alleviate the discreteness inherent in monitoring the number of defectives is to monitor the time (or number of good units) between observed defective units. Denote the time between observed defectives as X. In this way, we change the problem from one with discrete measurements to one with a continuous scale. Note that we also avoid the difficult problem of subgroup definition. This approach makes most sense for processes subject to 100% inspection.

Using the time between defectives as a test statistic, we may employ either a Shewhart type chart, or some sequential procedure, such as an EWMA chart or

CUSUM chart (Nelson, 1994). This approach was first suggested by Montgomery (1991), and further explored by Nelson (1994), and McCool and Joyner-Motley (1998). Nelson (1994) suggests an individual chart of $X^{.2777}$ (X to the power 0.2777) to monitor the time between defectives. McCool et al. (1998) consider a number of different possible test statistics and control charts. In particular, they suggest that an exponentially weighted moving average chart (EWMA) of $X^{.2777}$ or log(X) would be appropriate.

Unfortunately, there remain inherent difficulties with this approach in the PPM environment. First, it is expensive to perform the 100% inspection of units required to determine the time between defectives. If 100% inspection is not used the second difficulty is exacerbated. Second, a good estimate of in-control mean time between defectives is needed to set appropriate control limits. For PPM processes, the time between defectives is long, and thus the amount of time (or number of units) required to gather enough information to allow reasonably precise estimation of the mean time between defectives may be too long to be practical. For example, Nelson (1994) suggests that two dozen values of the time between defectives, while the process is in control, are required to reasonably estimate the mean time between failures. When the process produces 5 PPM defective, the expected number of parts between defectives is 200,000. Thus to get 24 values of X requires about 4.8 million units.

Even if we were able to estimate the in-control mean time between defectives accurately the control charts are still not very effective. For example, we may consider some typical results from McCool et al. (1998) for the 5 PPM process. They give the average run length (ARL) for a control chart based on the time between defectives. When the defective rate has increased to 500 PPM the ARL is given as 168.85. However, there are still, on average, 2000 units between failures, therefore on average the chart will pick up a change to 500 PPM defective only after 377,700 units have been inspected. Looking for increases in quality is even worse because the average time between defectives increases. Say the defective rate is reduced to .5 PPM then the ARL is 2.07 and this corresponds to, on average, over 4 million parts!

This problem is not avoided by using a sequential procedure such as an EWMA chart. The performance of an EWMA chart in detecting small shifts will be somewhat superior to the Shewhart chart, but the initial implementation of the EWMA control chart still requires an initial estimate for the in-control defective rate which is not available without massive production volume.

3 A Possible Remedy

The major difficulty in monitoring a PPM process using either the number of defectives in a subgroup or the time between defectives is the small amount of information per unit inspected that such data provides. One solution to this problem is to find an explanatory variate or combination of explanatory variates that are (strongly) related to the defective rate. We assume that the identified explanatory variate is not the underlying continuous measurement that defines defectives and non-defectives, but rather some other product or process characteristic. Naturally, the identified explanatory variate will not be a perfect predictor of whether a unit is defective or not. However, changes in the explanatory variate should be related to the defective rate.

If such an explanatory variate X is found, we can monitor the process using this continuous variate rather than based on the defective rate or time between defectives. This approach avoids the discreteness difficulty in the original problem. The problem is how to identify the important explanatory variable(*s*) when there are so few defective units. The key is to focus on the defectives that do occur. The defectives are compared to good units on as many process and product characteristics, denoted $X_1, \ldots, X_k$, as possible. This approach was promoted by Dorian Shainin (see Bhote and Bhote, 2000) and is the same as the idea underlying case/control studies that are widely used to identify risk factors for rare diseases in human populations (Schlesselman, 1982).

To identify important explanatory variates, we model the relationship between the defective rate p and the explanatory variates using a logistic regression model, i.e.

$$\text{logit}\,(p) = \text{logit}(p) = \log\left(p/(1-p)\right)$$

$$= g(X) = \beta_0 + \beta_1 X_1 + \ldots + \beta_k X_k \qquad (1)$$

Note that, in this context, often the best model would include non-linear terms, such as $(X_i - t_i)^2$, where t_i is the target value for characteristic X_i. With quadratic terms, any deviation of X_i from its target value will increase the defective rate. Of course interaction terms are also possible. The parameters in the above model $\beta_1,\ldots,\beta_k$ can be estimated using a sample of defective units and a sample of good units using standard approaches (Hosmer et al. 1989). The assumption is that each of these samples is representative of the two types. Based on such data, we can estimate a function that differs from $g(X)$ only in the intercept term. That is, we can estimate

$$\hat{\alpha}_0 + \hat{\beta}_1 X_1 + \ldots + \hat{\beta}_k X_k \qquad (2)$$

The intercept term β_0 in $g(X)$ is not estimable without knowledge of the sampling fraction of good and bad units. In fitting this logistic model, the goal is to find explanatory variates whose corresponding model parameters (β_l) are not zero.

3.1 Required Sample Sizes

An important question relates to the number of good (controls) and defective (cases) units that are needed or desirable for the analysis. Clearly when more defectives (and non-defectives) are used in the analysis significant explanatory variates can be identified more easily. However, the number of units required will depend on the magnitude of the effect of the particular explanatory variate.

We explore this issue in more detail by considering a single explanatory variate and assuming the distribution of the explanatory variate is Normal in both the cases and controls. In most work on sample size requirements in case control studies in the medical literature it is assumed that the explanatory variate is binary (Schlesselman, 1982). However, Lubin et al. (1988) consider a continuous explanatory variate. Assuming a single explanatory variate the sample size required for a logistic regression model to identify an explanatory variate as significant is equivalent to a comparison of the mean levels of the explanatory variate for good and defective units. We let μ_i and σ_i be the mean and standard deviation of the explanatory variate for good units ($i=0$) and for defectives ($i=1$). We find the sample size needed for a size α-level test with power $1-\beta$.

Following Lubin et al. (1988), the required number of defective units (cases), assuming k not defective units (controls) for each case is

$$m = \frac{k+1}{k} \frac{\left[Z_\alpha \sigma_x + Z_\beta \left\{ \left(k\sigma_1^2 + \sigma_0^2 \right) / (k+1) \right\}^{1/2} \right]^2}{\left(\mu_1 - \mu_0 \right)^2} \tag{3}$$

where $\sigma_x^2 = \left(\sigma_1^2 + k\sigma_0^2 \right) / (k+1) + \left(\mu_1 - \mu_0 \right)^2 k / (k+1)^2$, and Z_α and Z_β satisfy the equations $\Pr(Z > Z_\alpha) = \alpha$; $\Pr(Z > Z_\beta) = \beta$, where Z is a standardized Normal random variable.

To explore equation (3) we set $\alpha=0.05$ and $\beta=0.1$, and without loss of generality consider the case where for controls the distribution of the explanatory variate is standardized to have mean zero and standard deviation equal to one, i.e. $\mu_0=0$ and $\sigma_0=1$. This leads to three unknowns, the mean and standard deviation for defectives and the number of controls per case we choose. Only the number of controls per case can be chosen by the investigator, so these results should be consider simply guidelines. First, we consider the situation where we have a single control for each case. Figure 1 shows contours of constant m as a function of μ_1

and σ_1. We see that, as expected, the number of cases required decreases as the mean of the explanatory variate for cases increases.

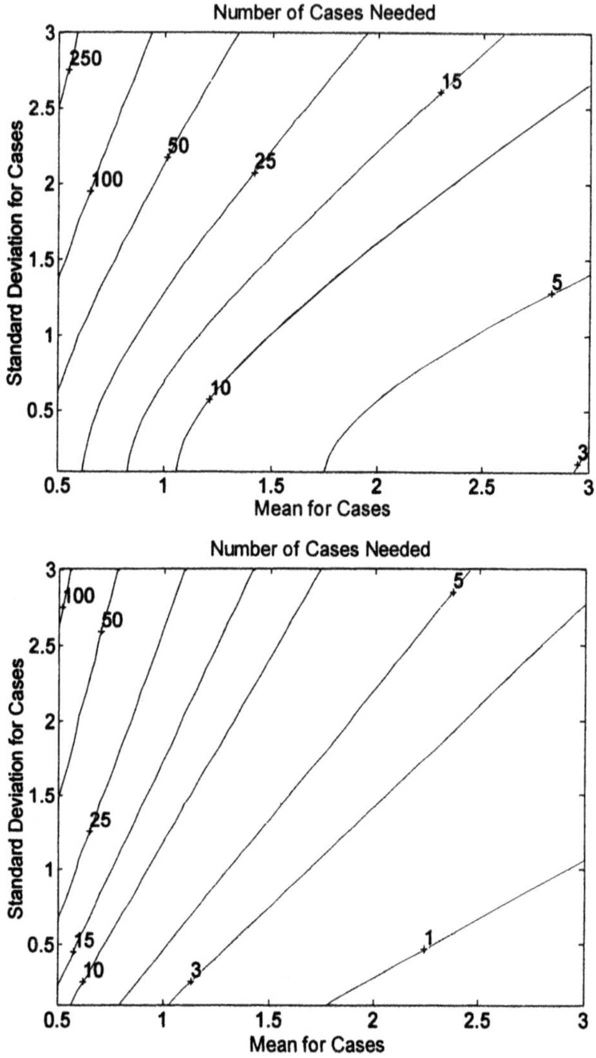

Fig. 1. Contours of the number of defective units needed
one control for each case on top, large number of controls for each case on the bottom

To give a better idea of the effectiveness of increasing the number of controls for each case Figure 2 shows the ratio of the required sample sizes from the two plots in Figure 1, i.e. Figure 2 plots m_1/m_∞, where m_k give the required number of cases when there are k not defective units (controls) for each case. It is interesting that the improvement possible through the use of multiple controls is much greater when the explanatory variate is continuous than when the explanatory is binary. Following Schlesselman (1982) using a large number of controls when the explanatory variate is binary can decrease the number of cases needed by at most a factor of 2. In the industrial PPM context increasing the number of controls (non defective units) for each case (defective units) is typically inexpensive.

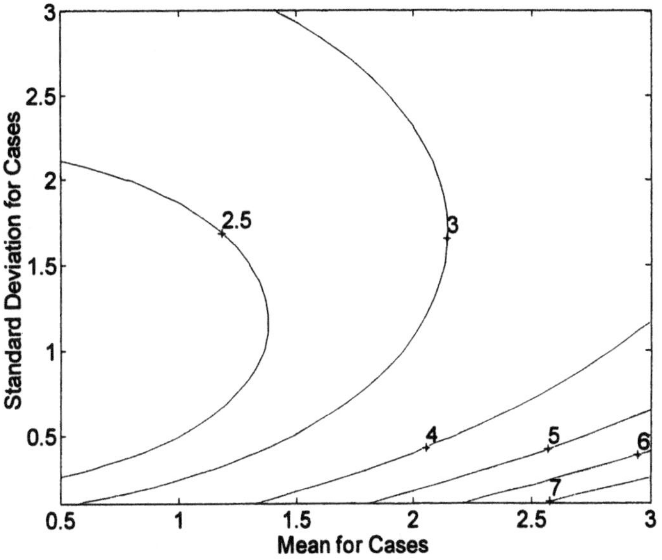

Fig. 2. Contours of the ratio of the required sample size for one control vs. many controls

4 Example

Exhaust valve seats are force fitted by insertion into the head of an engine. If the valve seat is not installed correctly, it can lead to a catastrophic engine failure. The quality of the fit is judged by visual inspection using feeler gauges. Given the high volume (four seats per head, two heads per engine, 1500 engines per day), 100% inspection is very costly and likely to be ineffective, especially since the expected defective rate is low, less than 50 PPM.

A sample of 25 defective seat insertions was collected over time. Pareto analysis showed that there was no evidence that the poorly fitted seats depended on location in the head. Since no head had more than one defective seat, the remaining three seats on the head were used as controls.

Eleven measurable, potentially important explanatory variates were identified. These are X_1, X_2, X_3, X_4, X_5, measurements of force, work and distance taken during the automated insertion process, X_6, X_7, X_8, X_9, dimensional and physical characteristics of the valve seat and X_{10} and X_{11} dimensional characteristics of the pocket in the head into which the seat is inserted.

The explanatory variates X_6 through X_{11} were measured on seats after insertion and there was suspicion that their values may have been distorted by the insertion process. Nevertheless these variates were included in the analysis. The plan was to include those characteristics identified as being important in the monitoring procedure. However, in that case, they would be measured before the insertion process.

A logistic regression model was fit to the 100 observations. Three explanatory variates, X_2, X_4, and X_9 were identified as important. Since X_2 and X_4 were measured on every insertion, an automated CUSUM chart based on a linear combination of X_2 and X_4 was constructed using the software available in the insertion process. The characteristic of the valve seat X_9 was monitored separately using an average and range chart based on subgroups of 5 parts collected with a regular frequency.

5 Discussion

In applying the proposed remedy it is important that the identified explanatory variates used in the linear combination (2) are truly related to the defect rate. If they are not we will be monitoring a linear combination that may well vary in ways unrelated to the defective rate. In this case, the introduction of spurious explanatory variates will "muddy the waters."

A typical reaction in industry to very rare defects is to do a complete "postmortem," including a detailed look for the probably causes, on each individual defective observed. While such an approach can yield improvements it is flawed unless we link the results from all the "postmortems" in an attempt to find some commonality. This idea of looking for commonalities across many defects is one of the principles underlying the proposed approach.

Finally, it should be noted that a designed experiment is not likely to be a viable alternative approach to process improvement in the context described in this article. With a binary response (defective / not defective) either very large sample sizes would be needed, or the factor levels would have to be set so extremely that the conclusions from the experiment would likely not be relevant to the current operating conditions of the process.

6 Summary

Monitoring of the defective rate of a process that produces defectives measured in parts per million (PPM) is often desirable. The use of control charts to monitor the defective rate of such processes based on the number of defectives or the time between defectives is shown to be infeasible. The sample sizes required to set up the control charts are much too large in most practical situations. As an alternative, we suggest a focus on the few defectives that are produced. In particular, we suggest a case/control type comparison of defectives and non-defectives based on as many of their other attributes as possible. If we can find some explanatory variable or combination of variables that is associated with defectives (or non defectives) we may be able to determine a continuous variate to monitor.

Acknowledgments

This research was supported, in part, by the Natural Sciences and Engineering Research Council (NSERC) of Canada.

References

1. Bhote, K.R. and Bhote, A.K. (2000), *World Class Quality*, 2nd edition, AMACOM, American Management Association, New York.
2. Geyer, P.L., Steiner, S.H. and Wesolowsky, G.O. (1996) "Optimal SPRT and CUSUM Procedures with Compressed Limit Gauges," *IIE Transactions*, 28, pp. 489-496.
3. Hosmer, D.W. Lemeshow, S. (1989), *Applied Logistic Regression*, Wiley and Sons, New York.
4. Lubin JH, Gail MH, Ershow AG (1988), "Sample Size and Power for Case-Control Studies When Exposures are Continuous," *Statistics in Medicine*, 7, 363-376.
5. McCool, J. I., and Joyner-Motley, T. (1998), "Control Charts Applicable When the Fraction Nonconforming is Small," *Journal of Quality Technology*, 30, 240-247.
6. Montgomery, D.C. (1991), *Introduction to Statistical Quality Control*, Second Edition, John Wiley and Sons, New York.
7. Nelson, L. S., (1994), "A Control Chart for Parts-Per-Million Nonconforming Items," *Journal of Quality Technology*, 26, 239-240.
8. Ryan, T.P. (1989), *Statistical Methods for Quality Improvement*, John Wiley and Sons, New York.
9. Schlesselman, J.J. (1982), *Case Control Studies*, Oxford University Press, Oxford.

On Monitoring Processes and Assessing Process Capability under a Hierarchical Model, Part 1

Poul Thyregod[1], Henrik Melgaard[2], and Jørgen Iwersen[2]

[1] Informatics and Mathematical Modelling (IMM), Technical University of Denmark, Richard Petersens Plads 321, DK 2800 Lyngby, Denmark
[2] Novo Nordisk A/S, Novo Alle, DK 2880 Bagsværd, Denmark

Abstract. In this paper we discuss the practical problems involved in following the rational subgrouping principle for control charting as many processes exhibit variation that exceeds the within subgroup variation. As an example we consider simple models for processes where the two extremes 'high frequency variation' and 'low frequency variation' both contribute to inherent process variation.

We emphasize the importance of distinguishing between charting aimed at identifying situations where minor process adjustments are needed to compensate for process disturbancies, and charting aimed at demonstrating a state of 'statistical control' with a predictable long-term distribution of process output. For the latter purpose we discuss the modelling of process variation by means of a hierarchical model for normally distributed measurements. The approach is illustrated by a case from the pharmaceutical industry. We finally discuss the assessment of process capability under a hierarchical model for process variation.

1 Introduction

In the summer of 1999, quality managers in the German automotive industry organized a workshop with the title "SPC on the test bench". The dire background for the workshop was that although the trend in current quality management standards (like e.g. QS 9000 [2]) is to put more emphasis on the use of statistical methods, and in particular on implementation of SPC, the experience in the industry was that use of current SPC tools was not adding value to the industry. The industry representatives reported on numerous cases where processes perpetually signalled an out-of-control situation although no cause could be assigned and, consequently no corrective action could be designated. A study of 1000 different processes in the industry had shown that about 90% of the processes exhibited variation of the charted averages that was significantly larger than what could be tolerated according to the calculated within subgroup variance. As such situations are not covered by the simple, first order approach to charting techniques described in industry standards, the experience from the practical implementation of SPC in the industry was that 'theory' is far from practice.

Figure 1 (from [3]) illustrates this observation. The figure shows an $\overline{X}-R$ chart for 200 consecutive samples of 0.1 m cuts of string. The cuts have been organized in subgroups of size 5, and standard 3-sigma limits were constructed based upon the measurements from these 40 subgroups. The process spread appears to be in control, but although the process mean appears to be fairly stable, the fluctuations of the subgroup averages is larger than allowed for by the within-group variation.

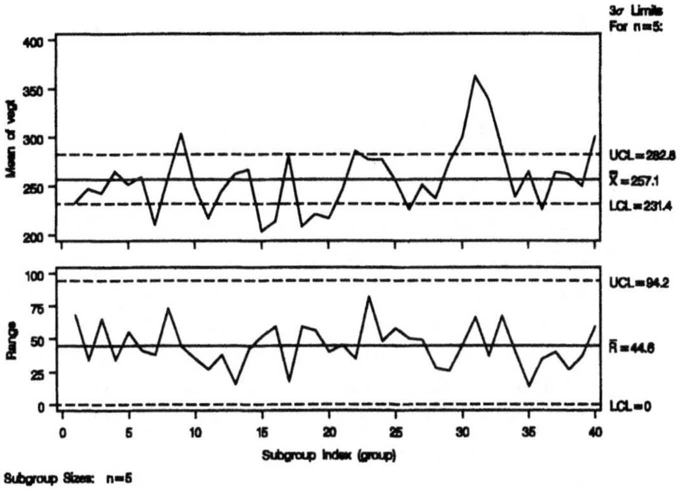

Fig. 1. Example of $\overline{X} - R$ chart for subgroups of size 5. Weight of successive samples of 0.1 m cuts of string

In the literature a number of authors have recognized the misleading nature of using within-subgroup variability to construct $\overline{X}$-chart control limits in the presence of inherent variability between subgroups, see e.g. the review by Woodall and Thomas [4] and the instructive examples given by Sparks and Field [5], but unfortunately this message does not play a prominent role in introductory textbooks, and misleads the user to put focus upon *variance* rather than *variation.*

In the paper we discuss various models for the transfer of variation in industrial processes, the relation to the designation of subgroups, and the consequence for the practical use of control charts and assessment of process capability. As a pragmatic, practical tool we suggest to base the charting on a hierarchical model reflecting between subgroup variation. In an accompanying paper [1], we discuss further details of such hierarchical models as well for the usual normal distribution model as for other 'natural' within-group distribution models.

As in all situations of statistical modelling, the in-control model is a matter of choice reflecting the purpose and intended use of the model. We extend the classical in-control model, which in the usual case is iid normal variates, to also include low frequency components of variation. In a traditional interpretation of the classical Shewhart approach this might be considered to be an out-of-control situation, but in some practical situations low frequency components are part of the normal operating conditions, and in these situations it is might well be argued that such variation should be part of the in-control model.

2 Cause systems, rational subgroups and process spread

2.1 Cause systems and rational subgroups

The basic idea in process monitoring and process improvement is to consider the process as resulting from a system of causes, and to understand the nature of variation and the transfer of variation in this system. This was already pointed out by Shewhart, and the socalled fishbone diagram introduced by Ishikawa is now a recognized tool for identifying potential causes of variation.

In the 1920's when Walter Shewhart developed the concepts and terms for what is now known as SPC, the modern theory of statistics was only in its beginning. The central limit theorem for independently acting causes had been established by Ljapunov. The theorem had a great impact for a theoretical probabilistic description of physical phenomena. Thus, statistical mechanics was thriving. The foundations of the theory of probability necessary for a formal probabilistic treatment of stochastic processes useful for modelling evolution in time (like diffusion processes) were, however, still under development by Kolmogorov and Khinchin.

In statistics, methods for statistical analysis of time series were very rudimentary, analysis of variance techniques were hardly formalised, and although Fisher in 1918 [6] did employ variance components analysis for his genetic research, applications of variance components models to industrial problems do not seem to have been published until 1931 in the textbook by Tippet [7]. Concepts, such as "blocks" related to design and analysis of experiments, and "clusters" related to sampling had not yet been formulated.

On that background, the fundamental concepts 'constant system of chance causes' and 'assignable causes' were introduced to distinguish *signal* from *noise* and to emphasize the basic message that physical laws are statistical in nature, and hence, variation exists in all processes. The ideal goal of control was to reduce output variation to resemble a pure noise process, i.e. a process subject to a constant system of chance causes.

The control chart together with the concept of a 'rational subgroup' were suggested as appropriate tools to demonstrate and monitor a state of statistical control. The fundamental idea being to use the variation within a "rational subgroup" to express the inherent process variation, i.e. the variation that may be ascribed to "a constant system of chance causes" in accordance with the central limit theorem.

Unfortunately, in practical applications, the within subgroup variation does not always reflect all inherent process variation, and therefore subgroup averages that deviate more from centerline than can be explained by the within group variation, do not necessarily signal that an 'assignable cause has occurred', but might just be a consequence of an inherent dynamical behaviour of the process that is not governed by the simple statistical theory of independently acting causes.

2.2 Slow, 'regular' variation

In order to illustrate the relation between the choice of sampling interval, inherent process dynamics and distribution of process output we shall consider the following simple model: Assume that samples of size n are taken at discrete times, equispaced at some unit interval, Δ_t.

In particular we want to illustrate the different effects of 'high frequency variation' reflected in the within subgroup variation and 'low frequency variation' reflected in the between subgroup variation. For simplicity we shall model the high frequency variation as independent 'noise' added to the process output. The inherent process dynamics is modelled by an AR(1) process.

Thus, we shall consider the following linear dynamical model for the output X_{ij} of the j'th item ($j = 1, 2, \ldots, n$) in the sample selected at time $t = i\Delta_t$:

$$\begin{aligned} X_{ij} &= \mu Z_i + E_{ij} \\ Z_i &= \phi Z_{i-1} + A_i \end{aligned} \tag{1}$$

with $0 < \phi < 1$ where E_{ij} are mutually independent and normally distributed with mean zero and variance $\mathsf{V}[E_{ij}] = \sigma_E^2$ and A_i are mutually independent and normally distributed with mean zero and variance $\mathsf{V}[A_i] = \sigma_A^2$, and also independent of the E's.

We emphasize that the E's do not represent measurement error, but are considered to contribute to the variation of true process output.

The total variation of process output is a combination of the 'low frequency' variation, $\sigma_A^2/(1-\phi^2)$ and the 'high frequency' variation, σ_E^2

$$\mathsf{V}[X_{ij}] = \mathsf{V}[Z_i] + \mathsf{V}[E_{ij}] = \frac{\sigma_A^2}{1-\phi^2} + \sigma_E^2 \,. \tag{2}$$

Measurements within a subgroup are correlated with *intra-subgroup correlation*

$$\rho_w = \frac{\mathsf{COV}\,[X_{ij}, X_{ij'}]}{\sqrt{\mathsf{V}\,[X_{ij}]\,\mathsf{V}\,[X_{ij'}]}} = \frac{\sigma_A^2}{\sigma_A^2/(1-\phi^2)+\sigma_E^2} \,. \tag{3}$$

The autocovariance function for process output is

$$\gamma_k = \mathsf{COV}[X_{ij}, X_{i-k,j}] = \sigma_A^2 \phi^k \tag{4}$$

for $1 \leq k$.

Assume now that the process is only sampled at intervals m time units apart where m is an integer. Thus, the new sampling times are $t' = i' \times (m\Delta_t)$, with $i' = 1, 2, \ldots$. The slow variation is governed by $Z'_{i'} = Z_{i'm}$ which again behaves like an AR(1)-process, but now with the autoregressive parameter $\phi' = \phi^m$ and innovations A' with variance

$$\sigma_m^2 = \frac{1-\phi^{2m}}{1-\phi^2}\,\sigma_A^2 = (1+\phi^2+\cdots+\rho^{2(m-1)})\sigma_A^2\,. \tag{5}$$

The smaller the frequency of sampling (i.e. the larger the value of m), the larger is the innovation variance, σ_m^2. For $m > \ln(0.01)/\ln(\rho)$, the innovation variance amounts to 99% of the variance, $\sigma_A^2/(1-\phi^2)$ caused by the Z-process.

The autocovariance function for the sampled process output under this scheme is

$$\gamma'_{k'} = \mathsf{COV}[X_{i',j}, X_{i'-k',j}] = \sigma_m^2 \phi^{k'm} = \frac{1-\phi^{2m}}{1-\phi^2}\,\phi^{k'm}\sigma_A^2 \tag{6}$$

for $1 \leq k'$. The lag-k autocovariance decreases with the thinning factor, m, and with the autogressive parameter, ϕ. Thus, for sufficiently large values of m, the lag-1 autovariance of the Z' process will be negligble, and successive subgroup averages will behave like independent observations.

Table 1 shows the inflation, σ_m^2/σ_A^2 from (5), of the innovation variance and the lag-1 autocovariance, γ'_1/σ_A^2 from (6), of the Z' process for various values of ϕ and m. Because of the the 'high frequency' noise term, E_{ij}, the autocorrelation in the series of subgroup averages, $\overline{x}_{k'}$ will be even smaller than indicated by $\gamma'_{k'}$.

Table 1. Inflation of innovation variance, and lag-1 autocovariance for a series thinned by a factor m

	$m = 5$		$m = 10$		$m = 15$	
ϕ	σ_m^2/σ_A^2	γ'_1/σ_A^2	σ_m^2/σ_A^2	γ'_1/σ_A^2	σ_m^2/σ_A^2	γ'_1/σ_A^2
0.9	3.43	2.02	4.62	1.61	5.04	1.03
0.8	2.48	0.81	2.75	0.29	2.77	0.10
0.7	1.91	0.32	1.96	0.06	1.96	0.01
0.6	1.55	0.12	1.56	0.01	1.56	0.00
0.5	1.33	0.04	1.33	0.00	1.33	0.00
0.4	1.19	0.01	1.19	0.00	1.19	0.00
0.3	1.10	0.00	1.10	0.00	1.10	0.00
0.2	1.04	0.00	1.04	0.00	1.04	0.00
0.1	1.01	0.00	1.01	0.00	1.01	0.00

We note that variation of process output corresponding to this thinned process is given as

$$\mathsf{V}[X_{i'j}] = \mathsf{V}[Z'_{i'}] + \mathsf{V}[E_{i'j}]$$

$$= \frac{\sigma_m^2}{1-\phi^{2m}} + \sigma_E^2 = \frac{\sigma_A^2}{1-\phi^2} + \sigma_E^2 \,. \tag{7}$$

As the use of a larger sampling interval essentially is equivalent to a further subsampling of the process, the variation of process output reflected by the samples from the larger interval is the same as (2), and, similarly, the intra-subgroup correlation is unchanged.

It may be shown (see e.g. [8]) that when the original process Z_i may be described by an ARMA(p,q) process then the process $Z'_{i'} = Z_{i'm}$ resulting from sampling at intervals m time units apart may be described by an ARMA(p,Q) process with $Q = [p + (q-1)/m]$, where $[x]$ means the integer part of x.

Thus, as already noted in the early days of industrial statistics [9], taking samples further apart diminishes the correlation between successive samples. Figure 2 shows a chart for individuals for a 'thinned ' version of the data presented in Figure 1. Only every fifth measurement of the original series has been used for the chart. Although there is still some correlation present, the chart does indicate that the process is in control. (The example should only be considered as an illustration of the effect of thinning measurements from a process with autocorrelation. The full story is that the measurements were part of a process study aiming at obtaining insight in the process dynamics in order to select an appropiate sampling unit and sampling interval. As the string was produced of natural fibres with individual fiber length of up to one meter, successive cuts of 0.1 meter contained almost the same fibres, and as weight variation was mainly due to variation between the fibres, a strong correlation was not surprising. In order to achieve independence between successive cuts, the weight of cuts of length 5 meter were used for charting. Sensitivity could then be tuned by the number of such cuts in a subgroup.)

The literature on control charting techniques offers a variety of techniques for charting processes with inherent dynamics deviating from the simple white-noise assumption underlying the classical Shewhart approach. When designating a sampling interval it is, however, important to consider the purpose of the charting and the nature of the 'assignable causes' that the chart should demonstrate.

2.3 The purpose of charting

Sometimes charting is used to signal the need for making such process adjustments as change of tools, or corrections for variation in input material or other process disturbancies. Even in highly automated processes with feedback adjustment of input compensatory variables there may sometimes be a need for more 'soft' procedures for process adjustments, or for monitoring the operation of the control algorithm and its effectiveness in performing process adjustments. Box and Luceño [10] discuss the choice of sampling interval under 'engineering process control' under consideration to the anticipated process disturbances.

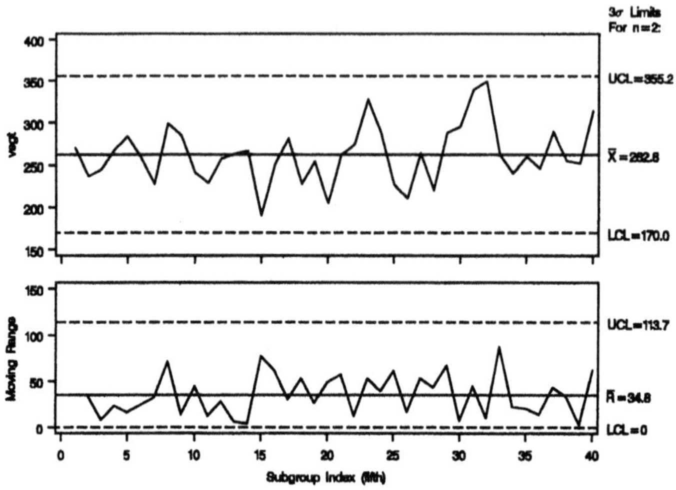

Fig. 2. Example of chart for individuals for the thinned version of the series shown in Figure 1

However, in many industrial applications, charting is used as a means to supervise the process and document the stability of process output. Assignable causes are unforeseen events that could make the product unfit for use. Usually such causes can not be handled by the operator but requires 'corrective actions' at a higher level. When an 'out-of-control' situation is signaled, provisions in the quality assurance procedures specify that an investigation shall be instituted, and, following the review by the investigation team, a corrective action should be taken to prevent this cause from ever occurring again. Most quality management standards require that out-of-control signals are recorded, the investigation is reported, and the effect of the corrective action is documented. This holds in particular for highly regulated industries like the pharmaceutical industry. Quality audit teams are usually very concerned that effective corrective actions have been implemented. Therefore, it is of great importance that the application of control charts reflects the intended purpose. If, too often, false alarms are received, confidence in the use of the control charts is lost, not only by the operators, but – as in the case of the German automotive industry – possibly also by auditing organizations and the quality management.

2.4 Slow, 'irregular' variation

The two purposes considered above represent two extremes in the whole spectrum of potential assignable causes that could be the target for the charting of the process. In any case, it is important that designation of subgroups, sampling interval and control limits reflect the purpose of the charting and the inherent process variation.

However, in practical applications it is often seen that the fundamental assumption that the within subgroup variation captures all 'common cause' variation is violated. Most often practical considerations suggest that the items in a subgroup are selected within a short time-interval thereby disregarding an inherent process dynamics. In other cases inherent variations, e.g. in raw material result in small shifts of process level at irregular time instants, not covered by the variation within a "rational subgroup".

For a large number of practical situations the simple model does not apply. For production processes run to run, or batch-to-batch variation may be considered as part of the natural variation. For laboratory controls, day-to-day variation may be considered a natural part of the process variation. Hence these sources of variation must be included as part of the process being in statistical control.

Although some autocorrelation in process output might be established empirically, such correlation might be result of a number of 'common causes', or factors, each with its own dynamic pattern of variation, like changes of laboratory controls, change of tools, changes in the different components of raw material. Although most of these factors are at a constant level within a subgroup, the aggregated effect of the change of levels between subgroups does not lend itself to be modelled just by the autocorrelation in the series, and sometimes it is more appropriate – under consideration to the purpose of the charting – to model the variation between subgroups as a variance component.

Thus, as suggested by Woodall and Thomas [4] the use of variance components may sometimes be more appropriate than a (more or less) equivalent correlated subsample representation. Woodall and Thomas give two major reasons for their preference, viz. that process improvement is easily expressed with variance components as reductions in σ_B^2 and σ_W^2 and, secondly that the use of variance components is much easier to extend to more than two components of common cause variability.

One further reason should also be emphasized. In the first, retrospective, stage of control limit determination when trial control limits are produced, the objective is to identify limits for the variation in the charted variable that can be expected when the process is running under normal (stable) conditions corresponding to the current state of control for the process. This stage is often referred to as stage 1. As pointed out by Ryan [11], in this Stage 1 the objective should be "to identify the points that have come from a common population, using whatever approach that will facilitate this identification". Ryan emphasizes the importance of the robustness of the charting approach in this stage. In other words, as the limits established in this stage should reasonable well represent the normal pattern of variation in future recordings, it is important that all 'common cause' variation is represented in this stage and taken into account when the limits are established. Thus, when the within subgroup variation fails to capture all common cause variation, it should be ensured that the control limits nevertheless do represent all inherent process variation. Similar considerations are relevant when designating charts to be used in regulated industries to document the stability of processes.

We find that a hierarchical model with random subgroup effects (also termed components of variance) could well be used to produce such robust limits. As indicated in section 2.2, by a proper choice of sampling interval such an approach may also fairly well represent the effect of autocorrelation.

Similar considerations are well appreciated in statistical models for the design and analysis of experiments. Here, the concept of 'blocking factors' to represent variation in operating conditions is an important tool to model the aggregated effect of factors that are not completely controlled, and observations are collected in 'clusters', each block being a cluster. The extension of the so-called general linear model to 'mixed models', and the availability of statistical software for the analysis under such models allows for assessment of the effect of blocking factors, also in unbalanced situation where the levels of these factor change at irregular intervals. Thus, utilizing the information in a retrospective analysis of past data in this way, the Stage 1 monitoring of the process using aggregated variance components is not in conflict with the quest for continual process improvement. When individual variance components have been identified, these components may be monitored individually as described by Yashchin [12], or the effect between runs may be eliminated by means of orthogonal constrasts in order to detect systematic changes in within-run conditions as described by Heredia-Langner et al. [13].

The statistical literature is concerned with type I and type II errors. We believe, however, that in practical statistical work, two other types should also be of concern:

Type A To consider an effect to be of practical significance, only because a statistical test has demonstrated significance

Type B Fail to consider an effect to be relevant, only because a statistical test has failed to demonstrate significance

Considerations of the errors of type A has led to the design of so-called 'acceptance control charts', see Freund [14].

An error of type B occurs when application of a standard test fails to demonstrate a significant between subgroup variation, and it is therefore decided to use a traditional charting procedure with limits based only upon the within group variation. Even when a statistical test has failed to demonstrate a significant between subgroup variation, such variation might still be part of inherent process variation, and use of such limits can lead to perpetual, unneccessary out of control signals as more data are collected and added to the control chart.

Therefore, a good understanding of inherent process variation and potential inherent between subgroup variation is of concern. In the simple normal case the statistical analysis might be performed by means of a single random factor analysis of variance. The random factor being interpreted as the low frequency variation and the residual variation as the high frequency variation of the data. A statistical test may reveal if the low frequency variation in the data is statistically significant. Sound engineering judgement should be used to combine the evidence produced by the statistical test with engineering knowledge of the process to decide upon the appropriate charting procedure. A significant low frequency variation may be ignored in the final model if this source of variation is not considered to be a part of the in-control state. On the other hand the random effect may be included based on process knowledge even if the statistical test failed to demonstrate significance on the current data. In some cases process knowledge alone might warrant inclusion of the random effect, even without statistical testing.

3 The variance component model for between subgroup variation

3.1 The model

The usual variance component model corresponding to the one-way classification by subgroups is

$$X_{ij} = \mu + A_i + E_{ij} \tag{8}$$

where E_{ij} are mutually independent and normally distributed with mean zero and variance $\mathsf{V}[E_{ij}] = \sigma_W^2$ and the *random effects*, A_i, are mutually independent and normally distributed with mean zero and variance $\mathsf{V}[A_i] = \sigma_B^2$, and also independent of the E's.

The total variation of process output is

$$\mathsf{V}[X_{ij}] = \mathsf{V}[A_i] + \mathsf{V}[E_{ij}] = \sigma_B^2 + \sigma_W^2 \,. \tag{9}$$

Borrowing the term from other fields of quality engineering, we shall denote the ratio

$$\gamma = \frac{\mathsf{V}[A_i]}{\mathsf{V}[E_{ij}]} = \frac{\sigma_B^2}{\sigma_W^2} \tag{10}$$

between the variance of 'low-frequency' process disturbancies and 'high-frequency' process disturbancies the *signal to noise ratio.*

Measurements within a subgroup are correlated with *intra-subgroup correlation*

$$\rho_w = \frac{\mathsf{COV}\,[X_{ij}, X_{ij'}]}{\sqrt{\mathsf{V}\,[X_{ij}]\,\mathsf{V}\,[X_{ij'}]}} = \frac{\sigma_B^2}{\sigma_B^2 + \sigma_W^2} = \frac{\gamma}{1+\gamma} \,. \tag{11}$$

In current statistical literature on variance components models – especially for genetic studies – this *intraclass correlation* is often used as a summarizing measure. An analogous measure, the *intracluster correlation*, is used in theories of sampling, when items are selected in *clusters*.

The subgroup averages, $\bar{x}_i$ are independent and identically distributed with variance

$$\sigma_a^2 = \mathsf{V}\left[\overline{X}_i\right] = \sigma_B^2 + \frac{\sigma_W^2}{n} \tag{12}$$

$$= \sigma_W^2\left(\gamma + \frac{1}{n}\right) \tag{13}$$

$$= \sigma_W^2\left(\frac{\rho_w}{1-\rho_w} + \frac{1}{n}\right) \tag{14}$$

$$\tag{15}$$

with γ and ρ_w defined in (10) and (11), respectively.

3.2 Estimation of the parameters

Consider the usual ANOVA-decomposition of the total variation into within subgroup and between subgroup variation

$$\sum_{i=1}^{N}\sum_{j=1}^{n}(x_{ij}-\bar{x})^2 = \sum_{i=1}^{N}\sum_{j=1}^{n}(x_{ij}-\bar{x}_i)^2 + \sum_{i=1}^{N} n(\bar{x}_i-\bar{x})^2\,, \tag{16}$$

and let

$$S_3^2 = \sum_i\sum_j (x_{ij}-\bar{x}_i)^2/[N(k-1)] \tag{17}$$

$$S_2^2 = \sum_i (\bar{x}_i-\bar{x})^2/(N-1) \tag{18}$$

represent the within-group empirical variance, and the empirical variance between group averages, respectively. The value of S_2^2 may be obtained from the usual mean square error between subgroups, in the ANOVA table, by dividing by the subgroup size, n.

As

$$\mathsf{E}[S_2^2] = \sigma_B^2 + \sigma_W^2/n \tag{19}$$

$$\mathsf{E}[S_3^2] = \sigma_W^2$$

the direct *method of moments* estimates of σ_W^2 and σ_B^2 are found to be

$$\tilde{\sigma}_W^2 = s_3^2 \tag{20}$$

$$\tilde{\sigma}_B^2 = s_2^2 - s_3^2/n \tag{21}$$

The estimates (20) and (21) are unbiased estimates of σ_W^2 and σ_B^2. Moreover, the empirical variance, s_2^2, of the sample averages, is an unbiased estimate of σ_a^2.

The distributional properties of s_2^2 and s_3^2 follow from the properties of the χ^2-distribution. Thus, the subgroup averages, $\overline{x}_i$ may be monitored by means of an $\overline{X}$-chart with limits directly determined from the estimate s_2^2 of σ_a^2, eg.

$$\mathrm{CL} = \overline{x} \pm 3s_2 \tag{22}$$

The discussion above referred to a situation where all subgroup sizes were equal. When subgroup sizes vary, various complications arise.

Assume that the size of subgroup i is n_i, $i = 1, 2, \ldots, N$. The decomposition (16) is now

$$\sum_{i=1}^{N}\sum_{j=1}^{n_i}(x_{ij} - \overline{x})^2 = \sum_{i=1}^{N}\sum_{j=1}^{n_i}(x_{ij} - \overline{x}_i)^2 + \sum_{i=1}^{N} n_i(\overline{x}_i - \overline{x})^2 , \tag{23}$$

with $\overline{x} = \sum_i n_i\overline{x}_i / \sum_i n_i$.

Consider the contribution to (23) from the variation between groups,

$$\mathrm{MSE}_B = \sum_{i=1}^{N} n_i(\overline{x}_i - \overline{x})^2/(N-1) . \tag{24}$$

As

$$\mathsf{E}[\mathrm{MSE}_B] = (\sigma_W^2 + n_0\sigma_B^2) \tag{25}$$

with

$$n_0 = \frac{\left(\sum_i n_i\right)^2 - \sum n_i^2}{(k-1)\sum_i n_i} \tag{26}$$

denoting an 'effective average sample size', the method of moments leads to the estimator

$$\tilde{\sigma}_B^2 = (\mathrm{MSE}_B - s_3^2)/n_0 \tag{27}$$

Individual 3-sigma control limits for $\overline{x}_i$ are then determined as

$$\mathrm{CL} = \overline{x} \pm 3\sqrt{\tilde{\sigma}_B^2 + s_3^2/n_i} \tag{28}$$

Introducing the simple arithmetic average of the sample sizes, $\overline{n} = \sum_i n_i/k$, the weighted average sample size, n_0 may be expressed as

$$n_0 = \overline{n}\left[1 - \frac{N}{N-1}\sum_{i=1}^{N}\left(\frac{n_i}{\sum_i n_i} - \frac{1}{N}\right)^2\right] , \tag{29}$$

where

$$\sum_{i=1}^{N}\left(\frac{n_i}{\sum_i n_i} - \frac{1}{N}\right)^2$$

measures the deviation between the actual distribution of the total sample $\sum_i n_i$ and the uniform allocation to groups. The larger the deviation from a uniform distribution, the larger a correction to the arithetic average is needed to obtain the 'effective sample size'.

The monograph by Rao [15] describes estimation, hypothesis testing and confidence interval procedures for balanced, as well as unbalanced mixed models. The implications for the estimation of process capability will be discussed in section 5.

4 Example

We consider laboratory data concerning a Kjeldahl analysis. The nitrogen content of dry insulin is determined by the Kjeldahl method, which consists of three steps, digestion, distillation and titration. *Blinds* are empty digestion tubes treated as samples. A number of quality parameters are monitored for this method, one of them are the blinds. The data considered are blind samples, 60 measurements, three per day during a 6 months period. It is of interest to assess whether the process may be considered to be in control. In a first attempt a standard $\overline{X}$, s - chart is applied to the data. The charts are shown in Figures 3 and 4

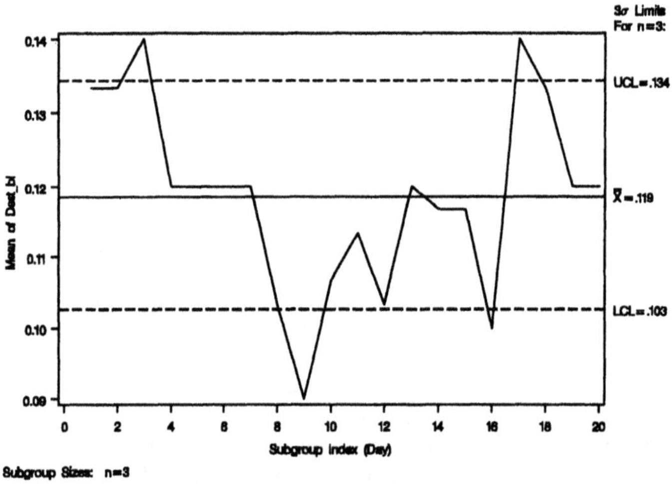

Fig. 3. x-bar chart for Kjeldahl analysis

Four observations are outside the limits of the $\overline{X}$ chart, indicating the process is not in-control.

The average is necessarily influenced by the inherent day-to-day variation, hence the random error within day is not sufficient to describe the random error of the average. We have to estimate the day-to-day variation as well as the random error, and account for both of them when the $\overline{X}$-chart is constructed.

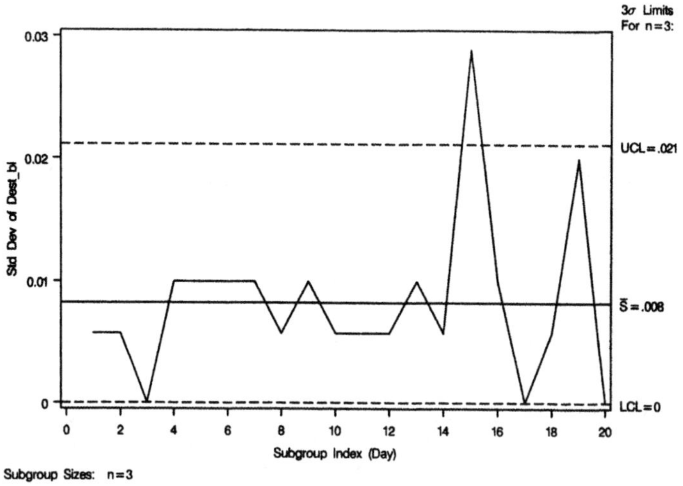

Fig. 4. s-chart for Kjeldahl analysis

When day-to-day variation is present, the limits of the $\overline{X}$- chart are given by :

$$\begin{aligned} \text{UCL} &= \overline{\overline{x}} + 3s \\ \text{CL} &= \overline{\overline{x}} \\ \text{LCL} &= \overline{\overline{x}} - 3s \end{aligned}$$

In the example the estimates are given by:

$$s = \sqrt{\text{Var(day-to-day)} + \text{Var(Random Error)}/n} \tag{30}$$

We found

$$\begin{aligned} s &= \sqrt{\text{Var(day-to-day)} + \text{Var(Random Error)}/n} \\ &= \sqrt{0.000142 + 0.00011/3} = 0.0134 \end{aligned} \tag{31}$$

Hence :

$$\begin{aligned} \text{UCL} &= 0.1185 + 3 \times 0.0134 = 0.1586 \\ \text{CL} &= 0.1185 \\ \text{LCL} &= 0.1185 - 3 \times 0.0134 = 0.0784 \end{aligned} \tag{32}$$

Note that the s-chart is unchanged. When these adjusted limits are used in the charts, we will not conclude that the process is out-of-control

5 Process capability indices

Process capability indices are used quite widely in quality assurance work. They are regarded as convenient indicators of the 'capability' of a process to

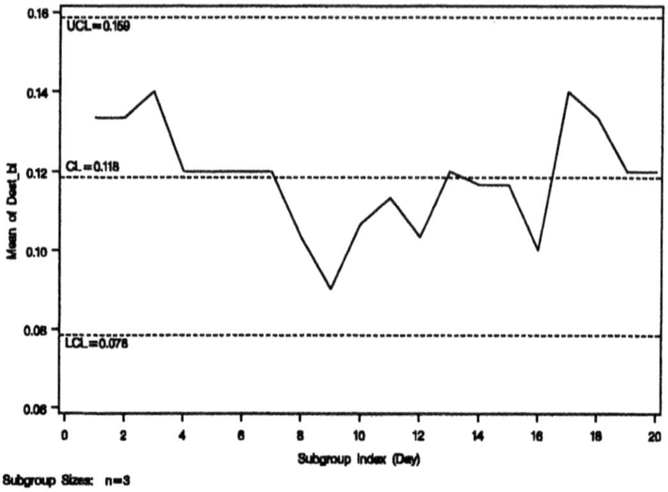

Fig. 5. adjusted x-bar chart for Kjeldahl analysis

produce items with a specified measurable characteristic between specified limits (LSL and USL). The basic concept was suggested by Juran who introduced the 'capability rate' of a process as a crude indicator of this capability. Juran's capability rate is currently denoted C_p defined as

$$C_p = \frac{USL - LSL}{6\sigma} \tag{33}$$

with σ denoting the standard deviation of output.

As indicated by Nelson [16], an important consideration is the procedure for estimating the standard deviation of the process for use in calculating a capability index. The phrase 'standard deviation of the process" refers to the standard deviation of the individual items produced by the process.

The following three methods – quoted from Nelson – are intuitive candidates for obtaining an estimate of the standard deviation of a process (assuming that the process is deemed to be in control)

1. Use the individual observations from which the control chart was made
2. Use the subgroup averages that constitute the $\overline{X}$-chart. If the subgroup averages are each of size n and the standard error of these subgroup averages is s_a, then the estimate of the standard deviation of the individuals is $s_a/\sqrt{n}$.
3. Use the average range (or average standard deviation) from the control chart

When the rational subgrouping has succeeded such that all 'common cause' variation is reflected in the within subgroup variation, all three methods would give essentially the same result (although, possibly, the confidence intervals might be of different width). However, when the fundamental assumption concerning the rational subgrouping is not fulfilled, and some remaining common cause variation is not captured by the within group variation, the resulting estimates will differ – and not represent the standard deviation of process output.

The three methods reflect the decomposition of the total variation in the samples in within and between subgroup variation (16). The corresponding expected values under the component of variance model,

$$\begin{aligned} \mathsf{E}[S_1^2] &= \sigma_W^2 + \frac{n(N-1)}{nN-1}\,\sigma_B^2 \simeq \sigma_W^2 + \sigma_B^2 \\ \mathsf{E}[S_2^2] &= \sigma_B^2 + \sigma_W^2/n \\ \mathsf{E}[S_3^2] &= \sigma_W^2\,, \end{aligned} \tag{34}$$

show that – although the most popular – method 3 could be very misleading as the low frequency process variation has been eliminated in this estimate that only measures the high-frequency process variation. The many signals from the corresponding control chart also indicate that is not appropriate to perform a standard capability analysis. When all items in a subgroup are selected within a short period of time, the estimate might be used to assess the so-called machine capability.

Method 2 provides the estimate $s_a^2 = s_2^2$ of the variance associated with the subgroup means, and although $s_a/\sqrt{n}$ might be suited as a crude plug-in in standard formulae for control limits for means, neither s_a^2 nor s_a^2/n provides the proper balance between σ_W^2 and σ_B^2 to reflect the total variation in process output.

Method 1 directly measures the total variation of process output, and – disregarding inferential problems arising from the distributional properties of S_1^2 – seems the most promising candidate for a crude estimate of process capability in a balanced situation. Approximate confidence intervals related to this estimate may be established by means of Satterthwaite's approximation, [17]. The validity of the approximation has been discussed in [18], [19] and [20].

For unbalanced data, the method of moments estimates could be improved upon by resorting to an estimation procedure that respects the unbalance in greater detail, like the residual maximum likelihood method (REML). For a discussion of the merits of various estimation methods see the monograph by Rao [15].

We have only considered the situation with normally distributed item characteristics. However, similar considerations are appropriate also under other distributional models. The important issue is still to distinguish between the distribution of the charted quantity used for establishing control limits (method 2) and the distribution of process output used for assessing process capability (method 1). Faddoul et al. [21] discuss the assessment of process capability under a bimodal distribution of process output resulting from a contaminated proces, and Shore [22] and Zhang [23] discuss the estimation of capability indexes when process data are autocorrelated.

In Part two we discuss the construction of stage 1 control limits under hierarchical models for non-normal distributions.

References

1. Thyregod, P., Iwersen, J. and Melgaard, H. (2003) On monitoring processes and assessing process capability under a hierarchical model, Part 2. To appear in *Frontiers in Statistical Quality Control 7*, H.-J.Lenz, P.-Th.Wilrich (Eds.), Physica Verlag, 2003
2. Quality System Requirements, QS - 9000, 3rd edition, Automotive Industry Action Group, Soutfield, Michigan
3. Sørensen, O. B. (editor) (1954) Statistisk kvalitetskontrol III Eksempler på Anvendelse. Dansk Ingeniørforenings Industrisektion, Teknisk Forlag, Denmark (In danish)
4. Woodall, W. H., Thomas, E. W. (1995) Statistical Process Control With Several Components of Common Cause Variability. IIE Transactions **27** 757-764
5. Sparks, R. S., Field, J. F. B. (2000) Using Deming's Funnel Experiment to Demonstrate Effects of Violating Assumptions Underlying Shewhart's Control Charts. The American Statistician **54**, 291-302.
6. Fisher, R. A. (1918) The Correlation Between Relatives on the Supposition of Mendelian Inheritance. Transactions of The Royal Society, Edinburgh **52** 399-433
7. Tippet, L. H. C. (1931) The Methods of Statistics, 1st Ed. William and Norgate, London
8. MacGregor, J. F. (1976) Optimal Choice of the Sampling Interval for Discrete Process Control. Technometrics **18** 151-160
9. Bayley, G. V. and Hammersley, J .M. (1946) The "Effective" Number of Independent Observations in an Autocorrelated Time Series. Journ. Roy. Statist. Soc. Suppl. **VIII** 185-197
10. Box, G., Luceño, A. (1994) Selection of Sampling Interval and Action Limit for Discrete Feedback Adjustment. Technometrics **36** 369-378
11. Ryan, T. P. (1997) Efficient Estimation of Control Chart Parameters. In Lenz H.-J., Wilrich P.-Th. (Eds) Frontiers in Statistical Quality Control **5**, Physica-Verlag, Heidelberg, 90-101
12. Yashchin, E. (1994) Monitoring Variance Components. Technometrics **36** 379-393
13. Heredia-Langner, A., Montgomery, D. C., Runger, G. C., Borror, C. M. and Post, R. I. (2001) Performance of Customized Control Charts to Detect Process Disturbancies. Quality and Reliability Engineering International **17** 205-218
14. Freund, R. A. (1957) Acceptance Control Charts. Industrial Quality Control **XIV** No.4, 13-19, 22
15. Rao, P. S. R. S. (1997) Variance Components Estimation. Mixed Models Methodologies and Applications. Chapman & Hall, London
16. Nelson, L .S. (1999) The Calculation of Process Capability Indices. Journal of Quality Technology **31**, 249-250

17. Satterthwaite, F. E. (1946) An Approximate Distribution of Estimates of Variance Components. Biometrics Bulletin **2** 110-114
18. Seely, J. F. and Lee, Y. (1994) A Note on the Satterthwaite Confidence Interval for a Variance. Commun. Statist. - Theory Meth. **23** 859-869
19. Stern, S. E. and Welsh, A. H. (2000) Likelihood Inference for Small Variance Components. The Canadian Journal of Statistics **28** 517-532
20. Longford, N. T. (2000) On Estimating Standard Errors in Multilevel Analysis. The Statistician **49** 389-398
21. Faddoul, N. R., English, J. R. and Taylor, G. D. (1996) The impact of mixture distributions in classical process capability analysis. IIE Transactions **28** 957-966
22. Shore, H. (1997) Process Capability Analysis when Data are Autocorrelated. Quality Engineering **9** 615-626
23. Zhang, N. F. (1998) Estimating Process Capability Indexes for Autocorrelated Data. Journal of Applied Statistics **25** 559-574

On Monitoring Processes and Assessing Process Capability under a Hierarchical Model, Part 2

Poul Thyregod[1], Jørgen Iwersen[2], and Henrik Melgaard[2]

[1] Informatics and Mathematical Modelling (IMM), Technical University of Denmark, Richard Petersens Plads 321, DK 2800 Lyngby, Denmark
[2] Novo Nordisk A/S, Novo Alle, DK 2880 Bagsværd, Denmark

Abstract. In industrial SPC-applications, it is not always appropriate to use a normal distribution with constant variance to describe the within group distribution of measurements. The common statistical properties for the basic distributions of counts, fractions and within group variance may be described by treating them as exponential dispersion families characterized by the relation of the variance to the mean. In the paper we review the theory of exponential dispersion families and discuss the use of standard conjugate distributions to model the between group distribution of means of the charted variable. We introduce a parametrization of the between group distribution by the overall mean of the charted statistic and a 'signal to noise ratio' describing the overdispersion as a factor to the average within group variance. The discussion is exemplified by two cases from industry, one showing overdispersion of fractions, and the other showing nonhomogeneity of within-subgroup variances.

1 Introduction

Thyregod et al. [1] discussed the importance of using control limits that truly reflect inherent process variation under normal (controlled) operating conditions in order to avoid undesirable out-of-control signals resulting from an inherent process variation between subgroups that exceeds the variation accounted for by the within-subgroup variation. In that paper the use of a simple variance component model was suggested as a basis for the charting of subgroup averages in a situation with normally distributed measurements. Using the theory for general linear mixed models, the model is easily extended to cover also situations where variation between subgroups may be ascribed to different identifiable sources.

However, in practice there is a need to treat the problem of 'overdispersion' of group averages when compared with the within group variance also for non-nonnormal distributions. Under a fixed effect model for group means, the extension of general linear models to the so-called generalized linear models (see McCullagh and Nelder [2]) has proven to be a valuable tool for analyzing the mean-value structure of observations from exponential dispersion families of distributions. The further development of generalized linear models to generalized linear mixed models, and hierarchical generalized linear models as described by Lee and Nelder [3] provides a framework for modeling random variation of subgroup means also for non-normal distributions.

In the paper we review the relevant theory for exponential dispersion families of distributions and generalized linear models, and introduce a simple hierarchical model to describe between subgroup variation in a Stage 1 retrospective analysis of historical data in order to describe the anticipated long term distribution of subgroup averages.

2 Exponential dispersion families of distributions

Among parametric statistical models, exponential families play a prime role because of their structure and their inferential properties. Exponential families of distributions were first highlighted by Fisher's [4] observation that non-trivial reduction by sufficiency is only possible in exponential families. Interest in exponential families continues to be strong, partially because of the development of methods for analysing generalized linear models. Among exponential families, the so-called *natural exponential families* with densities of the form

$$f_X(x;\theta) = c(x)\exp\{\theta x - \kappa(\theta)\} \tag{1}$$

play a fundamental role. (For $X \sim \mathrm{N}(\theta, 1)$ the density for the distribution of X is of the form (1) with $\kappa(\theta) = \theta^2/2$).

From the point of view of applications – in particular in relation to generalized linear models – the extension to exponential dispersion families provide a large class of probability distributions, discrete as well as continuous for 'common-cause' variation and makes it possible to obtain insight in the statistical properties of the models in a unified manner.

An *exponential dispersion family* of distributions for the variable Y has density (or probability function) of the form

$$f_Y(y;\theta,\lambda) = c(y,\lambda)\exp[\lambda\{\theta y - \kappa(\theta)\}]\ . \tag{2}$$

The parameter $\theta \in \Omega \subset \Re$ is called the *canonical parameter*, and $\lambda > 0$ is called the *precision parameter*. For most of the common distributions, the parameter space, Ω, is the entire real line. (For $Y \sim \mathrm{N}(\theta, \sigma^2)$ the density for the distribution of Y is of the form (2) with θ corresponding to the mean value, $\kappa(\theta) = \theta^2/2$ and $\lambda = 1/\sigma^2$).

The basic idea underlying this generalization is to separate the mean-value related distributional properties described by the *cumulant generator* $\kappa(\cdot)$ from such dispersion features as sample size, common variance, or common overdispersion that are not related to the mean value. These latter features are captured in the precision parameter, λ.

In some applications the precision parameter is a known number of observations as in the binomial distribution, or a known shape parameter (degrees of freedom) as in the gamma (or χ^2-) distribution. In other applications the precision parameter represents an unknown dispersion as in the normal distribution, or an *overdispersion* that is not related to the mean. In the latter cases it is usual practice (see e.g. [2]) to use the *dispersion parameter* $\sigma^2 = 1/\lambda$ to denote the reciprocal precision. To allow for an intuitive interpretation, we alternate between these two practices in the following presentation, and use σ^2 when an interpretation as common overdispersion is more appropriate than an interpretation as a precision weight. (In some statistical software systems the precision parameter is sometimes reparametrized to a so-called 'scale' parameter where the reparametrization depends on the distribution considered).

Formally, an exponential dispersion family is just an indexed set of natural exponential families, indexed by the precision parameter, λ, and therefore the family shares fundamental properties with exponential families.

The family is uniquely determined by the cumulant generator, $\kappa(\cdot)$, that essentially determines the moments (cumulants) of the distribution. In particular we have

$$\mathsf{E}\,[Y] = \kappa'(\theta) \tag{3}$$

$$\mathsf{V}\,[Y] = \frac{\kappa''(\theta)}{\lambda}\,, \tag{4}$$

and hence, the monotone function

$$\tau(\theta) = \kappa'(\theta) \tag{5}$$

defines a one to one mapping $\mu = \tau(\theta)$ of the parameter space, Ω, for the canonical parameter, θ, onto a subset, $\mathcal{M}$, of the real line, *the mean value space.* Roughly speaking, the mean value space is the convex hull of the support of the distribution. For some of the common distributions the mean value space is only a subset of the real line. For example, for the binomial distribution $0 < \mu < 1$.

Introducing the mean value parameter in (4) we may express the variance as a function, the *(unit) variance function*, of the mean value in the distribution as

$$V(\mu) = \kappa''(\tau^{-1}(\mu)) \tag{6}$$

Since a natural exponential family is uniquely determined (within the class of all natural exponential families) by its variance function (see e.g. Morris [5]), an exponential dispersion family is analogously uniquely determined by its variance function.

The variance function expresses the well-known feature for the commonly used one-parameter distributions that the variance in the distribution depends on the mean value. In SPC-applications the mean value is monitored using $\overline{X}$-charts, p-charts, u-charts, etc., and it is current practice to monitor the stability of the dispersion parameter, σ^2, only in the case of normally distributed data. The tacit assumption for data represented by counts and fractions is that within subgroup variation results from a constant system of chance causes and hence may be satisfactorily described by a Poisson or binomial distribution with no overdispersion within subgroups.

Morris [5] has identified the natural exponential families with at most quadratic variance functions. These include the normal, gamma, Poisson, binomial and negative binomial distributions. The inverse Gaussian distribution also deserves interest as a natural exponential family with variance function, $V(\mu) = \mu^3$, defined for $\mu > 0$.

The parametrization by the mean value, μ, facilitates the comparison of observations, y, with parameter values, μ, according to some model. As a means for comparing observations, y, with parameter values, μ, the *unit deviance* between an observation y and a parameter value μ may be defined as

$$d(y;\mu) = 2\int_{\mu}^{y} \frac{y-u}{V(u)}\,\mathrm{d}u \quad , \tag{7}$$

where $V(\cdot)$ denotes the variance function. Thus, the unit deviance is an expression of the deviation of y from μ, weighted by the reciprocal variance for each value between y and μ. In the case of a normal distribution where the variance does not change with the mean, i.e. $V(\mu) = 1$, the unit deviance is simply $d(y;\mu) = (y-\mu)^2$. For a gamma distribution, the coefficient of variation does not change with the mean, i.e. $V(\mu) = \mu^2$, the unit deviance is $d(y;\mu) = 2 \times \{y/\mu - \ln(y/\mu) - 1\}$.

Expanding $d(y;\mu)$ around $y = \mu$ one finds the approximation

$$d(y;\mu) \approx \frac{(y-\mu)^2}{V(\mu)} . \tag{8}$$

The density (2) may be expressed in terms of the mean value parameter, μ as

$$g_Y(y;\mu,\lambda) = a(y,\lambda)\ \exp\left\{-\frac{\lambda}{2}\,d(y;\mu)\right\} , \tag{9}$$

where $d(y;\mu)$ is the unit deviance (7). (See Jørgensen [6] for the measure-theoretical technicalities related to the representation for lattice distributions).

Thus, there are two equivalent representations for an exponential dispersion family. In one representation the family is characterised by the cumulant generator, $\kappa(\cdot)$ and parametrized by the natural parameter, $\theta \in \Omega$, and the precision parameter λ. In the other representation the family is characterised by the variance function, $V(\cdot)$ specifying the variance as function of the mean value parameter, $\mu \in \mathcal{M}$, and further parametrized by the precision parameter λ.

The two parametrizations (canonical parametrization and mean value parametrization) supplement each other. While the parametrization in terms of the canonical parameter, θ has the advantage that the parameter space is the real line and therefore well suited for linear operations, the parametrization in terms of the mean value parameter μ has the advantage that observations y may be directly compared with mean value parameters μ as they are expressed in the same units of measurement. It follows from (9) that the unit deviance (7) measures the 'closeness' of y and μ in terms of difference in log-likelihood.

The relation between the two parametrizations is determined as $\mu = \tau(\theta)$ with $\tau(\cdot)$ given by (5). The inverse mapping

$$\theta = \tau^{-1}(\mu) \tag{10}$$

is termed *the canonical link function*. It follows from (6) that $d\theta/d\mu = 1/V(\mu)$. In generalized linear models, a link function is any monotone mapping, $\eta = g(\mu)$ of the mean value space to (a subset) of the the real line used to form the *linear predictor*.

To emphasize that a family is characterized by the variance function, $V(\cdot)$, and the precision parameter, λ, the notation $Y \sim \mathrm{ED}(\mu, V(\mu)/\lambda)$ is sometimes used to indicate that Y has density (9) with $d(y;\mu)$ specified by the variance function, $V(\cdot)$, through (7). The notation is in analogy with the notation $\mathrm{N}(\mu, \sigma^2)$ to indicate normal distributions.

The notation highlights the importance of the variance function to describe the random variation around the mean, μ. The usual interpretation

$$\text{observed value} = \text{'true value'} + \text{error} \tag{11}$$

is not very suitable for non-normal distributions where the errors (deviations from model) are not necessarily identically distributed. Thus, in a Poisson regression, deviations from the model line are not necessarily integer valued, neither do they have the same variance. What matters, however, is that a model has been specified for the mean value, μ, and the possible dispersion of observations corresponding to this value of μ is characterized by the variance function, $V(\mu)$.

Table 1 lists the variance function, $V(\mu)$, the unit deviance, $d(y;\mu)$, and the precision parameter, λ, for the distributions most often used in industrial applications. We have included the inverse Gaussian distribution (see Seshadri [7]) as this distribution is often used to describe such right-skewed distributions as appear in applications regarding reliability and strength.

The precision parameter, λ, plays an important role in the expression of the convolution properties of exponential dispersion families. Let $Y_1, \ldots, Y_n$ be independent random variables with $Y_i \sim \mathrm{ED}(\mu, V(\mu)/\lambda_i)$, where $\lambda_1, \ldots, \lambda_n$ are known constants. Let $\lambda_* = \sum \lambda_i$, and consider the precision-weighted average,

$$\overline{Y} = \frac{1}{\lambda_*} \sum_{i=1}^{n} \lambda_i Y_i \,. \tag{12}$$

Table 1. Mean value space, variance function and unit deviance for common distributions

Family	$\mathcal{M}$	$V(\mu)$	unit deviance $d(y,\mu)$	λ	θ
Normal	$(-\infty,\infty)$	1	$(y-\mu)^2$	$1/\sigma^2$	μ
Poisson	$(0,\infty)$	μ	$2\{y\ln(y/\mu)-(y-\mu)\}$	$-$ [a]	$\ln(\mu)$
Gamma	$(0,\infty)$	μ^2	$2\times\{y/\mu-\ln(y/\mu)-1\}$	α [b]	$1/\mu$
Bin	$(0,1)$	$\mu(1-\mu)$	$2\{y\ln(y/\mu)+ (1-y)\ln((1-y)/(1-\mu))\}$	n [c]	$\ln(\mu/(1-\mu))$
Neg Bin	$(0,1)$	$\mu(1+\mu)$	$2\{y\ln(y(1+\mu)/[(1+y)\mu]) +\ln((1+\mu)/(1+\mu))\}$	r [d]	$\ln(\mu)$
I Gauss	$(0,\infty)$	μ^3	$(y-\mu)^2/(y\times\mu^2)$		$1/\mu^2$

[a] The precision parameter λ can not be distinguished from the mean value.
[b] Gamma distribution with shape parameter α and scale parameter μ/α.
[c] $Y=Z/n$, where Z is the number of successes in n independent Bernoulli trials.
[d] $Y=Z/r$, where Z is the number of successes until the rth failure in independent Bernoulli trials.

Then $Y\sim \mathrm{ED}(\mu, V(\mu)/\lambda_*)$. Thus, for $Y_1,\ldots,Y_n$ independent Bernoulli distributed, $Y_i\sim \mathrm{ED}(\mu,\mu(1-\mu)/1)$, we have $\overline{Y}\sim \mathrm{ED}(\mu,\mu(1-\mu)/n)$ corresponding to a binomial distribution of $Z=n\times\overline{Y}$.

The formulation of an exponential dispersion family in terms of the precision parameter, λ, also covers situations where $\lambda_i=w_i/\sigma^2$ with w_i denoting a known 'weight' for the ith observation y_i and σ^2 denoting an unknown dispersion parameter.

The expression (9) shows that maximum-likelihood estimates for generalized linear models specifying a mean value structure, $\mu_i=\mu(\boldsymbol{\beta})$ for independent observations $y_1,\ldots,y_N$ with densities $g_Y(y_i;\mu_i,\lambda_i)$ are obtained by minimizing the deviance

$$D(\boldsymbol{y},\mu(\widehat{\boldsymbol{\beta}}))=\sum_{i=1}^{N}\lambda_i d(y_i;\widehat{\mu}_i)\,. \tag{13}$$

(See [2]). When the distribution of the Y_i's is a normal distribution (with variance function $V(\mu)=1$) the solution is the usual least squares solution.

As the precision parameter, λ, enters in the normalizing factor $a(y,\lambda)$ in (9), maximum likelihood estimation of the precision parameter in situations with unknown dispersion is more complicated. It may be shown, however, that λ and the canonical parameter, θ, are orthogonal (see Pace and Salvan [8]).

Although the above review of exponential dispersion families has mainly concerned the families of order 1 traditionally used in SPC, it should be noted that many of the properties described above also apply to multivariate exponential distribution families. For a general introduction to the statistical properties of exponential dispersion families see Jørgensen [6] and Pace and Salvan [8].

3 Tests for between-group variation

3.1 General

Now consider a SPC-situation with subgroups $i = 1, 2, \dots N$, and measurements Y_{ij}, $j = 1, \dots, n_i$ from subgroup i.

We assume that $Y_{ij} \sim \mathrm{ED}(\mu_i, V(\mu_i)/\lambda)$ with all Y_{ij} independent. Thus, we assume that measurements are identically distributed within a subgroup with subgroup mean μ_i, and within subgroup variance, $V(\mu_i)/\lambda$, depending on the subgroup mean.

Although the distribution of the individual measurements, y_{ij} within a subgroup might be used to monitor the distributional assumption, we will here only be concerned with monitoring the process mean by charting the subgroup averages. To simplify notation we let Y_i, $i = 1, 2, \dots N$ denote the subgroup averages,

$$Y_i = \sum_{j=1}^{n_i} Y_{ij} \Big/ n_i \tag{14}$$

It follows from the reproductivity properties of ED-families that the subgroup averages, $Y_i \sim \mathrm{ED}(\mu_i, V(\mu_i)/(n_i\lambda))$.

Under a fixed effects model for the group means, μ_i, the likelihood-ratio test for the hypothesis of homogeneity,(i.e. $\mu_1 = \cdots = \mu_N$), has the test statistic

$$D(y; \overline{y}) = 2\sum_{i=1}^{N} n_i \lambda d(y_i; \overline{y}) \tag{15}$$

with $\overline{y}$ denoting the usual (weighted) average of the subgroup averages, $\overline{y} = \sum_i n_i y_i / \sum_i n_i$, and $d(y_i; \overline{y})$ denoting the unit deviance (7) between the subgroup average, y_i, and the overall average, $\overline{y}$.

In the framework of a generalized linear model with fixed effects, this statistic is the so-called (scaled) residual deviance under the null model corresponding to no difference between groups.

Under the hypothesis of homogeneity, the asymptotic distribution of the test statistic (for large subgroup sizes) is a $\chi^2(N-1)$ distribution. Skovgaard [9] has reviewed various methods for improving the χ^2 approximation to the distribution of likelihood ratio test statistics.

The Pearson test-statistic is

$$X^2 = \sum_{i=1}^{N} n_i \frac{(y_i - \overline{y})^2}{V(\overline{y})/\lambda} \quad . \tag{16}$$

It follows from (8) that the Pearson test statistic is asymptotically equivalent to the likelihood-ratio statistic (15). The Pearson statistic provides a crude comparison of the between group sum of squares,

$$\mathrm{SSQ}_B = \sum_{i=1}^{N} n_i(y_i - \overline{y})^2 \tag{17}$$

with the estimated variance, $V(\overline{y})/(n_i\lambda)$, under the null model, estimated by means of the specified variance function, $V(\cdot)$. Therefore, the test for homogeneity, (16) is sometimes referred to as a test for overdispersion.

The test statistics (15) and (16) are provided by standard statistical software-packages having options for generalized linear models.

Under the assumption of an $\mathrm{ED}(\mu_i, V(\mu_i)/\lambda)$ distribution within groups, both test statistics (16) and (15) compare the variation between group averages y_i with the estimated within group variation. Thus, the tests are analogous to the usual F-test statistic for testing fixed effects in a one-way analysis of variance and they share the lack of sensitivity to detect moderate deviations from the null hypothesis as the effect of such deviations is spread over $N-1$ degrees of freedom. Hence, considering a large number of groups does not necessarily imply a good sensitivity towards deviations from the null hypothesis.

3.2 Binomial distribution

When the group means, y_i represents fractions of nonconforming items in the sample, it is usually assumed that $Z_i = n_i \times Y_i$ is Binomial (n_i, μ_i) with $0 < \mu_i < 1$, i.e. $Y_i \sim \mathrm{ED}(\mu_i, \mu_i(1-\mu_i)/n_i)$. Hence, the likelihood ratio test statistic for homogeneity is

$$D(\boldsymbol{y}, \overline{y}) = 2\sum_{i=1}^{N} n_i\{y_i \ln(y_i/\overline{y}) + (1-y_i)\ln\left((1-y_i)/(1-\overline{y})\right)\}\,. \tag{18}$$

The χ^2-approximation may be improved by replacing the terms corresponding to $y_i = 0$ and $y_i = 1$ with

$$d(y_i;\overline{y}) = \begin{cases} [\Phi^{-1}(\frac{1}{2}(1-\overline{y})^{n_i})]^2 & \text{for } y_i = 0 \\ [\Phi^{-1}(\frac{1}{2}\overline{y}^{n_i})]^2 & \text{for } y_i = 1 \end{cases} \tag{19}$$

with $\Phi^{-1}(\cdot)$ denoting the fractile function for the standardized normal distribution. See Hansen and Thyregod [10].

The Pearson-test statistic

$$X^2 = \sum_{i=1}^{N} n_i \frac{(y_i-\overline{y})^2}{\overline{y}(1-\overline{y})} = \sum_{i=1}^{N} \frac{(z_i - n_i\overline{y})^2}{n_i\overline{y}(1-\overline{y})}\,, \tag{20}$$

is sometimes referred to as the 'binomial variance test for homogeneity'. Tarone [11] has discussed the properties of this test and the optimality against various alternatives.

3.3 Within group variance for normally distributed measurements

We consider the standard SPC-model for normally distributed measurements.

Let x_{ij} denote the jth measurement in subgroup i, $i = 1, \ldots, N$, where the ith subgroup is of size n_i.

As we are concerned with monitoring the within-subgroup variation, we shall let

$$y_i = \sum_{j=1}^{n_i} (x_{ij} - \overline{x}_i)^2 \Big/ (n_i - 1) \tag{21}$$

denote the empirical variance, (s^2), in the i'th group. Moreover, to avoid confusion with the dispersion parameter, σ^2, for exponential dispersion models, we shall use the symbol τ^2 to denote process variance for the x-measurements.

As the distribution of the within-group empirical variance is a $\tau^2\chi^2(f_i)$-distribution which is a gamma-distribution with shape parameter $f/2$ and scale parameter $2\tau^2/f_i$, the distribution of Y_i belongs to an exponential dispersion family of distributions, $Y_i \sim \mathrm{ED}\big(\mu_i, \mu_i^2/(f_i/2)\big)$ with $f_i = n_i - 1$ and with μ_i denoting the mean (τ_i^2) of Y_i, i.e. the process-variance.

Under a fixed effects model the hypothesis of homogeneity of within-group variances may be assessed by the the likelihood-ratio test statistic, $D(\boldsymbol{y}; \overline{\boldsymbol{y}})$ in (15), where we have

$$D(\boldsymbol{y}, \overline{\boldsymbol{y}}) = \sum_{i=1}^{N} f_i \times \{y_i/\overline{y} - \ln(y_i/\overline{y}) - 1\} \tag{22}$$

Under the hypothesis of variance homogeneity the distribution is approximately a χ^2-distribution with $N-1$ degrees of freedom. Bartlett [12] proposed a correction to $D(\boldsymbol{y}; \overline{\boldsymbol{y}})$, known as the Bartlett test. The Bartlett correction is essentially a transformation

$$Z = \frac{f D(\boldsymbol{y}; \overline{\boldsymbol{y}})}{\mathsf{E}\left[D(\boldsymbol{y}; \overline{\boldsymbol{y}})\right]} \tag{23}$$

that renders the expectation of Z equal to the nominal asymptotic value (see [13]). This usually produces a more satisfactory match with the reference χ^2-distribution.

It is of interest for the quality practitioner that most statistical software with provisions for analyzing generalized linear models include options for analyzing gamma-distributed observations, and hence, allow for analyzing variance structure, as indicated above. Thus, although one still might utilize that the transformation $\ln(s^2)$ carries the $\chi^2(f)/f$ -distribution into an approximate normal distribution for $f > 5$ (see Bartlett and Kendall [14]), an approach by means of a generalized linear model seems more appropriate.

4 Modelling hierarchical variation

In stage 1 of a control limit determination, the objective is to identify a common population (distributional structure) that represents the transfer of variation in the process that may be ascribed to common causes. In the case of such overdispersion (i.e. larger variation between subgroup averages than may be explained by the within subgroup variation) that can not be identified as out of control situations but is considered as part of inherent process variation, this overdispersion shall be included in the description of the variation as discussed by Thyregod et al. [1].

Although the modelling by means of exponential dispersion families apparently suggests using the parameter, σ^2, to describe overdispersion, see e.g. McCullogh and Nelder [2], this approach is mainly appropriate when the within-subgroup exhibits overdispersion and is more suited for control chart applications in Stage 2 where the aim is to achieve tighter control of the process. Lee and Nelder [15] report an application where systematic factors influence as well the mean value as the overdispersion of process output, and Lu [16] discusses a test for such overdispersion in a binomial regression.

Following the approach in Part 1, we shall therefore choose to model the between subgroup variation using a random-effect model.

When the within subgroup distribution may be modelled by a normal distribution, it is often natural also to choose a normal distribution to model the distribution of between subgroup random effects. This choice is in harmony with the traditional modelling of variance components, the *general linear mixed models*. When the observations are counts (Poisson) or fractions (binomial) and therefore the within subgroup variation can not be satisfactorily described by a normal distribution, the choice of distributional model for the random variation between subgroups is not as obvious. A normal distribution is generally not well suited to describe the variation of a parameter that can only take values in the interval $0 \leq \mu \leq 1$.

In the framework of an exponential dispersion family, one might choose to consider instead the canonical parameter (with possible values on the entire real axis) and model the variation of the canonical parameter between subgroups with a normal distribution. For generalized linear models this approach is often termed generalized linear mixed models (GLMM) to indicate the hybride between the generalized linear models for nonnormal distributions and the 'mixed models' for GLM's for normally distributed data. The inferential problems using likelihood procedures are inherited from the mixed models. (For a discussion of the estimation of variance components in GLMM's see Lee and Nelder [3], [18] and Pawitan [19]).

However, in many cases, the canonical parameter does not have a simple physical interpretation, and therefore, an additive 'true value' + error, with a normally distributed 'error' on the canonical parameter to describe variation between subgroups is not very transparent. We shall therefore use the *standard conjugate* distribution for the mean parameter of the within group distribution to model the between group variation in standard situations where the between group distribution is considered to be unimodal. In the framework of generalized linear models such models are termed *hierarchical generalized linear models* to distinguish between this class of models and GLMM's. Lee and Nelder (op.cit.) discuss the estimation of variance components in hierarchical generalized linear models.

Consider an exponential dispersion family $\mathrm{ED}(\mu, V(\mu)/\lambda)$ with density (2) for $\theta \in \Omega$. Let $\mathcal{M} = \tau(\Omega)$ denote the mean value space for this family. Let $m \in \mathcal{M}$ and consider

$$g_\theta(\theta; m, \gamma) = \frac{1}{C(m, \gamma)} \exp\{[\theta m - \kappa(\theta)]/\gamma\} \tag{24}$$

with

$$C(m, \gamma) = \int_\Omega \exp\{[\theta m - \kappa(\theta)]/\gamma\} \, d\theta$$

for all (positive) values of γ for which the integral converges.

Then (24) defines the density function of a probability distribution for θ. This distribution is called the *standard conjugate distribution* for θ corresponding to (2). The concept has its roots in the context of Bayesian parametric inference to describe a family of distributions whose densities have the structure of the likelihood kernel (see Consonni and Veronese [20]).

When the variance function, $V(\mu)$ for (2) is at most quadratic, the parameters m and γ have a simple interpretation in terms of the mean value parameter, $\mu = \tau(\theta)$, viz.

$$m = \mathsf{E}[\mu] \tag{25}$$

$$\gamma = \frac{\mathsf{V}[\mu]}{\mathsf{E}[V(\mu)]}, \tag{26}$$

with $\mu = \mathsf{E}[Y|\theta]$, and with $V(\mu)$ denoting the variance function (6). (See [21]).

Borrowing the term from other fields of quality engineering, we shall refer to the parameter γ defined in (26) as the *signal to noise ratio*.

When the variance function for the exponential dispersion family (2) is at most quadratic, the standard conjugate distribution for μ coincides with the standard conjugate distribution (24) for θ. However, for the Inverse Gaussian distribution, the standard conjugate distribution for μ is improper (see [22]), and in that case (25) and (26) are undefined.

The parametrization of the natural conjugate distribution for μ by the parameters m and γ has the advantage that location and spread are described by separate parameters. Thus, letting $\gamma \to 0$, the distribution of μ will converge towards a degenerate distribution with all its mass in m.

Now consider a SPC-situation with subgroups $i = 1, 2, \ldots N$ and measurements Y_{ij}, $j = 1, \ldots, n_i$ from subgroup i. To simplify notation we let Y_i, $i = 1, 2, \ldots N$ denote the subgroup average, i.e., $Y_i = \sum_i Y_{ij}/n_i$. It is natural to assume that the Y_{ij} are conditionally independent given the subgroup mean, μ_i. We shall furthermore assume that variation is hierarchical such that subgroup means varies randomly between subgroups. Specifically, we shall consider the following hierarchical model:

- Conditional on the subgroup mean, μ_i, the measurements, Y_{ij} are iid with $Y_{ij} \sim \mathrm{ED}(\mu_i, V(\mu_i)/\lambda)$
- The subgroup means, μ_i are iid random variables following the natural conjugate distribution on μ.

It then follows that the marginal mean of Y_{ij} is

$$\mathsf{E}\left[Y_{ij}\right] = \mathsf{E}\left[\mathsf{E}\left[Y_{ij}|\mu\right]\right] = \mathsf{E}\left[\mu\right] = m \tag{27}$$

and furthermore the marginal variance of Y_{ij} is

$$\mathsf{V}\left[Y_{ij}\right] = \mathsf{E}\left[\mathsf{V}\left[Y_{ij}|\mu\right]\right] + \mathsf{V}\left[\mathsf{E}\left[Y_{ij}|\mu\right]\right] = \mathsf{E}\left[V(\mu)\right]/\lambda + \mathsf{V}\left[\mu\right] \tag{28}$$

Thus, the parameter, γ, in the distribution of the group means reflects the usual decomposition of total variation into within-subgroup and between-subgroup variation. As the variation, $V(\mu)$, within a specific subgroup depends on the subgroup mean, μ, the within-subgroup variation is understood as an average within-subgroup variation, $\mathsf{E}\left[V(\mu)\right]$.

The marginal distribution of the subgroup means, Y_i, has mean and variance

$$\mathsf{E}\left[Y_i\right] = \mathsf{E}\left[\mathsf{E}\left[Y_i|\mu\right]\right] = \mathsf{E}\left[\mu\right] = m \tag{29}$$

$$\mathsf{V}\left[Y_i\right] = \mathsf{E}\left[V(\mu)\right](\gamma + \frac{1}{\lambda n_i}) = \sigma^2 \mathsf{E}\left[V(\mu)\right](\gamma + \frac{1}{n_i})\,, \tag{30}$$

where we have introduced the dispersion parameter, $\sigma^2 = 1/\lambda$ for the within-group distribution.

Thus, the variance in the marginal distribution generally exceeds the average variance in the within group distributions (i.e. whenever $\mathsf{V}\left[\mu\right] > 0$]). This phenonemon is sometimes referred to as overdispersion.

Measurements, Y_{ij} and Y_{ik} in the same subgroup are correlated with *intragroup correlation*

$$\rho_w = \frac{\mathsf{COV}\left[Y_{ij}, Y_{ik}\right]}{\mathsf{V}\left[Y_{ij}\right]} = \frac{\mathsf{V}\left[\mu\right]}{\mathsf{E}\left[V(\mu)\right]/\lambda + \mathsf{V}\left[\mu\right]} = \frac{\gamma}{1/\lambda + \gamma} \tag{31}$$

Introducing the dispersion parameter, $\sigma^2 = 1/\lambda$ for the within-group distribution, the intragroup correlation (31) may be expressed in terms of the within group dispersion parameter, σ^2, and the between group dispersion parameter, γ,

$$\rho_w = \frac{\gamma}{\sigma^2 + \gamma} \tag{32}$$

and hence,

$$\gamma = \frac{\rho_w}{1 - \rho_w}\,\sigma^2 \tag{33}$$

The variance in the marginal distribution of group averages may thus be expressed in terms of either within group variance, σ_W^2 and intergroup correlation, or between group variance, σ_B^2 and intergroup correlation as

$$\mathsf{V}\,[Y_i] = \mathsf{E}\,[V(\mu)]\left(\frac{\rho_w}{1-\rho_w} + \frac{1}{n_i}\right) \tag{34}$$

$$= \mathsf{V}\,[\mu]\left(1 + \frac{1-\rho_w}{\rho_w}\,\frac{1}{n_i}\right) \tag{35}$$

Now, consider the between group sum of squares,

$$\mathrm{SSQ}_B = \sum_{i=1}^{N} n_i (Y_i - \overline{Y})^2 \tag{36}$$

with $\overline{Y}$ denoting the usual (weighted) average of the subgroup averages, $\overline{Y} = \sum_i n_i Y_i / \sum_i n_i$.

It may be shown that

$$\mathsf{E}\,[\mathrm{SSQ}_B] = (k-1)\mathsf{E}\,[V(\mu)](\sigma^2 + n_0\gamma)\ , \tag{37}$$

with

$$n_0 = \frac{\left(\sum_i n_i\right)^2 - \sum n_i^2}{(k-1)\sum_i n_i} \tag{38}$$

denoting an 'effective average sample size'. Thus, a natural candidate for a method of moments estimate of the between-group dispersion, γ, is

$$S_2^2 = \sum_{i=1}^{N} n_i (Y_i - \overline{Y})^2 \Big/ (N-1) \tag{39}$$

with expected value

$$\mathsf{E}\,[S_2^2] = \mathsf{E}\,[V(\mu)](\sigma^2 + n_0\gamma) = \sigma^2 \mathsf{E}\,[V(\mu)] + n_0 \mathsf{V}\,[\mu]\ . \tag{40}$$

5 Monitoring fraction nonconforming

5.1 The problem

Results from the visual inspection of filled vials for cosmetic errors are considered. An inherent batch-to-batch variation is considered to be part of the process. There are some differences in batch-size and hence the sample size, which is considered to be the whole batch. Data are given i Table 2.

Table 2. Number of items with cosmetic errors in samples of filled vials

Batch	items with cosmetic errors	sample size	Batch	items with cosmetic errors	sample size
1	468	227190	14	294	87380
2	832	210800	15	228	60900
3	468	213950	16	371	90530
4	1021	214190	17	353	119780
5	1273	215810	18	785	122150
6	740	215540	19	211	87420
7	1368	217380	20	684	120040
8	217	212860	21	171	113200
9	502	205080	22	203	92090
10	3274	209870	23	160	93870
11	665	239830	24	504	94950
12	303	83160	25	154	90200
13	110	85760	26	379	80950

There is a wish to monitor the process, i.e. establish limits for the process in control, based on historical data. The first attempt to make these limits, is a p-chart for the proportion of cosmetic errors. Figure 1 shows the p-chart with 3-sigma limits for the data in Table 2.

Most of the observations are outside the limits, even though the process was considered to be in control, operating under under normal, controlled conditions. A specifically large observation is seen for sample no. 10, which may disturb the picture and the calculation of the limits. Therefore this observation is excluded from the calculation of the limits.

Figure 2 shows the p-chart with recalculated limits after exclusion of sample 10. Still there are only a few observations within the limits of the chart.

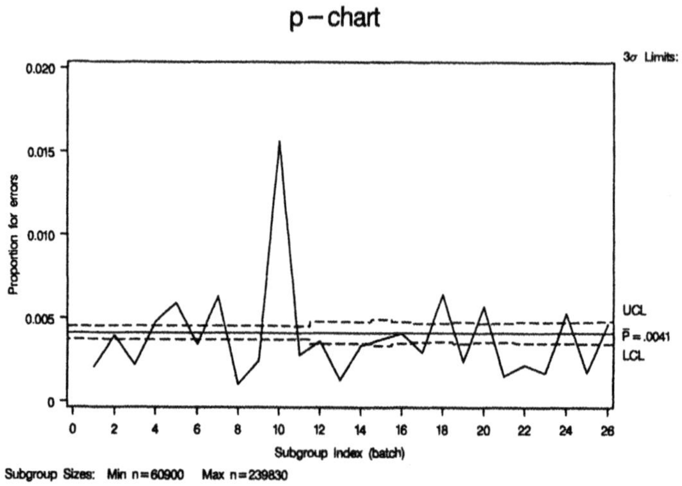

Fig. 1. p-chart for fraction of items with cosmetic errors for 26 succesive batches of a pharmaceutical product. Data from Table 2.

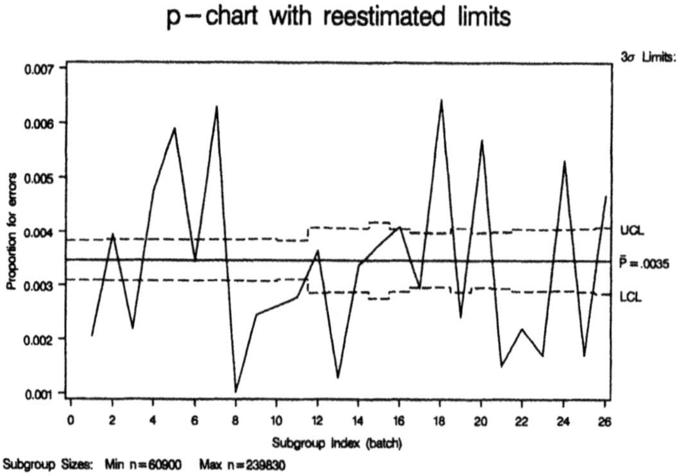

Fig. 2. p-chart for fraction of items with cosmetic errors for 25 batches of a pharmaceutical product. Limits reestimated after exclusion of sample 10

The assumption underlying the p-chart is that the subgroup fraction nonconforming, Y_i may be modelled as independent variables with $Y_i \sim \text{Bin}(n_i, p)/n_i$, or, expressed as an exponential dispersion family, $Y_i \sim \text{ED}(p, p(1-p)/n_i)$. Thus, from section 3.2 we may use the Pearson test statistic (20) to perform a numerical assessment of the homogeneity of the samples to support the visual impression from the charts. For all 26 samples we find the Pearson statistic (20) $X^2 = 9434.13$ with the corresponding p-value (from the $\chi^2(25)$-distribution) close to zero.

For the data shown in the re-calculated chart, Figure 2, the Pearson statistic is $X^2 = 2798.65$; the corresponding p-value (from the $\chi^2(24)$-distribution) is close to zero.

As the items with cosmetic errors were found to be randomly scattered throughout each batch there was no indication of a violation of the assumption of a constant probability of cosmetic errors throughout the production of a batch.

Because of the large sample sizes in this study, the power of the Pearson test is very good, even against small departures from the assumption of homogeneity, as indicated by the large values of the test statistic.

Thus, batch-to-batch variation is an inherent part of the process, and must be taken into account when calculating the limits of the control chart. Compared to the sampling variance of the binomial distribution there is overdispersion.

See Woodall [23] for a review and bibliography of control charts based on attribute data.

5.2 Statistical model for batch to batch variation

The natural conjugate distribution for μ is the Beta-distribution with density function

$$g(\mu) = \frac{1}{\mathrm{B}(\alpha, \beta)}\, \mu^{\alpha-1}(1-\mu)^{\beta-1} \quad \text{for } 0 < p < 1, \tag{41}$$

for $\alpha > 0$, $\beta > 0$ and with $\mathrm{B}(\alpha, \beta)$ denoting the beta function.

Skellam [24] showed that under this model for between group variation the marginal distribution of the group counts, $Z_i = n_i \times Y_i$ is the Polya distribution with probability frequency function

$$h(z) = \binom{n_i}{x} \frac{\Gamma(\alpha + z)}{\Gamma(\alpha)} \frac{\Gamma(\beta + n_i - z)}{\Gamma(\beta)} \frac{\Gamma(\alpha + \beta)}{\Gamma(\alpha + \beta + n_i)} \tag{42}$$

for $z = 0, 1, 2, \ldots, n_i$.

For obvious reasons the distribution is often termed beta-binomial distribution, or a mixed binomial distribution. Hald [25] showed that the distribution is the only distribution that is closed under hypergeometric sampling.

There is a long tradition in statistical litterature for using the beta-binomial distribution to describe overdispersion in supposedly binomial distributed samples. Ford [26] investigated samples from each of 229 production lots and found a satisfactory fit. Fitzner [27] has investigated 16 cases of lot-by-lot sampling from different industrial sectors and also found that the beta-binomial distribution provided a satisfactory description of the variation between lots.

The parametrization in terms of α and β is not very transparent for the practitioner. Introducing the parameters, m and γ (25) and (26) we find

$$m = \mathsf{E}[\mu] = \frac{\alpha}{\alpha + \beta} \tag{43}$$

$$\gamma = \frac{\mathsf{V}\,[\mu]}{\mathsf{E}\,[\mu(1-\mu)]} = \frac{1}{\alpha+\beta}\,, \tag{44}$$

where we have used that

$$\mathsf{E}\,[\mu] = \frac{\alpha}{\alpha+\beta} \tag{45}$$

$$\mathsf{V}\,[\mu] = \frac{\alpha\beta}{(\alpha+\beta)^2(\alpha+\beta+1)} \tag{46}$$

and hence,

$$\mathsf{E}\,[V(\mu)] = \mathsf{E}\,[\mu(1-\mu)] = m(1-m)\,\frac{\alpha+\beta}{\alpha+\beta+1} = \frac{m(1-m)}{1+\gamma}\,. \tag{47}$$

Griffiths [28] suggested the same parametrization for the beta-binomial although without the interpretation given above.

From (47) it is obvious $\mathsf{E}\,[V(\mu)] \leq V(\mathsf{E}\,[\mu]) = m(1-m)$ with equality only if $\gamma = 0$. By Jensen's inequality this is a direct consequence of the convexity of $V(\mu)$. However, (47) explicitly quantifies this effect.

For $\gamma \to 0$ (i.e. $\mathsf{V}[\mu] \to 0$), the beta-distribution (41) converges towards a one-point distribution in m, and, consequently, the probability frequency function (42) converges towards a binomial distribution with probability parameter m.

An explicit estimate for the parameters m and γ may be determined using the method of moments. From (40) we find the estimates

$$\widetilde{m} = \overline{y} \tag{48}$$

$$\widetilde{\gamma} = \frac{s_2^2 - \widetilde{m}(1-\widetilde{m})}{n_0\widetilde{m}(1-\widetilde{m}) - s_2^2}\,; \tag{49}$$

when the between group variation is less than $m(1-m)$ the solution is $\widetilde{\gamma} = 0$ indicating a one-point distribution of p concentrated at m, corresponding to perfect binomial situation.

The maximum likelihood estimates has to found by iteration. The maximum likelihood estimation of α and β in the balanced case ($n_1 = \cdots = n_N$) has been described, e.g.. by Skellam [24] and Hasselblad [29]. Griffiths [28] describes the estimation of the parameters m and γ, also only for the balanced case. The methods may easily be extended to the unbalanced case. In the balanced case, goodness of fit may be assessed by the usual Pearson statistic. Garren et al. [30] have reviewed various goodness-of-fit tests for the balanced case, and in [31] suggested a test that may be used in the unbalanced case.

When estimates of m and γ have been established, crude 3-sigma control limits for the group fraction nonconforming, y_i may then be determined from the general expression (30) as

$$\mathrm{CL} = \widetilde{m} \pm 3\sqrt{\frac{\widetilde{m}(1-\widetilde{m})}{1+\widetilde{\gamma}}\left(\widetilde{\gamma} + \frac{1}{n_i}\right)}\,, \tag{50}$$

or, the standardized values,

$$z_i = \frac{y_i - \widetilde{m}}{\sqrt{\dfrac{\widetilde{m}(1-\widetilde{m})}{1+\widetilde{\gamma}}\left(\widetilde{\gamma}+\dfrac{1}{n_i}\right)}}\,, \tag{51}$$

may be plotted on a chart with limits ± 3.

These limits may then be used to identify out-of-control situations in the historical data.

5.3 The control chart

For the full data set in Table 2 we found the maximum-likelihood estimates $\widehat{m} = 0.0039$ and $\widehat{\gamma} = 0.0014$. (The moment estimates determined from (48) and (49) were $\widetilde{m} = 0.0041$ and $\widetilde{\gamma} = 0.0026$). The control limits (50) corresponding to a sample size $n = 90000$ are found to be

$$\mathrm{CL} = 0.0039 \pm 3\sqrt{\frac{0.0039 \times 0.9961}{1+0.0014}\left(0.0014+\frac{1}{90000}\right)}\,. \tag{52}$$

Thus, the contribution, $\widehat{\gamma} = 0.0014$, to the limits reflecting the between group variation, exceeds the contribution, $1/90000 = 0.00001$ from the sample size by a factor 100, resulting in limits 10 times wider than the limits on the p-chart.

Figure 3 shows the standardized values (51) plotted on a chart with limits ± 3. Under this charting, only one sample, sample 10, indicates an out of control situation. It is concluded that this sample does not represent the common population of in-control conditions, and therefore, the limits are recalculated, excluding sample 10.

Excluding sample 10 from the data set in Table 2, we obtained the reestimated maximum-likelihood estimates $\widehat{m} = 0.00342$ and $\widehat{\gamma} = 0.00079$. (The moment estimates determined from (48) and (49) were $\widetilde{m} = 0.00347$ and $\widetilde{\gamma} = 0.00081$). We observe that exclusion of the 'outlier', sample 10, results in a much better agreement between maximum-likelihood and moment estimates. Moreover, we note that exclusion of sample 10 almost halves the estimate of the between subgroup variation, γ.

Figure 4 shows the standardized values (51) plotted on a chart with limits ± 3.

It is concluded that the hierarchical model provides a satisfactory description of the in-control distribution. Inherent batch-to-batch, or day-today variation has thus been included as part of the process, considered as being in statistical control. There is a direct interpretation of the parameters, in terms of mean level, m, and ratio, γ, of variation between subgroups to variation within subgroups.

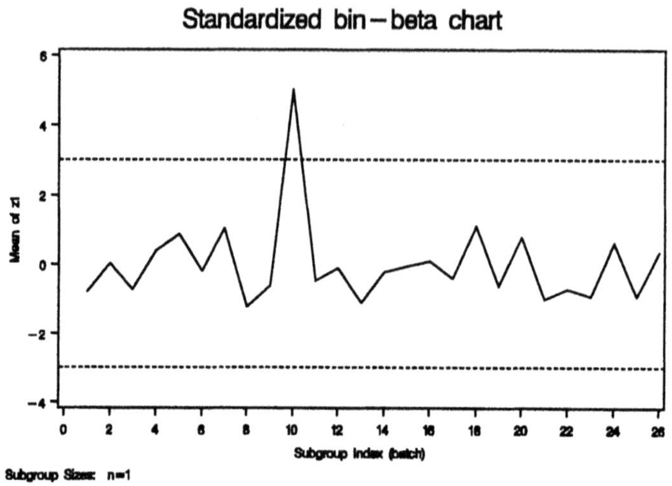

Fig. 3. Standardized bin-beta chart for fraction of items with cosmetic errors for 26 batches of a pharmaceutical product. Data from Table 2 .

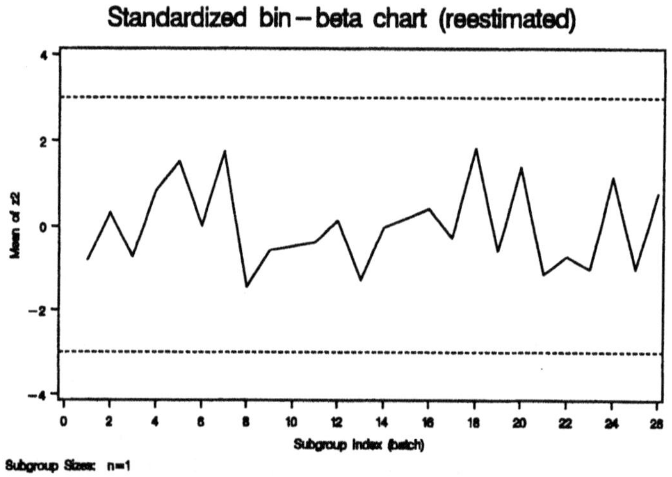

Fig. 4. Standardized bin-beta chart for fraction of items with cosmetic errors for 25 batches of a pharmaceutical product. Data from Table 2 (sample 10 excluded).

6 Monitoring within group variation

6.1 The problem

In the pharmaceutical industry requirements for batch release include requirements on content of active drug as well as batch homogeneity.

Different chemical analysis methods are used for assessment of content of active drug and batch homogeneity. Here we are only concerned with batch homogeneity.

For assessment of homogeneity, so-called homogeneity values are determined for each sample item. Figure 5 shows the distribution of homogeneity values from 10 consecutive batches. The within sample variances are given in Table 3.

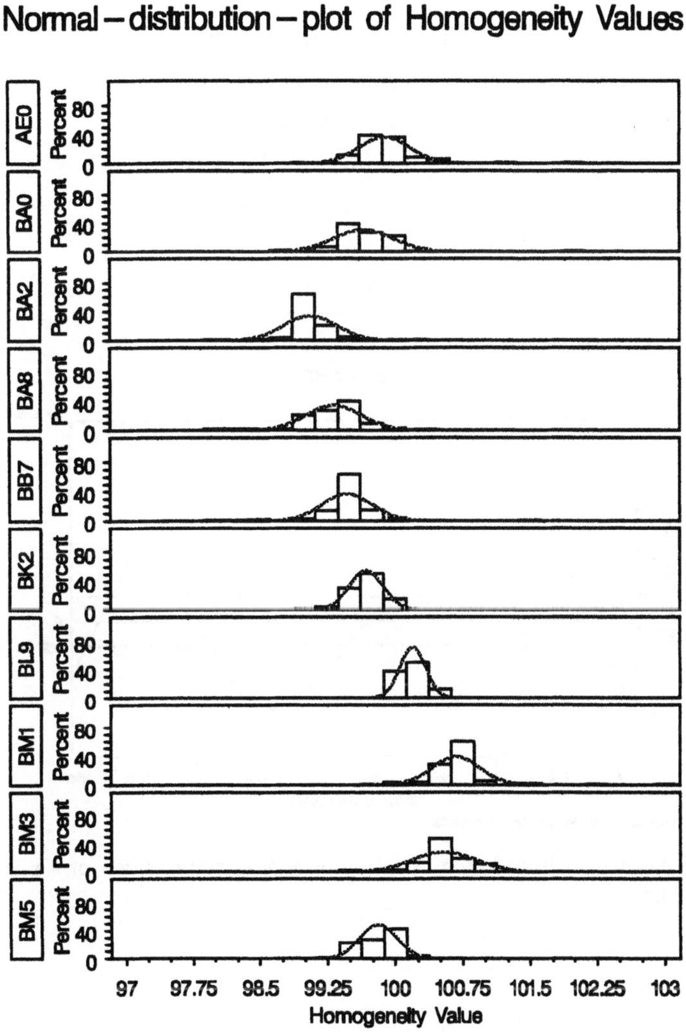

Fig. 5. Within group distribution of samples from 10 batches

Figure 6 shows a standard s-chart for the within group variation.

Tests for homogeneity of within-group variance under a 'fixed effect' model were discussed in section 3.3. We use the same notation as in section 3.3 and

Table 3. Empirical variances of homogeneity values in 10 batches of a pharmaceutical product

Batch	n	s^2	s
AE0	300	0.07601	0.2757
BA0	324	0.10821	0.3290
BA2	323	0.08805	0.2967
BA8	324	0.09116	0.3019
BB7	323	0.07384	0.2717
BK2	324	0.03334	0.1826
BL9	316	0.01971	0.1404
BM1	266	0.06743	0.2597
BM3	322	0.12747	0.3570
BM5	322	0.04184	0.2046

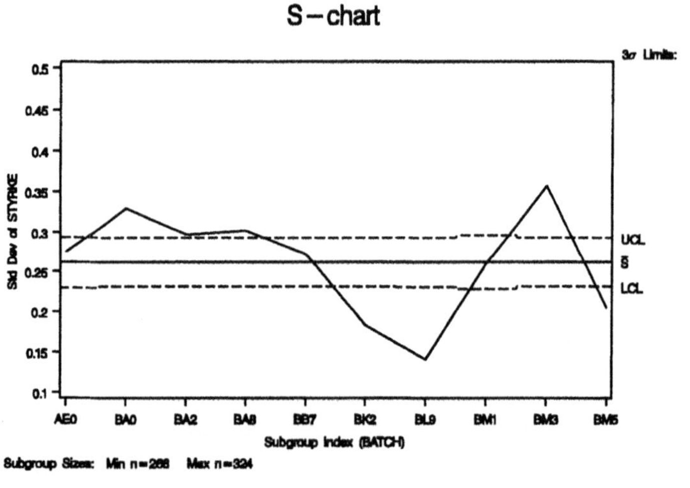

Fig. 6. S-chart for within batch variation of homogeneity values for a pharmaceutical product

let

$$y_i = \sum_{j=1}^{n_i} (x_{ij} - \overline{x}_i)^2 \Big/ (n_i - 1) \tag{53}$$

denote the empirical variance (s^2) in the ith group, and let τ_i^2 denote the corresponding process variance. Thus, $\mathsf{E}[Y_i] = \tau_i^2$.

The Bartlett test statistic for the data in Table 3 is $Z = 406.2$ with the corresponding p-value (from the $\chi^2(9)$-distribution) close to zero.

Due to the large number of observations in each sample, the power of the test is very high, even for small departures from homogeneity of variances.

6.2 Statistical model for batch to batch variation

The natural conjugate distribution for μ under this model is such that the distribution of $1/\mu$ is a gamma distribution with some shape parameter, α, and scale parameter $1/\beta$, i.e. with $\mathsf{E}[1/\mu] = \alpha/\beta$. It then follows that

$$\mathsf{E}[\mu] = \frac{\beta}{\alpha - 1} \tag{54}$$

$$\mathsf{V}[\mu] = \frac{\beta^2}{(\alpha-1)^2(\alpha-2)} = \frac{\mathsf{E}[\mu]^2}{\alpha-2} \tag{55}$$

$$\mathsf{E}[V(\mu)] = \mathsf{E}[\mu^2] = \mathsf{E}[\mu]^2 + \mathsf{V}[\mu] = \mathsf{E}[\mu]^2\,\frac{\alpha-1}{\alpha-2}\,, \tag{56}$$

Thus, assuming that $\alpha > 2$, we find

$$m = \beta/(\alpha - 1) \tag{57}$$

$$\gamma = \frac{\mathsf{V}[\mu]}{\mathsf{E}[\mu^2]} = \frac{1}{\alpha-2}\,. \tag{58}$$

Method of moment estimates of m and γ may be obtained from (40) with

$$s_2^2 = \sum_{i=1}^{N} f_i(y_i - \overline{y})^2 \Big/ (N-1) \tag{59}$$

denoting the between group sum of squares for the empirical variances, y_i.

The marginal distribution of y_i (the within group empirical variance) has density

$$h_i(y) = \frac{1}{\mathrm{B}(\alpha, f_i/2)}\,\frac{b^{\alpha} y^{f_i/2-1}}{(b+y)^{\alpha+f_i/2}} \quad \text{for } y > 0 \tag{60}$$

with $b = 2(\alpha - 1)m/f_i$. Under the model, the y_i's are independent, and, hence, (60) may be used to determine the maximum-likelihood estimates of m and $\gamma = 1/(\alpha - 2)$

For $\gamma \to 0$ (i.e. $\alpha \to \infty$, corresponding to $\mathsf{V}[\mu] \to 0$) will the distribution of μ converge towards a one-point distribution in m, and, consequently, the probability frequency function (60) converges towards a $m\chi^2(f_i)/f_i$ distribution.

The 3-sigma limits for the group-empirical variances may be determined from the general expression (30) as

$$\mathrm{CL} = m \pm 3m\sqrt{\frac{\alpha-1}{\alpha-2}\left(\frac{1}{\alpha-2} + \frac{1}{f_i}\right)} \tag{61}$$

For calculation of probability limits it could be utilized that the marginal distribution of y is such that the variable

$$F = \frac{m(\alpha - 1)}{\alpha Y_i} \tag{62}$$

follows an F-distribution with $(2\alpha, f_i)$ degrees of freedom.

We note that the parameter α in the distribution of the process variance, μ, plays a role as 'degrees of freedom', and therefore it might aid the interpretation of the between group distribution of process variance to express this distribution in terms of a χ^2-distributed variable. It follows from the relation between the χ^2- and the gamma distribution that the between group distribution of μ is such that

$$\frac{m}{\mu} \sim \frac{\chi^2(\nu)}{\nu - 2} \tag{63}$$

We emphasize that an ordinary $\overline{X}$-chart is not applicable for monitoring the subgroup averages. Under this model, the compound distribution of subgroup averages is a Pearson Type VII distribution. The distribution of the standardized values of the subgroup averages is a t-distribution.

6.3 The control chart

From the data in Table 3 we find the maximum-likelihood estimates, $\widehat{m} = 0.0806$ and $\widehat{\gamma} = 0.210$ (corresponding to $\widehat{\alpha} = 6.755$).

To illustrate the goodness of fit of the model, Figure 7 shows a histogram of the 10 reciprocal empirical variances from Table 3 with the estimated gamma-distribution of the reciprocal variances, $1/\mu$ (i.e. $1/\tau^2$). There appears to be a good agreement between the observed and the fitted distribution. Various standard tests for goodness of fit (Anderson-Darling, Cramer-von Mises, Kolmogorov-Smirnoff) all have p-values greater than 0.25. (By comparing with a gamma distribution instead of the marginal distribution corresponding to (60), we have disregarded the variation due to the $\chi^2(f_i)/f_i$ distribution overlayed the variation in μ. However, as the individual empirical variances, y_i are determined with more than 300 degrees of freedom, this variation is neglible).

Control limits for the within group variances, y_i may then be determined from (61). We find, using the maximum likelihood estimates above,

$$\text{CL} = 0.0806 \pm 3 \times 0.0806\sqrt{1.21\left(0.21 + \frac{1}{f_i}\right)} \tag{64}$$

When $f_i = 300$, the contribution, $\widehat{\gamma} = 0.21$ from the variation between groups exceeds the contribution $1/f_i = 0.003$ by a factor 70.

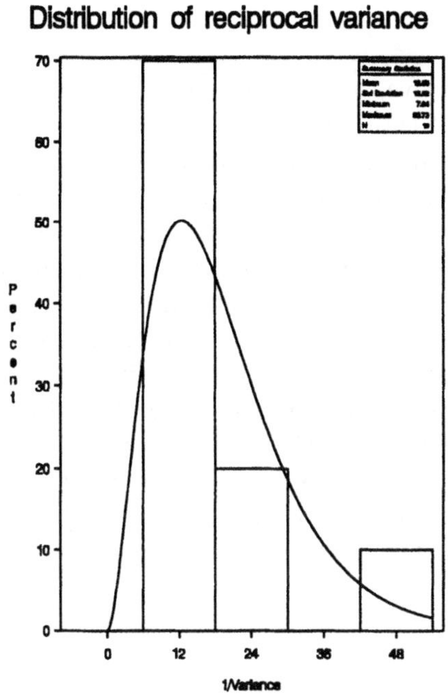

Fig. 7. Marginal distribution of reciprocal empirical within group variances

Figure 8 shows a s^2-chart for the data in Table 3 with control limits determined from (64) for $f_i = 300$. As the lower 3 sigma control limit is negative, only the upper limit, CL = 0.204 has been indicated. All points are seen to be within the control limits, and it is therefore believed that the hierarchical model describes natural process variation of batch homogeneity.

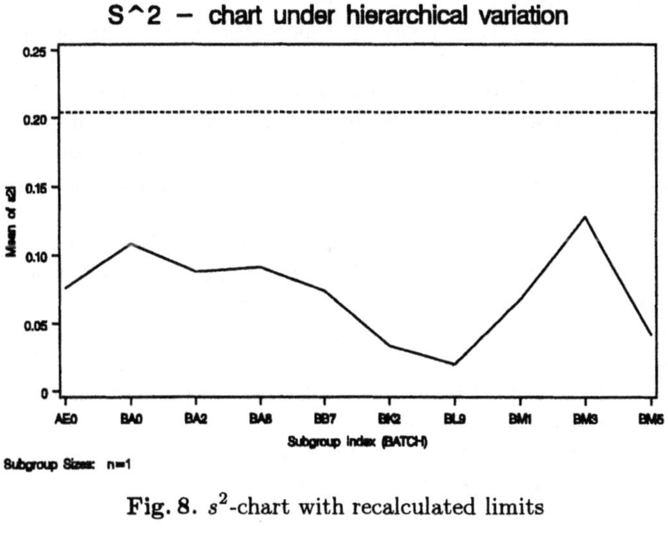

Fig. 8. s^2-chart with recalculated limits

7 Conclusion

In the first, retrospective, stage of control limit determination the objective is to identify limits for the variation in the charted variable that can be expected when the process is running under normal (stable) conditions corresponding to the current state of control for the process. In this Stage 1 it is therefore imperative that all common cause variation is reflected in the determination of the limits. Often, in industrial practice, some of the inherent common causes are only slowly varying and their variation is not reflected in the 'high frequency' within subgroup variation. Usually, all sources of slowly varying common cause variation have not been identified in this stage of the charting, and it is appropriate to model the aggregated effect of these sources as a random between subgroup effect. For non-normal distributions such random effect models are a special case of the hierarchical generalized models. We have discussed a parametrization of these models that allows for a direct assessment and interpretation of the overdispersion due to these slowly varying common causes.

References

1. Thyregod, P., Melgaard, H. and Iwersen, J. (2003) On monitoring processes and assessing process capability under a hierarchical model, Part 1. Frontiers
2. McCullagh, P. and Nelder, J. A. (1989) Generalized Linear Models, Second Edition. Chapman & Hall, London
3. Lee, Y. and Nelder, J. A. (1996) Hierarchical Generalized Linear Models (with discussion). Journ. Roy. Statist. Soc. B, **58**, 619-678
4. Fisher, R. A. (1934) Two new properties of Mathematical Likelihood. Proceedings of the Royal Society A **144** 285-307
5. Morris, C. N. (1982) Natural Exponential Families with Quadratic Variance Functions. Annals of Statistics **10** 65-80
6. Jørgensen, B. (1997) The Theory of Dispersion Models, Chapman & Hall, London
7. Seshadri, V. (1993) The Inverse Gaussian Distribution. Clarendon Press, Oxford.
8. Pace, L. and Salvan, A. (1997) Principles of Statistical Inference: From a Neo-Fisherian Perspective. World Scientific Publishing Co., Singapore
9. Skovgaard, I. M. (1989) A Review of Higher Order Likelihood Methods. Bullletin of the International Statistical Institute, **53**, vol. 3, 331-351
10. Hansen, C. K. and Thyregod, P. (2000) Analysis of Integrated Circuit Fault Data Using Generalised Linear Models. Quality and Reliability Engineering International **16** 173-186
11. Tarone, R. E. (1979 Testing the Goodness of Fit of the Binomial Distribution. Biometrika **66** 585-90
12. Bartlett, M. S. (1937) Properties of Sufficiency and Statistical Tests. Proceedings of Royal Society A **160** 268-282
13. Bartlett, M. S. (1954) A Note on the Multiplying Factors for Various χ^2 Approximations. Journ. Roy. Statist. Soc. B, **16**, 296-298

14. Bartlett, M. S. and Kendall, D. G. (1946) The Statistical Analysis of Variance-heterogeneity and the Logarithmic Transformation. Journ. Roy. Statist. Soc. Suppl. **VIII** 128-138
15. Lee, Y. and Nelder, J. A. (1998) Generalized Linear Models for the Analysis of Quality-improvement Experiments. The Canadian Journal of Statistics, **26** 95-105
16. Lu, W.-S. (1999) Testing Extra-binomial variations. Journal of Statistical Computation and Simulation **63** 93-103
17. Lee, Y. and Nelder, J. A. (2000) Two Ways of Modelling Overdispersion in Non-normal Data. Applied Statistics. Journal of the Royal Statistical Society C **49** 591-598
18. Lee, Y. and Nelder, J. A. (2001) Hierarchical Generalised Linear Models: A Synthesis of Generalised Linear Models, Random-Effect Models and Structured Dispersions. Biometrika, **88**, 987-1006
19. Pawitan, Y. (2001) Two-staged Estimation of Variance Components in Generalized Linear Mixed Models. Journal of Statistical Computation and Simulation **69** 1-17
20. Consonni, G. and Veronese, P. (1992) Conjugate priors for Exponential Families Having Quadratic Variance Functions. Journal of the American Statistical Association **87** 1123-1127
21. Müller-Funk, U. and Pukelsheim, F. (1987) How Regular are Conjugate Exponential Families ? Statistics & Probability Letters **7** 327-333
22. Gutiérez-Peña E. and Smith A.F.M (1995) Conjugate Parametrizations for Natural Exponential Families, Journ. Amer. Statist. Assoc., **90** 1347-1356
23. Woodall, W. H. (1997) Control Charts Based on Attribute Data: Bibliography and Review. Journal of Quality Technology **29** 172-183
24. Skellam, J. G. (1948) A Probability Distribution Derived from the Binomial Distribution by Regarding the Probability of Success as Variable Between the Sets of Trials. Journal of the Royal Statistical Society B **10** 257-261
25. Hald A. (1960) The compound Hypergeometric Distribution and a System of Single Sampling Inspection Plans Based on Prior Distributions and Costs. Technometrics, **2** 275-340, Discussion, 370-372
26. Ford, J.H. (1951) Examples of the process curve. M. Sc. Thesis, Imperial College, London.
27. Fitzner, D. (1979) Adaptive Systeme Einfacher Kostoptimaler Stichprobenpläne für die Gut-Schlecht-Prüfung. Physica-Verlag. Würzburg.
28. Griffiths, D. A. (1973) Maximum Likelihood Estimation for the Beta-binomial Distribution and an Application to the Household Distribution of the Total Number of Cases of a Disease. Biometrics **29** 637-648
29. Hasselblad, V., Stead, A. G. and Anderson, H. S. (1981): DISFIT: A Distribution Fitting System 1. Discrete Distributions, U.S. Environmental Protection Agency, EPA 600/281010.
30. Garren, S. T., Smith, R. L. and Piegorsch, W. W. (2000) On a Likelihood-based Goodness-of-fit Test of the Beta-Binomial Model. Biometrika **56** 947-949
31. Garren, S. T., Smith, R. L. and Piegorsch, W. W. (2001) Bootstrap Goodness-of-fit Test for the Beta-binomial Model. Journal of Applied Statistics **28** 561-571

SPC Monitoring and Variance Estimation

C. Alexopoulos[1], D. Goldsman[1], K.-L. Tsui[1], and W. Jiang[2]

[1] School of Industrial and Systems Engineering, Georgia Institute of Technology, Atlanta, GA 30332-0205, USA
[2] INSIGHT, AT&T Labs, Short Hills, NJ 07078, USA

Abstract. According to W. Shewhart, process variation can be classified into assignable cause and common cause variations. Assignable cause variation can be eliminated by statistical process control (SPC) methods through identification and elimination of the root cause of the process shift. Common cause variation is inherent in the process and is generally difficult to reduce by SPC methods. However, if the common cause variation can be modeled by an autocorrelated process and physical variables are available to adjust the output, the common cause variation can be reduced by automatic process control (APC) methods through feedback/feedforward controllers. Integration of SPC and APC methods can result in major improvements in industrial efficiency.

Most SPC monitoring methods and traditional APC process adjustment methods (such as those based on one-step-ahead minimum mean squared error predictors or proportional integral derivative controllers) involve one or both of the following two steps: (i) "whitening" the process by subtracting a predictor and (ii) monitoring the prediction errors with appropriate control limits. This paper reviews common process monitoring and adjustment methods for process control of autocorrelated data, including model-free methods based on batch means, and investigates the general relationships and properties of the underlying models.

1 Introduction

Automatic process control (APC), or engineering process control (EPC), is a popular strategy for process optimization and improvement. In APC schemes, the manufacturing process is described as an input-output system where the input variable(s) can be manipulated (or adjusted) to counteract the uncontrollable disturbances to maintain the process. The output of the process can be measurements of the final product or critical in-process variables that need to be controlled. In general, without any control actions (adjustment of inputs), the output may shift or drift away from the desired quality target due to disturbances (Box and Luceno 1997). These disturbances often are not white noise but exhibit a dependence on past values; hence they are autocorrelated. It is thus possible to anticipate the process behavior based on past observations and to control the process by adjusting the input variables.

The basic objective of APC is to minimize the average deviation of the process output from its target. Two APC schemes have been popular in the industry: the minimum mean squared error (MMSE) control and the proportional integral derivative (PID) control (Box and Luceno 1997). The MMSE controller is developed based on a known time series transfer function model between the process input and output (Box, Jenkins, and Reinsel 1994). The PID controller is formed by the summation of three control components: proportional, integral, and derivative.

Statistical process control (SPC) is an effective tool for achieving process stability and improving process capability through variation reduction. The basic idea of SPC is to quickly detect the occurrence of process shifts (referred as special cause variation) so that the process can be investigated and corrective action may be taken before quality deteriorates and defective units are produced. The first major SPC tool was developed in the 1920s by Walter Shewhart of Bell Telephone Laboratories and has received tremendous success in manufacturing applications (see Montgomery 1996; Montgomery and Woodall 1997; Woodall, Tsui, and Tucker 1997).

SPC techniques have gained widespread acceptance in the manufacturing industry for process variation reduction and quality improvement. However, traditional SPC charts are developed based on the assumption that process data are independently and identically distributed (iid), whereas serial correlation often exists in many industrial processes, especially in batch production processes. Traditional control charts have been shown to have poor performance in monitoring and controlling such processes (Harris and Ross 1991; Alwan 1992).

One common SPC strategy for monitoring autocorrelated processes is to modify the control limits of traditional charts and then to apply the modified charts to the original autocorrelated data. An example is the Exponentially Weighted Moving Average chart for stationary processes (EWMAST) studied by Zhang (1998). Jiang, Tsui, and Woodall (2000) extended this to a general class of control charts based on the autoregressive moving average transformation, the ARMA charts. They show that these charts could have good performance when certain chart parameters are chosen appropriately.

Another popular method is the use of Special-Cause Charts (SCC) proposed by Alwan and Roberts (1988). Their idea is to first "whiten" the autocorrelated data by subtracting their one-step-ahead minimum mean squared error (MMSE) prediction, and then monitor the residuals (or the prediction errors). When the model or prediction is accurate, the prediction errors are approximately uncorrelated. Thus traditional SPC techniques such as Shewhart charts can be applied to these prediction errors. The SCC method has attracted considerable attention and has been further studied by many authors. Among them, Harris and Ross (1991), Superville and Adams (1994), Runger and Willemain (1995), and Runger, Willemain and Prabhu (1995) investigate process monitoring based on the MMSE prediction errors for simple autoregressive (AR(1)) models; Wardell, Moskowitz, and Plante (1992, 1994) discuss the performance of SCC's for ARMA(1, 1) models; and Vander Wiel (1996) studies the performance of SCC's for integrated moving average (IMA(0, 1, 1)) models. Montgomery and Mastrangelo (1991) recommend the use of exponentially weighted moving average (EWMA) predictors as one-step-ahead predictors in the SCC method (hereafter called the M-M chart).

The general idea behind these charts is to apply some predictors to the correlated process and then monitor the corresponding prediction errors. A control chart based on this approach is thus called a forecast-based chart. In general, any predictor can be used in this approach. Jiang, Wu, Tsung, Nair, and Tsui (2002) propose the use of proportional-integrated-derivative (PID) predictors with subsequent monitoring of the prediction errors; these are called PID-based charts. The family of PID-based charts includes the SCC, EWMA, and M-M charts as special cases. Jiang et al. (2001) show that the predictors of the EWMA chart and M-M chart may sometimes be inefficient and the SCC may be too sensitive to model deviation. On the other hand, the performance of the PID-based chart can be predicted via chart parameters through measures of two "capability indices". As a result, for any given underlying process, one can tune the parameters of the PID-based chart to optimize its performance.

The third SPC approach is the use of batch means methods to monitor autocorrelated data. It is well known that the batch means method can reduce the correlation among data. Runger and Willemain (1995) consider monitoring classical (unweighted) and weighted batch means of autocorrelated data, where the weights depend on the underlying process model. The classical method determines a batch size based on a simplistic iterative procedure à la Fishman (1978) or Law and Carson (1979). The method of weighted batch means requires knowledge of the underlying process model, whereas the classical method is "model-free". Runger and Willemain show that the SCC chart is a special case of the weighted batch means method. They conclude that monitoring the batch means (weighted or classical) can be more efficient in detecting small process mean shifts than the forecast-based charts in some situations. In a follow-up study, Runger and Willemain (1996) compare the classical batch means method to the method of "spaced" batch means (with unused observations between batches) suggested by Alwan and Radson (1992) using an AR(1) process. Their study suggested that the classical method yields in-control average run lengths (ARL's) that are comparable to those of iid processes and is more sensitive than the spaced batch method in detecting small mean shifts.

Recent development in APC and SPC reveals an opportunity to combine the two techniques for process improvement. Although the two techniques have different origins and implementation strategies (APC is often applied in process industries and SPC is often applied in parts industries), they both aim at the same objective of reducing process variation. To avoid confusion, Box and Luceno (1997) refer APC activities as process adjustment and SPC activities as process monitoring. As explained by Shewhart (1931), the process variation can be classified into two categories, special cause and common cause variations. The special cause variation can be eliminated by implementation of SPC methods through identification and elimination of the root cause of the process changes. The common cause variation is inherent in the process and is generally difficult to reduce by SPC methods. However, if the common cause variation can be modeled as an autocorrelated process, it can be reduced by implementation of APC methods through feedback/feedforward control schemes.

As suggested by Box and Kramer (1992) and others, it is possible to reduce both the special cause and common cause variations by applying SPC methods to monitor the output of an APC-controlled process. In practice, when an APC control scheme is applied to reduce the systematic variation, it also unintentionally compensates for the (special cause) process shift at the same time. This makes it difficult to apply standard SPC methods to detect the process shift. However, Box and Kramer (1992) point out that it is important to identify this type of process shift so that the engineer can understand and eliminate the root cause and thus improve the long term performance of the process.

In most aforementioned work, it is assumed that the model for the process data and the model parameters are known exactly, which is unlikely to be true in reality. When the process model is unknown and the focus is on model-based methods, the model and the parameters first need to be estimated and only then can the monitoring statistic be computed. For both model-based and model-free methods, in order to implement the control chart the variance of the monitoring statistic first needs to be estimated, and then the control limits are determined based on some pre-specified requirements on the in-control and out-of-control ARL's. Although the variance estimation problem plays such an important role in implementing control charts in the presence of serial correlation, very little work has been done in the SPC research literature, especially for monitoring complex statistics such as those mentioned above (see Cryer and Ryan 1990; Adams and Tseng 1998). In addition, even though the variance of the monitoring statistic can be accurately estimated, it is very difficult to determine the number of sigma units in the control limits so that the requirements on the ARL's are satisfied, which is generally referred to as the design of control chart problem.

2 SPC Methods for Monitoring Autocorrelated Processes

2.1 ARMA Monitoring Methods

We first introduce the ARMA chart for monitoring the mean of iid processes. Suppose that we are monitoring an iid normal process, $a_1, a_2, \ldots$ with mean zero and variance σ_a^2, and we wish to detect shifts in the mean of the process. The monitoring statistic of an ARMA chart is defined to be the result of a *generalized* ARMA(1, 1) process applied to the iid process, i.e.,

$$Z_t = \theta_0 a_t - \theta a_{t-1} + \phi Z_{t-1} = \theta_0(a_t - \beta a_{t-1}) + \phi Z_{t-1},$$

where $\beta = \theta/\theta_0$ and θ_0 is chosen so that the sum of the coefficients is unity when Z_t is expressed in terms of the a_t's, i.e.,

$$\left.\frac{\theta_0 - \theta B}{1 - \phi B}\right|_{B=1} = \frac{\theta_0 - \theta}{1 - \phi} = 1,$$

where B is the backshift operator ($Ba_t = a_{t-1}$). Thus, $\theta_0 = 1 + \theta - \phi$. To guarantee that the monitoring process is reversible and stationary, we have the constraints that $|\beta| < 1$ and $|\phi| < 1$. The ARMA chart signals when $|Z_t| > L\sigma_Z$, where L is an appropriate constant. Note that the ARMA chart reduces to the EWMA chart when $\theta = 0$ and $\phi = 1 - \lambda$. Thus the ARMA chart can be considered as an extension of the EWMA chart.

Since $W_t = Z_t/\theta_0$ is a standard ARMA(1, 1) process, it is easy to show that the steady-state variance (or "variance parameter") and the first-lag autocovariance of the monitoring statistic are

$$\sigma_Z^2 = \left[\frac{2(\theta - \phi)(1 + \theta)}{1 + \phi} + 1\right]\sigma_a^2 \quad \text{and} \quad \gamma_1 = \frac{(1 + \theta)(1 - \phi)(\phi - \theta)}{1 + \phi}\sigma_a^2.$$

It can be shown that, for a given ϕ, the variance of the monitoring process is minimized when $\phi - \theta = 1 + \theta$ or $\theta = -(1 - \phi)/2$. Notice that this is the condition for a positive first-lag autocovariance γ_1. The autocorrelations of the monitoring process are $\rho_1 = \phi - \theta\theta_0/\sigma_Z^2$ and $\rho_k = \phi^{k-1}\rho_1$ $(k \geq 2)$.

Suppose there is a step shift μ in the mean of the original process a_t occurring at time t_0. Then the corresponding mean shift on the monitoring process becomes

$$\mu_{t_0} = \theta_0\mu, \qquad \mu_{t_0+k} = (\theta_0 - \theta)\mu + \phi\mu_{t_0+k-1} \quad (k \geq 1). \tag{1}$$

We now consider the application of the ARMA chart (with parameters ϕ and θ) to an ARMA(1, 1) process with parameters u and v. As shown in Jiang et al. (2000), the steady-state variance can be simplified in terms of the parameters of the process. Suppose the underlying process is

$$X_t - uX_{t-1} = a_t - va_{t-1}$$

with $|u| < 1$ and $|v| < 1$. It follows that the variance is

$$\sigma_X^2 = \frac{1 - 2uv + v^2}{1 - u^2}\sigma_a^2$$

while the lagged correlation coefficients are

$$\rho_1 = u - v\sigma_a^2/\sigma_X^2 \quad \text{and} \quad \rho_k = u\rho_{k-1} \ (k \geq 2).$$

By substituting these terms into the steady-state variance, we have

$$\sigma_Z^2 = \left\{\theta_0^2 + \frac{\alpha^2}{1 - \phi^2} + 2\left[\theta_0\alpha + \frac{\phi\alpha^2}{1 - \phi^2}\right]\frac{\rho_1}{1 - \phi u}\right\}\sigma_X^2.$$

In addition, it can be shown that application of the ARMA chart to an ARMA(1, 1) process will result in a generalized ARMA(2, 2) process. Whenever a step shift μ occurs at $t = t_0$ in the underlying process $\{X_t\}$, the shift pattern of the monitoring process $\{Z_t\}$ after the occurrence follows from equation (1). Solving this difference equation, the shift pattern following t_0 is

$$\mu_{t_0+k} = [1 + (\theta - \phi)\phi^k]\mu \quad (k \geq 0).$$

We observe that this shift pattern depends only on the shift size and the charting parameters ϕ and θ. However, since the autocorrelation structure of the ARMA chart on an ARMA(1, 1) process depends on the parameters of the charting process (ϕ and θ) as well as those of the original process (u and v), the performance of the ARMA chart depends on all four parameters. It is therefore hard to characterize the performance of the ARMA chart.

Jiang et al. (2000) show that the ARMA chart includes the SCC and EWMAST charts as special cases. The performance of the SCC and EWMAST charts has been studied extensively in the literature. Wardell, Moskowitz, and Plante (1992) show that the SCC chart performs better than the EWMAST chart in some cases, but worse in others. The relative performance depends critically on the parameters of the original process (u and v), and neither chart is uniformly better than the other. Since both the SCC and EWMAST charts are special cases of the ARMA chart, Jiang et al. (2000) propose an algorithm to choose ARMA chart parameters based on two capability ratios. Jiang et al. (2000) investigate the performance of the ARMA charts for monitoring the first-order ARMA process. Their simulation study compared the ARMA-based charts with the popular EWMAST and SCC charts. They found that the ARMA charts with appropriately chosen parameters outperform both EWMAST and SCC charts in most cases. This is not surprising as the ARMA charts include the EWMAST and SCC charts as special cases. With appropriate adjustment of chart parameters, one would expect that these new ARMA charts will exhibit better performance.

2.2 Forecast-Based Monitoring Methods

Forecast-based charts, such as the MMSE-based SCC and the EWMA-based M-M charts, apply traditional control charts to monitor the prediction error. Therefore, as proposed in Jiang et al. (2002), the PID-based charts apply traditional control charts to monitor the prediction error of the PID predictor. As shown in Tsung, Wu, and Nair (1998), the PID predictor is highly efficient and robust, but the MMSE predictor has a serious robustness problem and the EWMA predictor is inefficient in many situations. Thus, we would expect the PID-based chart to perform better than the existing charts if the parameters are appropriately chosen.

In a PID-based chart, any traditional control chart can be used to monitor the prediction error. For simplicity, one can apply the Shewhart chart to the prediction error e_t, which can be written as the difference between the underlying process $\{X_t\}$ and its PID predictor $\{\hat{X}_t\}$, i.e.,

$$e_t = X_t - \hat{X}_t = (1-\lambda_1)e_{t-1} - \lambda_2 e_{t-2} - \lambda_3 e_{t-3} + (X_t - X_{t-1}),$$

where $\lambda_1 = k_I + k_P + k_D$, $\lambda_2 = -(k_P + 2k_D)$, $\lambda_3 = k_D$, and k_P, k_I and k_D are the parameters of the PID predictor. We will use the parameters k_P, k_I, and k_D to specify the PID-based chart and reflect its connection with the PID control. It is reasonable to require that the prediction error process $\{e_t\}$ be stationary when the monitored process is stationary. Otherwise, the charting process $\{e_t\}$ would drift even when the monitored process has no mean shift.

It is important to note that the I control leads to the EWMA predictor, and the EMWA prediction-based chart is the M-M chart. There is no connection between the EWMA predictor and the EWMA chart. As shown in Jiang et al. (2002), the EWMA chart is the same as the P-based chart. As both the M-M chart and the EWMA chart have good performance in monitoring autocorrelated processes, one would expect that the PID-based chart with combined terms may combine the advantages of the pure P chart (the EWMA chart) and the pure I chart (the M-M chart).

Jiang et al. (2002) investigate the performance of P-, I-, D-, PI-, and PD-based charts under iid processes to understand how individual parameters affect the ARL performance. In summary, the three pure charts (P, I, and D) have different advantages when monitoring different types of mean shifts. The pure P chart is sensitive in detecting small shifts, the pure D chart is sensitive in detecting large shifts, and the pure I chart is sensitive in detecting oscillating shifts. For charts with combined terms (PI- and PD-based charts), the authors show that the advantages also combine. For constant mean shifts, the ARL performance of PI-based charts is sensitive to the parameter k_I, whereas the performance of PD-based charts is sensitive to the parameter k_P. For oscillating mean shifts, the parameter k_I tends to amplify the shifts and helps to detect them quickly. These properties show that PID-based charts have great potential for monitoring autocorrelated processes.

Jiang et al. (2002) also investigate the performance of the PID-based charts for monitoring a first-order ARMA process. In their simulation study, the PID-based charts are compared with the popular EWMAST and SCC charts. The authors found that the EWMAST outperforms the SCC chart for some ARMA processes and vice versa for other processes. As expected, the PID-based charts with appropriately chosen chart parameters perform competitively with the winner of the EWMAST and the SCC charts under various ARMA processes. In addition, the advantage of the PID charts is that their ARL performance can be predicted based on the two capabilities computed from the charting parameters.

Most of the methodology described in Sections 2.1 and 2.2 explicitly assumes that the model for the process data and the model parameters are known exactly—an assumption that is unlikely to be true in practice. This motivates the consideration of the model-free methods in the next section.

3 Batch-Means Monitoring Methods

In this section, we outline the proposed batch-means-based methodology for monitoring processes in the presence of serial correlation. Batch-means methods have two different applications in SPC process monitoring: (a) estimation of the variance of the monitoring statistics of some commonly used SPC charts and (b) creation of new monitoring statistics based on batch means. Below we first focus on batching the data, with the key contribution being the development of accurate and precise estimators for the variance of the resulting batch means.

To begin, let $\{X_t\}$ be a discrete stationary monitoring process with mean μ. Since the observations X_t are usually correlated in the problems under consideration here, we propose to construct control charts by forming nonoverlapping batches of size m each and using the respective batch means as approximately iid observations. Specifically, the ith batch consists of the observations $X_{(i-1)m+1}, \ldots, X_{im}$ and the ith batch mean is the sample average of the ith batch:

$$Y_i = \frac{1}{m} \sum_{j=1}^{m} X_{(i-1)m+j} \quad (j \geq 1).$$

Under some mild moment and mixing assumptions (e.g., phi-mixing, see Billingsley 1968), one can show that, as m becomes large, the batch means become approximately iid normal. However, to form control charts, one needs to estimate the variance $\sigma_Y^2 = \mathrm{Var}(Y_1)$, which is a difficult problem for serially correlated data.

The simulation literature reveals a variety of methods for estimating σ_Y^2. These methods are based on the existence of a positive, finite quantity σ^2 (often referred to as the *variance parameter*) so that a central limit theorem holds:

$$\frac{\bar{X}_t - \mu}{\sigma/\sqrt{t}} \Rightarrow N(0,1) \quad \text{as } t \to \infty,$$

where $\bar{X}_t = t^{-1} \sum_{i=1}^{t} X_i$ is the sample mean of $X_1, \ldots, X_t$ and $\Rightarrow$ denotes convergence in distribution. The estimation of σ_Y^2 involves two steps. The first step uses a sufficiently large set of historical data (say $X_1^*, \ldots, X_n^*$) and a method to form an estimate, say $\widehat{\sigma^2}$, of the variance parameter σ^2; and the second step estimates σ_Y^2 by $\widehat{\sigma_Y^2} = \widehat{\sigma^2}/m$.

The set of methods one can use to estimate the variance parameter σ^2 includes, among others, nonoverlapping batch means (NBM), overlapping batch means (OBM), and standardized time series (STS). We now discuss each in turn.

3.1 Nonoverlapping Batch Means Variance Estimator

The "classical" NBM method "splits" the data into b^* approximately independent non-overlapping batches each of size m^* ($b^*m^* \leq n^*$) and estimates the variance parameter σ^2 by

$$\widehat{\sigma^2}_{NBM} = \frac{m^*}{b^* - 1} \sum_{i=1}^{b^*} (Y_i^* - \bar{Y}_{b^*}^*)^2,$$

where the Y_i^*'s are the batch means from the data $X_1^*, \ldots, X_n^*$ and $\bar{Y}_{b^*}^*$ is their sample mean.

3.2 Overlapping Batch Means Variance Estimator

The method of OBM (see Meketon and Schmeiser 1984 or Sargent, Kang, and Goldsman 1992) splits the observations into overlapping batches, with the full realization that the resulting overlapping batch means are *not* independent (though they are identically distributed and asymptotically normal). Then OBM applies the "standard" sample-variance estimator, appropriately scaled, to these highly correlated overlapping batch means. This seemingly bizarre technique uses results from spectral theory to obtain an estimator that is provably superior to NBM, at least asymptotically, for certain serially correlated time series. It is worth mentioning that OBM estimators can be expressed as the averages of estimators based on nonoverlapping weighted batch means.

What do we mean by superior? We most often care about the bias and variance of an estimator for σ^2 [and the resulting mean squared error (MSE)]. Batching typically increases bias (bad), but decreases variance (good); its effect on MSE takes a bit more work to analyze. What is nice about the OBM estimator is that it has the same asymptotic bias as, but smaller variance than, the corresponding performance measures for NBM. In fact, the variance of the OBM estimator is only about 2/3 that of the NBM. Thus, OBM gives better (asymptotic) performance than NBM "for free".

3.3 Standardized Time Series Variance Estimators

This class of variance estimation techniques, originally proposed by Schruben (1983) and generally referred to as standardized time series methods, has enjoyed a great deal of interest over the years. The general idea is to use a Functional Central Limit Theorem to standardize an entire stationary time series into a limiting Brownian bridge process (instead of simply standardizing the sample mean to a limiting standard normal distribution via the usual Central Limit Theorem). *This technique is related to CUSUM methods, but has never been applied in the quality control literature for some reason.* This is unfortunate, since certain STS variance estimators have both very low bias and very low variance.

Now some specifics. The *standardized time series* of one batch of stationary X_i's is

$$T_{m^*}(t) \triangleq \frac{\lfloor m^* t \rfloor (\bar{X}_{m^*} - \bar{X}_{\lfloor m^* t \rfloor})}{\sigma \sqrt{m^*}} \quad \text{for } 0 \leq t \leq 1.$$

Under mild assumptions, it can be shown that

$$(\sqrt{m^*}(\bar{X}_{m^*} - \mu), \sigma T_{m^*}) \Rightarrow (\sigma \mathcal{W}(1), \sigma \mathcal{B}),$$

where $\mathcal{W}$ is a standard *Brownian motion* process, and $\mathcal{B}$ is a standard *Brownian bridge* process on $[0, 1]$, i.e., Brownian motion conditioned to equal 0 at time 1 (see Foley and Goldsman 2001; Glynn and Iglehart 1990; Schruben 1983). One then uses properties of the Brownian bridge to develop estimators for σ^2.

As the simplest example, let us consider the *weighted area* STS estimator for σ^2 (see Goldsman, Meketon, and Schruben 1990; Goldsman and Schruben 1990), formed by the square of the weighted area under the standardized times series

$$A(f; m^*) \triangleq \left[\frac{1}{m^*} \sum_{k=1}^{m^*} f\left(\frac{k}{m^*}\right) \sigma T_{m^*}\left(\frac{k}{m^*}\right) \right]^2$$

and its limiting functional

$$A(f) \triangleq \left[\int_0^1 f(t) \sigma \mathcal{B}(t)\, dt \right]^2,$$

where $f(t)$ is continuous on the interval $[0, 1]$ and normalized so that $\mathsf{Var}(\int_0^1 f(t)\mathcal{B}(t)\, dt) = 1$. Then $\int_0^1 f(t)\sigma\mathcal{B}(t)\, dt \sim \sigma \mathsf{Nor}(0, 1)$, and under mild conditions, we have $A(f; m^*) \Rightarrow A(f) \sim \sigma^2 \chi_1^2$. (Here, $\mathsf{Nor}(0, 1)$ denotes a standard normal random variable, $\sim$ means "is distributed as", and χ_ν^2 denotes a chi-square random variable with ν degrees of freedom.) Hence, $A(f; m^*)$ is an (unbiased) estimator of σ^2 based on one batch of observations.

This trick can be repeated on b^* subsequent nonoverlapping batches of observations, yielding approximately iid $\sigma^2 \chi_1^2$ random variables; these are then averaged to produce what is called the *weighted area* estimator for σ^2—asymptotically distributed as $\sigma^2 \chi_{b^*}^2 / b^*$. For certain choices of weighting functions, one can show that the STS weighted area estimator is less biased as an estimator of σ^2 than is the NBM estimator. Both the area and NBM estimators have about the same variance.

The area estimator is just the beginning. The STS plate is full of other estimators for σ^2, some of which enjoy excellent performance properties. We list some briefly, without going into the notational fog of the details.

- Cramér–von Mises (CvM) estimators. Instead of dealing with the square of the weighted area under the standardized time series, we use the area under the square. The resulting estimators have excellent bias properties, as well as lower variance than that of NBM. See Goldsman, et al. (1999) for more details.
- L_p-norm estimators. These estimators, based on the L_p norm of the standardized time series, can be regarded as generalizations of the area and CvM estimators. While we have not yet proven small-sample bias superiority over NBM, we have shown that the L_p-norm estimators possess very low variance vs. NBM. See Tokol, Goldsman, Ockerman, and Swain (1998) for the complete discussion.
- Overlapping versions of the above STS estimators. Here we see what happens when we form an STS estimator for each *overlapped* batch, and then take the average of all of these estimators. The resulting average maintains the low bias of its nonoverlapped counterpart, but has the added bonus of much lower variance, usually at least 50% (depending on the type of estimator, weight employed, etc.). See Alexopoulos, Goldsman, Tokol, and Argon (2002) for the latest developments.
- Jackknifed versions of the above.

Time and Space Requirements

For fixed batch size m^* and sample size n^*, the vast majority of the estimators in this section can be computed in $O(n^*)$ time, but the storage requirements depend on the type of the estimator. (The computational issues related to the overlapping versions of STS estimators are currently under investigation.)

The LBATCH and ABATCH procedures of Fishman and Yarberry (1997) deserve special attention since they dynamically increase the batch size and the number of batches based on the outcome of a hypothesis test for independence among the batch means. Both procedures require $O(n^*)$ time and $O(\log_2 n^*)$ space (Alexopoulos, Fishman, and Seila 1997). Although like complexities are known for static algorithms, the dynamic setting of the LBATCH and ABATCH rules offers an important additional advantage not present in the static approach. As the analysis evolves with increasing sample path length, it allows a user to assess how well the estimated variance of the sample mean stabilizes. This assessment is essential to gauge the quality of the variance estimate. These procedures are implemented in the LABATCH.2 package described in Fishman (1998).

4 Variance Estimation of Monitoring Statistics

How is σ_Y^2 estimated in the quality control literature? The short answer is: only via a 1970's-style implementation of the NBM method. Obviously, this limited body of work certainly needs to be enhanced in a number of ways. We feel that there are three pressing gaps in the literature.

Gap 1. *The level of mathematical rigor should be improved somewhat.* Typically, the batch-means-based research up to this point has been heuristic, being based entirely on limiting arguments. No mention has ever been made of finite-sample bias properties of the NBM variance estimator—which obviously affect procedure performance.

Gap 2. *One should use better estimators for the variance of the batch means.* The research up to this point only uses the NBM variance estimator (or trivial variants thereof). A number of other variance estimators are superior. For example, weighted area and CvM STS variance estimators have lower bias than does NBM. Further, the OBM and certain STS estimators have lower variance than does NBM—sometimes substantially lower—at no loss in bias superiority.

Gap 3. *The implementation methodology ought to be improved.* For instance, the existing control literature does not mention the problem of batch size determination. Further, it does not address the important question of updating variance estimators on the fly as more information becomes available. Current procedures usually assume that the variance estimator is known (a naive assumption), or has somehow been estimated a priori (by the NBM variance estimator).

Below we describe two alternative approaches:

Fixed Variance Estimator

This approach requires the existence of "sound" historical data. Specifically, it uses a method from Section 3 to obtain an estimate of σ_Y^2 and monitors the process $\{X_t\}$ by using batch means of observations. We are currently evaluating the variance estimation methods from Section 3 on the basis of industrial data and common theoretical benchmark models (e.g., AR and ARMA models).

Dynamic Update of Variance Estimator

This approach has two goals: (a) it accounts for the lack of historical data and (b) it attempts to improve the precision of the variance estimator by appending real-time data to the historical sample until a shift in the process is observed. For instance, suppose that no historical data are available and that the process is considered to be in control. One can proceed as follows:

(i) Use any of the NBM procedures in the LABATCH.2 implementation of Fishman (1998) until a proper batch size m is identified and an estimate of σ_Y^2 is obtained. The collected data forms the initial historical data set.

(ii) Form a control chart based on the estimate of σ_Y^2. Add each observation to the historical data set and update/store the variance estimates once a batch of real-time observations is complete. Continue until an out-of-control condition is detected. Then discard the last batch, identify/rectify the cause of this condition, and once the process is back in control use the estimate of σ_Y^2 that was computed before the process went out-of-control.

Note that the batch means methods discussed above focus on estimating the batch mean variance. In general, it is important to obtain variance estimates of commonly used monitoring statistics, such as the ARMA and PID charts discussed. One approach is to express the monitoring statistic as some form of batch means and apply a traditional batch means variance estimation method. An alternative approach is to extend existing batch means methods to develop accurate variance estimates of complex monitoring statistics.

5 New Monitoring Statistics Based on Batch Means

Runger and Willemain (R&W) (1996) studied the applicability of classical NBM as the monitoring statistic for an AR(1) process and determined the batch size based on an iterative procedure à la Fishman (1978) or Law and Carson (1979). The alternative approach in R&W (1996) assume an ARMA model and uses weighted nonoverlapping batch means as the monitoring statistic based on the recommendations in Bischak, Kelton, and Pollock (1983). (Of course, the latter approach has limited applicability as it assumes a particular model.) There are several advantages of using batch means as monitoring statistics. First, one can assume that the batch means are uncorrelated and then use existing charting techniques on them. Second, charting the batch means instead of the original data may have a better performance in detecting shifts in the process. Finally, accurate and precise variance estimates of batch means are available, making it easier to determine the control limits of batch means than those of other complex monitoring statistics. As shown by R&W, simple unweighted batch means and weighted batch means can be effective in detecting small process shifts when the batch mean size is appropriately chosen. For monitoring AR(1) processes, R&W (1995) propose to determine the optimal batch size based on simulation results of certain processes. In general, though, it is not clear how the optimal batch size can be determined for AR or ARMA processes. This problem can be addressed by the methods in Section 3. In particular, one should compare the optimal batch sizes required for detecting small and large shifts.

As mentioned in Section 3, in addition to the classical NBM methods discussed in R&W (1995, 1996), many other methods have been developed for variance estimation. There is no reason why these alternative methods cannot be used as monitoring statistics in SPC charting. For instance, the OBM method in Section 3.2 is an interesting candidate. As this approach reuses the original observations in each batch mean, the total number of observations before signaling out-of-control can be minimized if the monitoring statistic has enough power. The disadvantage of this approach is that the correlation structure of the monitoring process becomes more complex. It is interesting to investigate the use of interval-overlapping and other batch means as the monitoring statistics in SPC charts, and to develop accurate and precise variance estimators of the proposed monitoring statistics to determine appropriate control limits.

References

1. Adams, B.M., I.T. Tseng. 1998. Robustness of forecast-based monitoring systems. *Journal of Quality Technology* **30** 328–339.
2. Alexopoulos, C., G.S. Fishman, A.F. Seila. 1997. Computational experience with the batch means method. In *Proceedings of the 1997 Winter Simulation Conference*, 194-201. IEEE, Piscataway, NJ.
3. Alexopoulos, C., D. Goldsman, G. Tokol. 2001. Properties of batched quadratic-form variance parameter estimators for simulations. *INFORMS Journal on Computing* **13** 149–156.
4. Alexopoulos, C., D. Goldsman, G. Tokol, N. Argon. 2001. Overlapping variance estimators for simulations. Technical Report, School of Industrial and Systems Engineering, Georgia Institute of Technology, Atlanta, GA.
5. Alexopoulos, C., A.F. Seila. 1998. Output data analysis. In *Handbook of Simulation*, J. Banks (ed.), 225–272. John Wiley and Sons, New York.
6. Alwan, L.C. 1992. Effects of autocorrelation on control chart performance. *Communications in Statistics – Theory and Methods* **41(4)** 1025–1049.
7. Alwan, L.C., D. Radson. 1992. Time-series investigation of subsample mean charts. *IIE Transactions* **24** 66–80.
8. Alwan, L.C, H.V. Roberts. 1988. Time-series modeling for statistical process control. *Journal of Business and Economic Statistics* **6** 87–95.
9. Billingsley, P. 1968. *Convergence of Probability Measures*. John Wiley and Sons, New York.
10. Bischak, D.P, W.D. Kelton, S.M. Pollock. 1983. Weighted batch means for confidence intervals in steady-state simulations. *Management Science* **39** 1002–1019.
11. Box, G.E.P, G.M. Jenkins, G.C. Reinsel. 1994. *Time Series Analysis, Forecasting and Control*, 3rd Ed. Prentice-Hall, Englewood Cliffs, NJ.
12. Box, G.E.P., T. Kramer. 1992. Statistical Process Monitoring and Feedback Adjustment - A Discussion, *Technometrics* **34** 251–285.
13. Box, G. E. P., A. Luceno. 1997. *Statistical Control by Monitoring and Feedback Adjustment*. John Wiley and Sons, New York.
14. Cryer, J.D., T.P. Ryan. 1990. The estimation of sigma for an X chart: $\overline{MR}/d_2$ or S/d_4? *Journal of Quality Technology* **22** 187–192.
15. Fishman, G.S. 1978. Grouping observations in digital simulation. *Management Science* **24** 510–521.
16. Fishman, G.S. 1998. LABATCH.2 for analyzing sample path data. Technical Report UNC/OR TR-97/04, Department of Operations Research, University of North Carolina, Chapel Hill, NC.

17. Fishman, G.S., L.S. Yarberry. 1997. An implementation of the batch means method. *INFORMS Journal on Computing* **9** 296–310.
18. Foley, R.D., D. Goldsman. 2001. Confidence intervals using orthonormally weighted standardized times series. To appear in *ACM Transactions on Modeling and Computer Simulation.*
19. Glynn, P.W., D.L. Iglehart. 1990. Simulation output analysis using standardized time series. *Mathematics of Operations Research* **15** 1–16.
20. Glynn, P.W., W. Whitt. 1991. Estimating the asymptotic variance with batch means. *Operations Research Letters* **10** 431–435.
21. Goldsman, D., K. Kang, A.F. Seila. 1999. Cramér–von Mises variance estimators for simulations. *Operations Research* **47** 299–309.
22. Goldsman, D., M.S. Meketon, L.W. Schruben. 1990. Properties of standardized time series weighted area variance estimators. *Management Science* **36**, 602–612.
23. Goldsman, D., L.W. Schruben. 1990. New confidence interval estimators using standardized time series. *Management Science* **36** 393–397.
24. Harris, T.J., W.M. Ross. 1991. Statistical process control for correlated observations. *The Canadian Journal of Chemical Engineering* **69** 48–57.
25. Jiang, W., K.-L. Tsui, W.H. Woodall. 2000. A new SPC monitoring method: The ARMA chart. *Technometrics* **42** 399–410.
26. Jiang, W., H. Wu, F. Tsung, V.N. Nair, K.-L. Tsui. 2002. Proportional integral derivative control charts for process monitoring. *Technometrics* **44** 205–214.
27. Law, A.M., J.S. Carson. 1979. A sequential procedure for determining the length of a steady-state simulation. *Operations Research* **27** 1011–1025.
28. Meketon, M.S., B.W. Schmeiser. 1984. Overlapping batch means: Something for nothing? In *Proceedings of the 1984 Winter Simulation Conference*, 227–230. IEEE, Piscataway, NJ.
29. Montgomery, D.C. 1996. *Introduction to Statistical Quality Control*, 3rd edition. John Wiley and Sons, New York.
30. Montgomery, D.C., C.M. Mastrangelo. 1991. Some statistical process control methods for autocorrelated data, *Journal of Quality Technology* **23** 179–204.
31. Montgomery, D.C., W.H. Woodall (eds.). 1997. A discussion on statistically-based process monitoring and control. *Journal of Quality Technology* **29** 121–162.
32. Pedrosa, A.C., B.W. Schmeiser. 1994. Estimating the variance of the sample mean: Optimal batch size estimation and 1–2–1 overlapping batch means. Technical Report SMS94–3, School of Industrial Engineering, Purdue University, West Lafayette, IN.
33. Runger, G.C., T.R. Willemain. 1995. Model-based and model-free control of autocorrelated processes. *Journal of Quality Technology* **27** 283–292.
34. Runger, G.C., T.R. Willemain. 1996. Batch means charts for autocorrelated data. *IIE Transactions* **28** 483–487.
35. Runger, G.C., T.R. Willemain, S.S. Prabhu. 1995. Average run lengths for CUSUM control charts applied to residuals . *Communications in Statistics – Theory and Methods* **25** 273–282.
36. Sargent, R.G., K. Kang, D. Goldsman. 1992. An investigation of finite-sample behavior of confidence interval estimators. *Operations Research* **40** 898–913.
37. Schmeiser, B.W. 1982. Batch size effects in the analysis of simulation output. *Operations Research* **30** 556–568.
38. Schruben, L.W. 1983. Confidence interval estimation using standardized time series. *Operations Research* **31** 1090–1108.

39. Sherman, M., D. Goldsman. 2001. Large-sample normality of the batch means variance estimator. To appear in *Operations Research Letters.*
40. Shewhart, W.A. 1931. *Economic Control of Quality of Manufactured Product.* Van Nostrand, New York.
41. Song, W.-M., B.W. Schmeiser. 1993. Variance of the sample mean: Properties and graphs of quadratic-form estimators. *Operations Research* **41** 501–517.
42. Song, W.-M., B.W. Schmeiser. 1995. Optimal mean-squared-error batch sizes. *Management Science* **41** 110–123.
43. Superville, C.R., B.M. Adams. 1994. An evaluation of forecast-based quality control schemes. *Communications in Statistics – Simulation and Computation* **23** 645–661.
44. Tokol, G., D. Goldsman, D.H. Ockerman, J.J. Swain. 1998. Standardized time series L_p-norm variance estimators for simulations. *Management Science* **44** 234–245.
45. Tsung, F., H. Wu, V.N. Nair. 1998. On efficiency and robustness of discrete proportional integral control schemes. *Technometrics* **40** 214–222.
46. Vander Wiel, S.A. 1996. Monitoring processes that wander using integrated moving average models. *Technometrics* **38** 139–151.
47. Wardell, D.G., H. Moskowitz, R.D. Plante. 1992. Control charts in the presence of data correlation. *Management Science* **38** 1084–1105.
48. Wardell, D.G., H. Moskowitz, R.D. Plante. 1994. Run-length distributions of special-cause control charts for correlated observations. *Technometrics* **36** 3–17.
49. Woodall, W.H. 1999. Control charting based on attribute data: Bibliography and Review. *Journal of Quality Technology* **29** 172–183.
50. Woodall, W.H., D.C. Montgomery. 1999. Research issues and ideas in statistical process control. To appear in *Journal of Quality Technology.*
51. Woodall, W.H., K.-L. Tsui, G.R. Tucker. 1997. A Review of statistical and fuzzy quality control charts based on categorical data. In *Frontiers in Statistical Quality Control*, 83–89. Physica-Verlag, Heidelberg, Germany.

Control Charts for Time Series: A Review

Sven Knoth and Wolfgang Schmid

Europe-University, Dep. of Statistics, Postfach 1786, 15207 Frankfurt(Oder), Germany

Abstract. A state of the art survey in Statistical Process Control (SPC) for dependent data will be given. First papers about the influence on and the modification of standard SPC schemes are written some decades ago. After that, in the end of the 1980s and during the 1990s the consideration of dependence in the field of SPC became very popular. It turned out, that falsely assuming independence mostly leads to improper SPC schemes. Therefore, one has to be aware of dependence.

Roughly speaking we can distinguish two different types of SPC schemes for dependent data, residual and modified schemes. The first ones are based on the standard schemes that are applied to the (empirical) residuals of a suitable time series model. For getting the latter ones the stopping rule of the scheme is modified. Furthermore, some special issues of CUSUM schemes in case of dependent data are considered.

In case of the most simple model of dependent data – the autoregressive process of order one – we compare modified and residual EWMA and CUSUM schemes.

We present some approximation methods for computation of SPC schemes characteristics. Furthermore, some theoretical relations between the schemes for dependent data and for independent data will be sketched.

1 Introduction

The objective of process monitoring, i. e. of Statistical Process Control (SPC), is to detect changes in a process that may result from uncontrollable and unpredictable causes at unknown times. The most important tools for monitoring data are control charts. In principle, a control chart is a decision rule which signals that the process has changed if the control statistic lies outside the control limits. The first control charts were introduced by Shewhart (1926, 1931). Since this pioneering work many other control schemes have been introduced, such as the cumulative sum (CUSUM) chart of Page (1954) and the exponentially weighted moving average (EWMA) chart of Roberts (1959).

Up to the end of the 1980s one of the basic assumptions of nearly all papers was that the underlying observations are independent over time. Only some papers dealt with the dependence problem. Goldsmith and Whitfield (1961) investigated the influence of correlation on the CUSUM chart by Monte Carlo simulations. In the same year Page (1961) mentioned: *"The effects of correlated observations have been examined in one case by Goldsmith and Whitfield; wider study is desirable."* The problem was addressed but one had to wait over 10 years more for the next papers. Then, Bagshaw and Johnson (1974,75,77) approximated the CUSUM statistic for correlated data by Brownian motions. Rowlands (1976) studied the performance of CUSUM schemes for AR(1) and ARMA(1,1) data (cf. Rowlands and Wetherill (1991)). Furthermore, Nikiforov (1975,79,83,84) wrote a lot about CUSUM charts for ARIMA data. Finally, Vasilopoulos and Stamboulis (1978) proposed modified alarm limits for the Shewhart chart in case of dependent data.

After the more theoretical oriented period of dealing with dependence of the data, in recent years it has been pointed out that quite often the independence assumption is incorrect. Wetherill (1977) reminded in his monography, that observations from modern industrial processes are often autocorrelated. Then, in the end of the 1980s treating the dependence problem became very popular. Alwan (1989) analyzed 235 datasets which were frequently used as (standard) SPC examples. It turned out that in about 85 % the control limits were misplaced. In nearly the half of these cases neglected correlation was responsible. Now, the problem was realized as severe.

Besides assessing these strong effects of correlation on the control charts as quality control tools in industrial statistics, one has to consider correlation as a typical pattern of data which we can find in medicine, environmental and financial statistics, and in automatic control in engineering. Here time series models are standard.

In the following, we present modifications of SPC schemes such as Shewhart chart, CUSUM procedure, and EWMA scheme in order to take into account the dependence.

First, we introduce the utilized time series models in Section 2. Then, it follows in Section 3 a short description of the (standard) control charts in the independent case. Moreover, we demonstrate the performance of the control charts designed for independent data and applied to AR(1) data.

In Section 4 and 5 we introduce the residuals and modified charts, respectively. There are two different types of SPC schemes for dependent data, residual and modified schemes. The first ones (e. g., Alwan and Roberts (1988), Wardell et al. (1994a,b)) are based on the standard schemes that are applied to the (empirical) residuals of a suitable time series model. It seems so, that Harrison and Davies (1964) were the first who applied control charts (CUSUM) to residuals (prediction errors). For getting the modified charts (e. g., Vasilopoulos and Stamboulis (1978), Schmid (1995, 1997a,b)) the stopping rule (variance term, control limit) of the scheme is modified. Furthermore, some special issues of CUSUM schemes (*keyword* likelihood ratio approach, e. g., Nikiforov (1975 and later), Schmid (1997b)) in case of dependent data are considered.

In Section 6 a comparison study is performed for modified and residual EWMA and CUSUM schemes for AR(1) data.

We restrict ourselves on the above charts because they are the most common charts for applications. Moreover, up to now it is difficult to find results for other SPC schemes in case of dependent data. In the following we use the average run length (ARL) as performance measure. The ARL denotes the average number of observations (or samples) from the beginning of the observation until the scheme signals. Thereby, it is assumed that the parameters are constant over time, i. e. a possible change takes place before the first observation was taken. In case of an undisturbed process the ARL should be large, otherwise small. Naturally, there are other measures too, but then the problem becomes much more complicate in presence of dependence.

2 Modeling

The aim of statistical process control (SPC) consists in detecting structural deviations in a process over time. It is examined whether the observations can be considered as realizations of a given target process $\{Y_t\}$. The procedure is a sequential one, i. e. a change should be detected immediately after its occurrence. The observed or actual process is denoted by $\{X_t\}$. In this paper it is always assumed that at each time point t exactly one observation is available. Thus the sample size is equal to 1. The extension to a larger sample size is straightforward.

For a long time it was always assumed that the random variables $\{Y_t\}$ are independent and identically distributed (iid). It has turned out that in applications this condition is frequently not fulfilled. The data have a memory. Within time series analysis many suitable models have been introduced to describe such a behavior. A good introduction to this topic is given, e. g., by Brockwell and Davis (1991) and Shumway and Stoffer (2000). In the next Subsection we present some basic results of this area which will be of importance in the following.

To assess the performance of a control procedure it is necessary to know the relationship between the target process $\{Y_t\}$ and the observed process $\{X_t\}$. Such (change-point) models are considered in the second part of this Section.

2.1 ARMA Processes

A time series is defined to be a collection of observations x_t, $t \in T_0$ indexed according to the order they are obtained in time. T_0 is equal to the set of time points of the observations. The observations are considered to be realizations of a stochastic process $\{X_t : t \in T\}$ with $T_0 \subset T$. In the following it is mainly assumed that $T = \mathbb{Z}$. As usually we use the term time series for both the process and a particular realization. No notational distinction is made between the two concepts.

A very important class of time series analysis are stationary processes. A stochastic process $\{X_t\}$ is called to be (weakly) stationary if

$$E(X_t) = \mu \quad \forall t \in \mathbb{Z}, \qquad Cov(X_t, X_{t+h}) = \gamma_h \quad \forall t, t+h \in \mathbb{Z} .$$

For stationary processes the mean function is constant over time and the covariance function between two time points depends only on their time spacing. Important characteristics of a stationary process in the time domain are the mean μ, the autocovariance function γ_h, and the autocorrelation function $\rho_h = \gamma_h/\gamma_0$.

The most popular family of time-correlated models are autoregressive moving average processes (ARMA). A stochastic process $\{X_t\}$ is called an ARMA process of order (p,q) with mean μ if it is a solution of the stochastic difference equation

$$X_t = \mu + \sum_{i=1}^{p} \alpha_i (X_{t-i} - \mu) + \varepsilon_t + \sum_{j=1}^{q} \beta_j \varepsilon_{t-j} . \tag{1}$$

Here it is assumed that $\{\varepsilon_t\}$ is a white noise process, i.e. that

$$E(\varepsilon_t) = 0, \quad Var(\varepsilon_t) = \sigma^2, \quad Cov(\varepsilon_t, \varepsilon_s) = 0 \quad \forall t \neq s .$$

In the case $p = 0$ the process $\{X_t\}$ is said to be a moving average process (MA(q)). For $q = 0$ an autoregressive process is given (AR(p)).

Now, of course, not all solutions of the equation (1) are stationary. It is possible to derive conditions on the coefficients such that there exists a stationary solution of the stochastic difference equation. This holds if all roots of the polynomial $1 - \sum_{i=1}^{p} \alpha_i z^i$ lie outside the unit circle (Brockwell and Davis (1991, p. 85)), i.e.

$$1 - \sum_{i=1}^{p} \alpha_i z^i \neq 0 \quad \forall z \in \mathbb{C}, |z| \leq 1 . \tag{2}$$

For an AR(1) process this condition reduces to $|\alpha_1| < 1$. For an AR(2) process the condition (2) is fulfilled if the coefficients α_1 and α_2 lie within the triangle which is built by $\alpha_1 + \alpha_2 < 1$, $\alpha_2 - \alpha_1 < 1$, and $|\alpha_2| < 1$. In the following we always assume that (2) holds.

If the coefficients of the ARMA process satisfy the condition (2) then there is exactly one stationary solution of the stochastic difference equation and it is given by

$$X_t = \mu + \sum_{i=0}^{\infty} s_i \varepsilon_{t-i} . \tag{3}$$

The coefficients $\{s_i\}$ can be determined by expanding the polynomial fraction $\left(1 + \sum_{j=1}^{q} \beta_j z^j\right) / \left(1 - \sum_{i=1}^{p} \alpha_i z^i\right)$ in a power series about the point $z = 0$. Then s_i is equal to the ith coefficient of the power series. If additionally $\varepsilon_t \sim \mathcal{N}(0, \sigma^2)$ then $\{X_t\}$ is a Gaussian process. This means that the joint distribution function of every finite collection of the X_t's is a multivariate normal distribution.

Using the representation (3) of a stationary ARMA process its autocovariance function can be calculated (cf. Brockwell and Davis (1991, ch. 3)). For an AR(1) process it follows that

$$\gamma_0 = \frac{\sigma^2}{1-\alpha_1^2} \quad , \gamma_i = \alpha_1^{|i|}\,\gamma_0\,, \quad \rho_i = \alpha_1^{|i|}\,, \quad i \in \mathbb{Z}$$

and for an AR(2) process it holds that

$$\gamma_0 = \frac{\sigma^2\,(1-\alpha_2)}{(1+\alpha_2)\big((1-\alpha_2)^2-\alpha_1^2\big)}\,, \quad \gamma_1 = \frac{\sigma^2\,\alpha_1}{(1+\alpha_2)\big((1-\alpha_2)^2-\alpha_1^2\big)}\,,$$

$$\gamma_i = \alpha_1\,\gamma_{i-1} + \alpha_2\,\gamma_{i-2}\,, \quad i \geq 2\,.$$

Many papers about time series investigate the problem of forecasting future developments. It is desirable to have theoretically good predictors whose explicit calculation is simple. The theory of best linear predictors turns out to be simple, elegant and useful in practice (cf. Brockwell and Davis (1991, ch. 5)). If $\{X_t\}$ is a stationary process then the best (in the L^2 sense) linear predictor of X_t in terms of $X_1, \ldots, X_{t-1}$ is given by

$$\hat{X}_t = \mu + \sum_{i=1}^{t-1} \phi_{ti}\,(X_{t-i} - \mu)\,. \tag{4}$$

The coefficients $\phi_{t1}, \ldots, \phi_{tt-1}$ are real numbers which are obtained by solving the minimization problem

$$E\left(X_t - \mu - \sum_{i=1}^{t-1} b_{ti}\,(X_{t-i}-\mu)\right)^2 \quad \rightarrow \quad \min_{b_{t1},\ldots,b_{tt-1}}\,.$$

It immediately follows that these weights fulfill the linear system of equations

$$\Gamma_t\,\boldsymbol{\phi}_t = \boldsymbol{\gamma}_t$$

where $\Gamma_t = (\gamma_{|i-j|})_{i,j=1,\ldots,t-1}$, $\boldsymbol{\gamma}_t = (\gamma_1, \ldots, \gamma_{t-1})'$ and $\boldsymbol{\phi}_t = (\phi_{t1}, \ldots, \phi_{tt-1})'$. The prediction error is given by $v_t = E\big(X_t - \hat{X}_t\big)^2 = \gamma_0 - \boldsymbol{\gamma}_t'\,\Gamma_t^{-1}\,\boldsymbol{\gamma}_t$ (Brockwell and Davis (1991, p. 168)). Recursive algorithms for determining the coefficients $\phi_{t1}, \ldots, \phi_{tt-1}$ can be found in Brockwell and Davis (1991, ch. 5).

For instance, it holds for an AR(p) process that for $t \geq p+1$

$$\hat{X}_t = \mu + \sum_{i=1}^{p} \alpha_i\,(X_{t-i} - \mu)\,.$$

The calculation of the predictors for the first p values shows to be more complicate but it always holds that $\hat{X}_1 = \mu$.

2.2 Modeling the Observed Process

By analyzing datasets we are frequently faced with structural changes of the underlying process like mean shifts, scale changes, outliers, drifts etc. Up to now in published literature about statistical process control for dependent data mostly changes in the mean and the variance of a time series have been investigated (there are some exceptions in literature on signal detection, see, e. g., the survey of Basseville (1988)). It is interesting that other types of changes like, e. g., changes in the correlation coefficients have been treated within econometrics (e. g., Johnston and DiNardo (1997)). However, in this context it is usually demanded that the time point of a possible change is known.

In this paper we restrict ourselves to deviations from the mean. A very general relationship between the target process $\{Y_t\}$ and the observed process $\{X_t\}$ is described by the model

$$X_t = Y_t + \sqrt{\gamma_0}\, f_t(\boldsymbol{a}, m)\, I_{\{m,m+1,\dots\}}(t)\,, \tag{5}$$

where I_A denotes the indicator function of the set A and $\gamma_0 = Var(Y_t)$. Consequently it holds that $X_t = Y_t$ for $t < m$. The influence of the change is described by the deterministic function $f_t(\boldsymbol{a}, m)$ which depends on m, the time index t, and the parameter vector $\boldsymbol{a}$. In the case $f_t(\boldsymbol{a}, m) = 0$ for all t it is said that the process $\{X_t\}$ is in control. The observed process and the target process coincide. Especially, it holds that $E(X_t) = E(Y_t)$ for all t. Sometimes $\{E(Y_t)\}$ is called the target sequence. If $E(Y_t) = \mu$ then μ is the target value. However, if $f_m(\boldsymbol{a}, m) \neq 0$, then there is a change at position m. The process $\{X_t\}$ is said to be out of control for $t \geq m$. In the case $f_t(\boldsymbol{a}, m) = a$ a shift in the mean is observed at position m. This model is the most widely discussed one in literature. If $\{Y_t\}$ is stationary then the process changes to another level for the mean. Setting $f_t(\boldsymbol{a}, m) = a\,(t - m + 1)$ a linear drift is present, while for $f_t(\boldsymbol{a}, m) = a_1\,(t - m + 1) + a_2\,(t - m + 1)^2$ a quadratic drift can be observed. Furthermore the choice $f_t(\boldsymbol{a}, m) = a\, I_{\{m\}}(t)$ leads to an outlier of type I (cf. Fox (1972)).

3 Standard Control Charts

The foundations of SPC were laid by Shewhart in the 1920s and 1930s. He introduced different control charts for monitoring the mean and the variance (e. g., Shewhart (1926,1931)). The Shewhart control charts do not have a memory. More precisely, the decision whether a change happened or not, is based only on the recent observation (sample). This is the main difference between Shewhart charts and the schemes of Page (1954) and Roberts (1959). Here, the influence of past data is controlled by an additional smoothing parameter.

For all these approaches it was always assumed that the data are independent. Therefore, we call these schemes standard control charts. Because these charts are well-known we only want to sketch them. For more details we confer to Montgomery (1991). Again, we restrict ourselves to the case with one single observation at each time point. We consider the following simplified change point model

$$X_t = Y_t + \sqrt{\gamma_0}\, a\, I_{\{m,m+1,\ldots\}}(t)\,, \quad t \geq 1\,. \tag{6}$$

The target value is given as $\mu = E(Y_t)$. We assume that μ and $\gamma_0 = Var(Y_t)$ are known. Usually, they are estimated by using a prerun (for $\{Y_t\}$). The process $\{X_t\}$ is in control if $a = 0$. Then $E(X_t) = E(Y_t) = \mu$ is valid for all $t \geq 1$. Otherwise, the process is out of control (for $t \geq m$).

3.1 EWMA Charts

The EWMA chart was introduced by Roberts (1959). It is based on the sequence

$$Z_t = (1-\lambda)\, Z_{t-1} + \lambda\, X_t\,, \quad t \geq 1\,. \tag{7}$$

The starting value Z_0 is set equal to z_0. Conventionally, the target value (of the mean) is chosen, i.e. $z_0 = E(Y_t) = \mu$. The parameter λ is named "smoothing parameter". It controls the influence of past data. For $\lambda = 1$ the EWMA chart becomes the Shewhart chart.

Solving (7) leads to

$$Z_t = z_0\,(1-\lambda)^t + \lambda \sum_{i=1}^{t} (1-\lambda)^{t-i}\, X_i\,.$$

This presentation shows that the smaller the smoothing parameter λ, the smaller the weight of the current observation and thus the larger the weight of the past data. The influence of the previous observations decreases exponentially (geometrically). For that reason the scheme is denoted as EWMA - exponentially weighted moving average, or in the past GMA - geometric moving average.

With $\sigma^2 = Var(Y_t)$ we obtain (cf. (6))

$$E(Z_t) = \mu + (1-\lambda)^t\,(z_0 - \mu) + a\sigma\left(1 - (1-\lambda)^{t-m+1}\right) I_{\{m,m+1,\ldots\}}(t)\,,$$

$$Var(Z_t) = \frac{\lambda}{2-\lambda}\left(1 - (1-\lambda)^{2t}\right)\sigma^2\,.$$

Thus,

$$\lim_{t\to\infty} E(Z_t) \underset{m<\infty}{=} \mu + a\sigma \quad , \quad \lim_{t\to\infty} Var(Z_t) = \frac{\lambda}{2-\lambda}\,\sigma^2\,.$$

In applications control charts based on the asymptotic variance prevailed. Their advantage is that the control limits are constant over time. One concludes that the process is out of control at time t if

$$Z_t > UCL = \mu + c\sigma\sqrt{\lambda/(2-\lambda)} \quad \text{or} \quad Z_t < LCL = \mu - c\sigma\sqrt{\lambda/(2-\lambda)}\,.$$

UCL and LCL denotes the upper and lower control limit, respectively. The critical value $c > 0$ is the second design parameter of the EWMA scheme (besides λ). The EWMA run length is given by

$$N_e = \inf\left\{t \in \mathbb{N} : Z_t > UCL \lor Z_t < LCL\right\}.$$

The most popular performance measure for control charts is the average run length (ARL). It is assumed that the process is out of control already at the beginning. This means that $m = 1$. Now, given a smoothing constant λ the critical value c is determined in order to get a prespecified in-control ARL, say ξ. Thus, one concludes on average after ξ observations that the process is out of control, although the process is in control. c is obtained as a solution of that implicit equation. In industrial statistics $\xi = 500$ is a common choice (or $\xi = 370$ as a result of the popular Shewhart $\bar{X}$ chart with 3σ limits), while in finance $\xi = 60$ might be an appropriate choice.

The computation of the ARL (as a function of the parameters λ and c) is quite complicate. Contrary to the Shewhart chart an explicit formula does not exist. Therefore, one has to use approximation techniques. The Markov chain approximation due to Brook and Evans (1972, for EWMA cf. Lucas/Saccucci (1990)) and the numerical solution of an integral equation (Crowder (1986)) are very popular techniques. Waldmann (1986a) presented a very accurate, but only seldom applied approximation method for computing the distribution of the run length of the EWMA scheme.

There remains the question about the choice of the parameter λ. Lucas and Saccucci (1990) determined for known a (out-of-control shift) the optimal λ. They minimized the out-of-control ARL. In their paper a lot of tables for independent and normally distributed data are given.

3.2 CUSUM Charts

CUSUM control charts were introduced by Page (1954). They are connected to the sequential probability ratio test (SPRT). Here, we derive CUSUM directly from the related SPRT.

Let $Y_t \sim \mathcal{N}(\mu, \sigma^2)$ for $t \geq 1$. Because of (6) it follows for $m = 1$ that $X_t \sim \mathcal{N}(\mu + a\,\sigma, \sigma^2)$. We denote with f_a the density of X_t. First, we consider the simple testing problem $H_0 : a = 0$ against $H_1 : a = a_0$ with known $a_0 > 0$. The SPRT says that sampling is stopped at time t if

$$\sum_{i=1}^{t} T_i \notin [A, B]$$

$$\text{with} \quad T_i = a_0 \frac{X_i - (\mu + a_0\,\sigma/2)}{\sigma}.$$

If $\sum T_i > B$ then H_0 is rejected. Otherwise, if $\sum T_i < A$ then H_0 is accepted. The (random) sample number is given by

$$N = \inf\left\{ t \in \mathbb{N} : \sum_{i=1}^{t} T_i \notin [A, B] \right\}.$$

Based on the SPRT we get the standard CUSUM chart by setting $A = 0$, $h = B/a_0$, and $k = a_0/2$. Thus,

$$N = \inf\left\{ t \in \mathbb{N} : \sum_{i=1}^{t} (X_i - \mu - k\,\sigma) \notin [0, h\,\sigma] \right\}.$$

Now the CUSUM scheme can be obtained as follows. The SPRT is continued as long as H_0 is not rejected. The first rejection of the null hypothesis H_0 leads to a signal of the CUSUM chart. The first signal of the SPRT is given at

$$N_1 = \inf\left\{ t \in \mathbb{N} : \sum_{i=1}^{t} (X_t - \mu - k\,\sigma) \notin [0, h\,\sigma] \right\}.$$

If $\sum_{i=1}^{N_1} (X_i - \mu - k\,\sigma) > h\,\sigma$ then the SPRT rejects H_0. The CUSUM decision rule signals that the mean has changed from μ to $\mu + a\,\sigma$. Otherwise, if the first SPRT ends with $\sum_{i=1}^{N_1} (X_i - \mu - k\,\sigma) < 0$, then there is no reason to signal a parameter change and a new SPRT is started at time point $N_1 + 1$. Thereafter, the second stopping of the SPRT is at

$$N_2 = \inf\left\{ t > N_1 : \sum_{i=N_1+1}^{t} (X_i - \mu - k\,\sigma) \notin [0, h\,\sigma] \right\}.$$

The procedure is continued until the SPRT rejects H_0 for the first time. In other words, we add consecutively $X_i - \mu - k\,\sigma$. It must be taken into account that the part of the sum which leads to the stop below 0 has to be deleted. Hence, it is sufficient to consider

$$\begin{aligned} S_0^+ &= 0\,, \\ S_t^+ &= \max\left\{ 0, S_{t-1}^+ + (X_t - \mu - k\,\sigma) \right\} \quad \text{for } t \geq 1\,. \end{aligned}$$

The run length of the CUSUM chart is given by

$$N_c^+ = \inf\left\{ t \in \mathbb{N} : S_t^+ > h\,\sigma \right\}.$$

In the same way we can derive a decision rule for $a_0 < 0$. It is based on

$$\begin{aligned} S_0^- &= 0\,, \\ S_t^- &= \min\left\{ 0, S_{t-1}^- + (X_t - \mu + k\,\sigma) \right\} \quad \text{for } t \geq 1\,. \end{aligned}$$

Let N_c^- denote the corresponding stopping time.

If the sign of the change is unknown then both one-sided CUSUM charts are combined. The two-sided CUSUM chart gives a signal at time t if $S_t^+ > h\sigma$ or $S_t^- < -h\sigma$. We denote the run length of the two-sided scheme by $N_c = \min\{N_c^+, N_c^-\}$.

Again, the critical value h is chosen such that the in-control ARL is equal to a given quantity ξ. The parameter k is called reference value. Usually, a_0, i.e. k, is not known. If one is interested in detecting large changes then a_0 and k should be taken large. Otherwise, small shifts are detected faster if a_0 is small, too. Several simulations showed that for detecting a shift of size $a_0\,\sigma$ the best choice seems to be $k = |a_0|/2$. This is not surprising because of the optimality of the related SPRT for testing μ against $\mu + a_0\,\sigma$. This means that we have a hint how to determine the parameter k for the CUSUM chart. This is an advantage of the CUSUM chart because for the EWMA chart a similar formula is not known (there are some suitable approximations, see Srivastava and Wu (1997)).

As for the EWMA chart we can compute the ARL as a function of the chart parameters by using approximation techniques. We mention the Markov chain approach due to Brook and Evans (1972) which was originally developed for the CUSUM chart, the numerical solution of the ARL integral equation (e.g., Goel and Wu (1971), Vance (1986)), and the Waldmann approach (1986b). Starting with Roberts (1966) a lot of comparison studies were performed (cf. Lucas and Saccucci (1990), Srivastava (1994)). It turned out that for large changes, i.e. large a, the Shewhart charts have the best performance. For smaller shifts control charts with a memory like, e.g., the EWMA and the CUSUM chart lead to a smaller out-of-control ARL. There are only slight differences between them. Besides the ARL criterion some other measures for evaluating control charts (or change point detection schemes) are used. Considering a worst case and a min-max criterion Moustakides (1986) and Ritov (1990), respectively, showed that CUSUM is optimal. For more details on the CUSUM chart we confer to Hawkins and Olwell (1998).

3.3 Performance in the Presence of Autocorrelation

Frequently, SPC data are dependent. Therefore the question arises whether the standard control charts can be applied without any modification. In this Subsection we want to show the influence of correlation on the performance of the Shewhart, EWMA, and CUSUM chart.

We assume that the target process $\{Y_t\}$ is a stationary AR(1) process with mean 0. The white noise $\{\varepsilon_t\}$ is assumed to be normally distributed. Hence, the target value is equal to 0 and the variance of the target process is $\gamma_0 = \sigma^2/(1-\alpha_1^2)$.

In the following we assume that γ_0 is known. Consequently, the SPC user knows the variance of the target process. This assumption is not unusual. E.g., by using a (in-control) prerun it is possible to estimate the process variance. A natural estimator is the sample variance. This estimator is consistent and, unfortunately, biased. In Kramer and Schmid (2000) the influence of estimating unknown process parameters (see also Lu and Reynolds (1999)) on the ARL of Shewhart charts is investigated. They compare different variance estimators. Especially, they give recommendations for the sample size of the prerun. Here, we assume that the SPC user correctly determined the variance. That is, within the scope of our study we are only interested in the effects of neglecting correlation on the ARL.

Let ξ, i. e. the in-control ARL, be equal to 500. The smoothing parameter λ of the EWMA scheme and the reference value k of the CUSUM chart are determined such that a change of size $\sqrt{\gamma_0}$ (one unit measured in standard deviations, i. e. $a = 1$) is detected as fast as possible. This leads to $\lambda = 0.1$ (cf. Lucas and Saccucci (1990)) and $k = 0.5$ (see above). Now, suppose that the analyst falsely assumes that the data are independent. Then he will take the critical values from standard tables for independent normally distributed data. This procedure leads to the following stopping times N_s, N_e and N_c (Shewhart, EWMA, CUSUM)

$$N_s = \inf\left\{ t \in \mathbb{N} : |X_t| \geq 3.0902 \sqrt{\gamma_0} \right\},$$

$$N_e = \inf\left\{ t \in \mathbb{N} : |Z_t| \geq 2.8143 \sqrt{\lambda/(2-\lambda)} \sqrt{\gamma_0} \right\}$$
$$\text{with} \quad Z_0 = 0 \,,\ \lambda = 0.1 \,,$$

$$N_c = \inf\left\{ t \in \mathbb{N} : \max\{-S_t^-, S_t^+\} \geq 5.0707 \sqrt{\gamma_0} \right\}$$
$$\text{with} \quad S_0^- = S_0^+ = 0 \,,\ k = 0.5 \,.$$

Let $E_a(N)$ denote the ARL of the control chart with stopping time N if in (6) (with $\gamma_0 = \sigma^2/(1-\alpha_1^2)$) a shift of size $a\sqrt{\gamma_0}$ at position $m = 1$ is present. Here we want to focus on the cases $a = 0$ and $a = 1$. Thus, for independent variables the in-control ARL $E_0(N)$ is equal to $\xi = 500$. In the out-of-control case the ARL is given by

$$E_1(N_s) \;=\; 54.6 \,,\ E_1(N_e) \;=\; 10.3 \,,\ E_1(N_c) \;=\; 10.5 \,.$$

In Figure 1 we see the influence of the correlation on the ARL of the Shewhart chart. Remember, that $\alpha_1 = Corr(Y_t, Y_{t-1})$. The above graph shows the in-control ARL as a function of α_1. The in-control ARL is symmetric in α_1 around 0. Further, it is relatively robust for $|\alpha_1| \leq 0.5$. E. g., for $\alpha_1 = 0.4$ the relative deviation from the target ARL $\xi = 500$ is only 3.08 % and for $\alpha_1 = 0.5$ it is equal to 6.24 %. If the absolute value of α_1 is large then strong deviations from ξ can be noticed. A completely different behavior can be observed for the EWMA and the CUSUM chart (see Figure 2). Both schemes react very sensitively even to small correlations. For negative values of α_1 the in-control ARL is larger than assumed, for positive correlation the

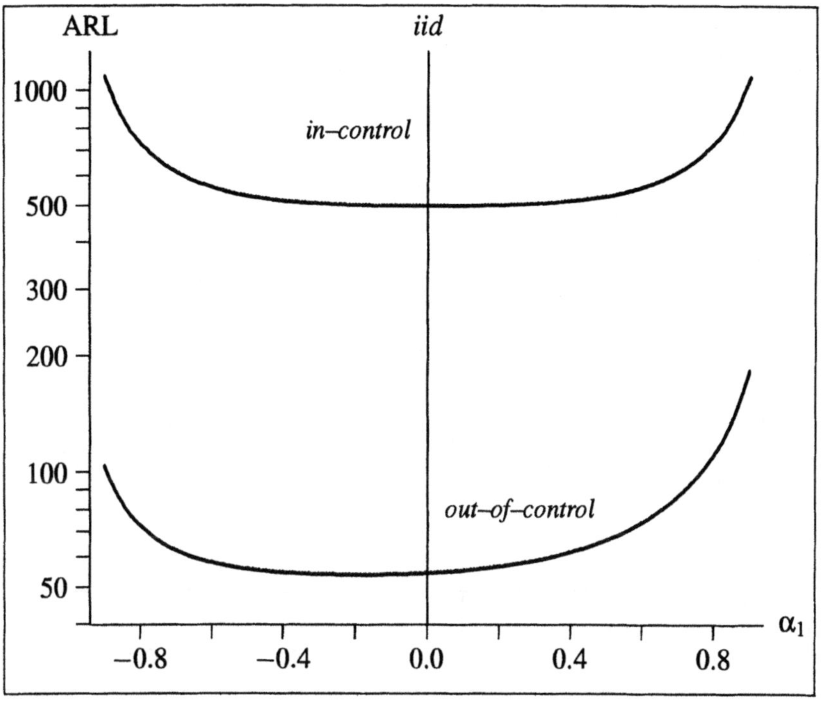

Fig. 1. ARL of the standard Shewhart chart in the presence of autocorrelation (model (6) with $m = 1$; $a = 0$: upper graph, $a = 1$: lower graph)

ARL is smaller. Already for $\alpha_1 = -0.1$ the relative deviation of $\xi = 500$ is equal to 100.6 % for the EWMA chart and even 123.9 % for the CUSUM chart. Although for positive correlations the deviations are not so large as for negative ones, nevertheless they can be dramatical (43.4 % and 47.6 %). Moreover, it can be stated that the CUSUM chart is more sensitive than the EWMA chart.

In the out-of-control situation the Shewhart chart is less robust with respect to correlation than in the in-control case. For $\alpha_1 = 0.5$ the deviation from the ARL of independent variables (i. e., $\alpha_1 = 0$) is equal to 21.9 %. The deviations are larger for positive than for negative autocorrelations. Again, the EWMA and the CUSUM chart behave different. It is interesting that they are more robust in the out-of-control state than in the in-control state. For $\alpha_1 = 0.5$ the deviation is merely 2.0 % (10.12 instead of 10.33) and 4.3 % (10.07 instead of 10.52), respectively. Note, that the graphs of the EWMA and the CUSUM schemes for $a = 1$ nearly coincide.

At first sight it is surprising that the Shewhart chart is more robust than the EWMA chart. In both cases the control statistic is compared with a target value. It is examined whether the deviation is larger than a multiple of the standard deviation of the statistic. Remember, that we assumed that

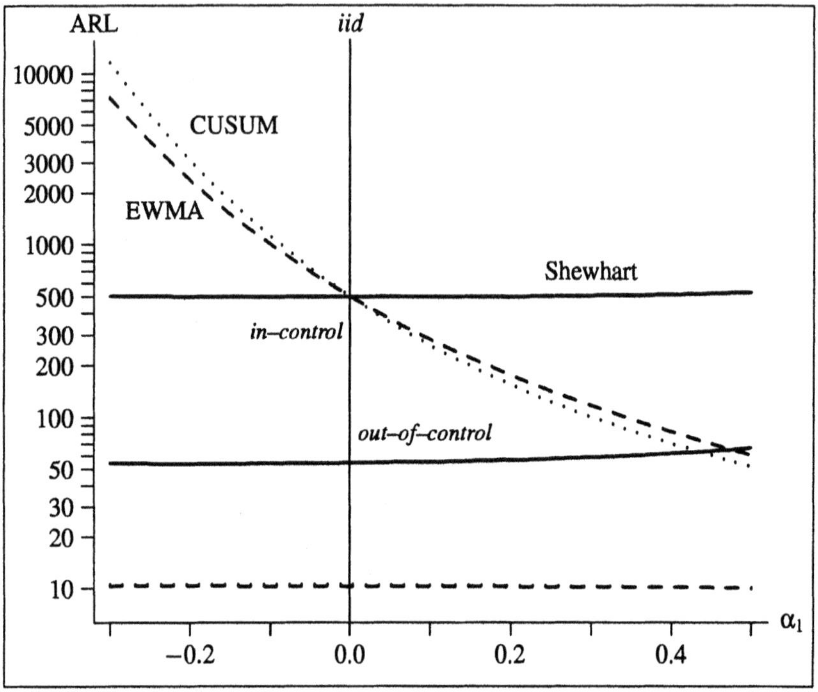

Fig. 2. ARL of the standard Shewhart (solid line), EWMA (dashed line) and CUSUM (dotted line) chart in the presence of autocorrelation (model (6) with $m = 1$; $a = 0$: upper graphs, $a = 1$: lower graphs)

the standard deviation of the target process is known. Consequently, we use the correct standard deviation for the Shewhart chart while the true variance of the EWMA statistic is not regarded (cf. Section 5.2). For that reason the EWMA chart may fail.

4 Residual Charts

Residual charts are a special type of control charts for time series. They have been intensively discussed in the last years (e. g., Alwan and Roberts (1988), Harris and Ross (1991), Montgomery and Mastrangelo (1991), Wardell et al. (1994a,b), Lu and Reynolds (1999)). Residual charts are based on the idea to filter the original data. The transformation is chosen such that the transformed quantities (residuals) are independent and identically distributed. Then all of the classical control schemes can be applied to the residuals. Suppose that (6) holds. The residuals $\hat{\varepsilon}_t = X_t - \hat{X}_t$, $t \geq 1$ are based on the comparison of the observation with a predictor $\hat{X}_t$ of X_t. Usually the best linear predictor of X_t (see Subsection 2.1) is taken. This quantity is calculated under the assumption that no shift in the mean is present ($a = 0$ in (6)). If

$\{X_t\}$ is stationary then $\hat{X}_t = \mu + \sum_{i=1}^{t-1} \phi_{ti}\,(X_{t-i} - \mu)$ (cf. (4)). Consequently it holds for $t \geq 1$ that

$$\hat{\varepsilon}_t = X_t - \mu - \sum_{i=1}^{t-1} \phi_{ti}\,(X_{t-i} - \mu)\,.$$

If $\{Y_t\}$ is a Gaussian process then the random variables $\{\hat{\varepsilon}_t\}$ are independent in the in-control state, but they are not identically distributed. For that reason the residuals are normalized by the standard deviation $\sqrt{v_t}$ (cf. Subsection 2.1). The resulting variables $\hat{\Delta}_t = \hat{\varepsilon}_t/\sqrt{v_t}$, $t \geq 1$ are independent and identically distributed. It has to be emphasized that in the out-of-control state at least the second property does not hold any longer.

For an AR(1) process with mean μ it follows that $\hat{X}_t = \mu + \alpha_1\,(X_{t-1} - \mu)$ for $t \geq 2$ and $\hat{X}_1 = \mu$. Consequently it holds that $\hat{\varepsilon}_t = X_t - \mu - \alpha_1\,(X_{t-1} - \mu)$ for $t \geq 2$ and $\hat{\varepsilon}_1 = X_1 - \mu$. Because $Var(\hat{\varepsilon}_1) = \sigma^2/(1 - \alpha_1^2) \geq \sigma^2 = Var(\hat{\varepsilon}_t)$ for $t \geq 2$ the residuals have to be normalized. We get

$$\hat{\Delta}_t = (X_t - \mu - \alpha_1\,(X_{t-1} - \mu))/\sigma \quad \text{for } t \geq 2 \text{ and}$$
$$\hat{\Delta}_1 = \sqrt{1 - \alpha_1^2}\,(X_1 - \mu)/\sigma\,.$$

Now in the in-control state the variables $\hat{\Delta}_t$, $t \geq 1$ are independent and identically distributed. All control charts of Section 3 can be directly applied to the normalized residuals $\{\hat{\Delta}_t\}$. If the process is out of control from the beginning then $E(X_t) = \mu + a\,\sqrt{\gamma_0}$ for $t \geq 1$ and

$$E(\hat{\Delta}_t) = a\sqrt{\frac{1-\alpha_1}{1+\alpha_1}} \quad \text{for} \quad t \geq 2, \qquad E(\hat{\Delta}_1) = a\,. \tag{8}$$

If $\alpha_1 \geq 0$ then $E(\hat{\Delta}_t) \leq a$ and for $\alpha_1 \leq 0$ we get $E(\hat{\Delta}_t) \geq a$. Consequently the effect of the transformation is that for positive coefficients the influence of the change $a\,\sqrt{\gamma_0}$ decreases while it increases for negative coefficients.

In many papers the starting problem is not treated. But this procedure is not satisfactory because then the knowledge of values X_t, $t \leq 0$ is tacitly demanded. Taking $\hat{\varepsilon}_1 = X_1 - \alpha_1 X_0$ it is assumed that X_0 is known. With respect to the ARL this means a conditional ARL is determined. A thorough discussion of the starting problem is given by Kramer and Schmid (1997).

As an extension of the residual charts the so-called ARMA charts (Jiang et al. (2000)) can be considered. Here the general intention is to transform the data in order to detect a change faster.

5 Modified Control Charts

Residual charts are based on a transformation of the observations. This is a completely different attempt than for modified schemes. The idea behind this type of charts is to extend the control design of the classical charts directly to the case of correlated data. Thus, this method seems to be more devoted to practitioners.

5.1 Modified Shewhart Charts

Let $\{Y_t\}$ be a stationary process with mean μ and autocovariance function $\{\gamma_h\}$. The relationship between the target process and the observed process is again described by model (6) with $m = 1$. Then it is concluded that the process is out of control at time t if

$$|X_t - \mu| > c\,\sqrt{\gamma_0}\,.$$

In the following it is assumed that both μ and γ_0 are known. The decision rule is based on the comparison of the observation at time t with the target value μ. If the distance between both quantities is larger than a certain multiple of the standard deviation then the process is declared to be out of control. This is the same idea as for the standard Shewhart charts. The only difference is the use of the variance of the time series (and a possibly different critical value c). The run length of this chart is denoted by

$$N_s = \inf\left\{t \in \mathbb{N} : |X_t - \mu| > c\sqrt{\gamma_0}\right\}.$$

This control chart was proposed by Vasilopoulos and Stamboulis (1978). A thorough discussion is given in Schmid (1995).

The control statistics of the modified schemes are not independent. Therefore the analytical analysis of these charts is more difficult than in the case of independent samples. In the next theorem the run length of the control chart for correlated data is compared with that for independent observations. We use the symbol P_{iid} to indicate that the process $\{Y_t\}$ is assumed to be independent and identically distributed.

Theorem 1: Let $\{Y_t\}$ be a Gaussian process with mean μ and variance $\gamma_0 > 0$. Let $k \in \mathbb{N}$. Then it holds in the in-control state that

$$P(N_s > k) \;\geq\; P_{iid}(N_s > k) \;=\; \left(2\,\Phi(c) - 1\right)^k. \tag{9}$$

Proof: see Schmid (1995).

In Theorem 1 the event is considered that an in-control process is assessed to be in control at least up to time k. On the right hand side of (9) this probability is determined for iid variables with mean μ and variance γ_0. The theorem says that for correlated data the probability of the event is greater or equal than for independent ones. Consequently, if a practitioner falsely assumes that the underlying data are independent and identically distributed, then he will determine c according to $c_{iid} = \Phi^{-1}\left(1 - 1/(2\xi)\right)$. Theorem 1 says that the true value of c is less or equal to c_{iid}.

Theorem 2: Let $\{Y_t\}$ be a stationary AR(1) process. Assume that the variables $\{\varepsilon_t\}$ are independent and normally distributed with mean 0 and variance σ^2. Then it holds that in the in-control state $P(N_s > k)$ is a nondecreasing function in $|\alpha_1|$.

Proof: see Schmid (1995).

A discussion of the behavior in the out-of-control state is given by Schmid (1995). It is proved for a large family of distribution functions that given the model (5) the out-of-control ARL is smaller or equal to the in-control ARL.

Now $E(X_t - \mu)/\sqrt{\gamma_0} = a$. For an AR(1) process it holds that for a positive autocorrelation coefficient $E(\hat{\Delta}_t) \leq a$ for $t \geq 1$ (cf. (8)). This shows that the shift has a greater effect on the (modified) Shewhart chart than on the residual chart. This leads to the conjecture that the modified Shewhart chart will provide better results than the classical Shewhart scheme applied to the residuals.

Explicit formulas of the ARL for correlated data are at the moment only available for exchangeable variables (Schmid (1995)). For AR(1) processes an approximation of the ARL can be determined by the Markov chain approach of Brook and Evans (1972). For more complex processes, however, estimators for the ARL can only be obtained by simulations.

5.2 Modified EWMA Charts

The modified EWMA chart was introduced by Schmid (1997a). It is based on the recursion

$$Z_t = (1-\lambda)\, Z_{t-1} + \lambda\, X_t$$

for $t \geq 1$. Let $Z_0 = z_0$ and $\lambda \in (0,1]$. The target process $\{Y_t\}$ is assumed to be a stationary process with mean μ and autocovariance function $\{\gamma_v\}$.

Lemma 1: Let $\{Y_t\}$ be a stationary process with mean μ and autocovariance function $\{\gamma_v\}$. Furthermore suppose that (5) holds.

a) If $f_t(a, m) = a\, g_t(m)$ then it holds for $t \geq m$ that

$$E(Z_t) = \mu + (1-\lambda)^t\,(z_0 - \mu) + \lambda \sum_{j=0}^{t-m} (1-\lambda)^j\, g_{t-j}(m)\,.$$

b)

$$\sigma^2_{e,t} := Var(Z_t) = \lambda^2 \sum_{|i| \leq t-1} \gamma_i \sum_{j=\max\{0,-i\}}^{\min\{t-1,t-1-i\}} (1-\lambda)^{2j+i}\,.$$

If $\{\gamma_v\}$ is absolutely summable then it holds that

$$\sigma^2_{e,t} \to \sigma^2_e := \frac{\lambda}{2-\lambda} \sum_{j=-\infty}^{\infty} \gamma_j\,(1-\lambda)^{|j|}$$

as $t \to \infty$.

c)

$$Cov(Z_t, Z_{t+h}) = \lambda^2 \sum_{i=-(t+h-1)}^{t-1} \gamma_i \sum_{\tau=\max\{0,-h-i\}}^{\min\{t-1-i,t-1\}} (1-\lambda)^{2\tau+i+h}$$

Proof: The results can be obtained by straightforward calculations.

For instance, it follows that for an AR(1) process

$$\sigma_e^2 = \sigma^2 \frac{\lambda}{2-\lambda} \frac{1+\alpha_1 (1-\lambda)}{1-\alpha_1 (1-\lambda)} .$$

Putting $\alpha_1 = 0$ the variance for independent variables is obtained. If $\{Y_t\}$ has a spectral density f then we get

$$\sigma_e^2 = \lambda^2 \int_{-\pi}^{\pi} \frac{f(x)}{1-2(1-\lambda)\cos x + (1-\lambda)^2} \, dx .$$

It is concluded that the process is out of control at time t if

$$|Z_t - \mu| > c\sigma_e .$$

This decision criterion is a direct extension of the design of the standard EWMA chart. To simplify the scheme the asymptotic variance is used again. If $\gamma_v = 0$ for $v \neq 0$ then the classical chart of Roberts is present. Choosing $\lambda = 1$ the chart reduces to the modified Shewhart chart.

A comparison of the run length of the EWMA chart for correlated and for independent variables is given by Schmid and Schöne (1997). They deal with a one-sided control chart whose run length is given by

$$N_e^u = \inf \left\{ t \in \mathbb{N} : Z_t - \mu > c\sqrt{Var(Z_t)} \right\} .$$

Note that here the variance at time t is used and not the asymptotic variance. The next theorem can be considered to be a generalization of Theorem 1, however, for the one-sided problem.

Theorem 3: Let $\{Y_t\}$ be a stationary Gaussian process with mean μ and autocovariance function $\{\gamma_v\}$ with $\gamma_v \geq 0$ for all v. Let $z_0 = \mu$. Then it holds for $k \in \mathbb{Z}$ that

$$P(N_e^u > k) \geq P_{iid}(N_e^u > k) .$$

Proof: see Schmid and Schöne (1997).

In Theorem 3 the index *iid* means that the target process $\{Y_t\}$ is assumed to be independent and identically distributed with mean μ and variance γ_0. Note that this does not imply that the EWMA statistics $\{Z_t\}$ are independent. Even for independent variables $\{Y_t\}$ they are correlated (cf. Lemma 1, c)). Furthermore, Theorem 3 is no longer valid if the autocovariances may be negative (see Schmid and Schöne (1997)).

In Theorem 3 the ARL of the EWMA chart in the case of correlated observations is compared with its ARL for independent variables. This result was generalized in Schöne et al. (1999). They proved that the ARL shows a monotonicity behavior in the autocovariances.

Let $\{Y_t\}$ be a Gaussian process with mean μ and autocovariance function $\gamma = \{\gamma_v\}$. In order to express the dependence upon the autocovariances we use the notation P_γ, E_γ and Var_γ.

Theorem 4: Let $\{Y_t\}$ be a stationary Gaussian process with mean μ and autocovariance functions $\{\gamma_v\}$ and $\{\delta_v\}$, respectively. Assume that

$$\gamma_0 > 0,\ \delta_0 > 0,\ \gamma_h \geq 0,\ \delta_h \geq 0 \quad \text{for} \quad 1 \leq h \leq k-1\,,$$

and

$$\gamma_i\,\delta_j \quad \geq \quad \gamma_j\,\delta_i \quad \text{for} \quad 0 \leq j < i \leq k-1\,.$$

Then it holds for all real numbers $c_1, \ldots, c_k$ that

$$P_\gamma\left(Z_t - E_\gamma(Z_t) \leq c_t\sqrt{Var_\gamma(Z_t)}\ ,\ t = 1, \ldots, k\right)$$
$$\geq P_\delta\left(Z_t - E_\delta(Z_t) \leq c_t\sqrt{Var_\delta(Z_t)}\ ,\ t = 1, \ldots, k\right)\,.$$

Proof: see Schöne et al. (1999).

The design of the modified control charts is simpler than that of the residual charts. The main problem consists in the determination of the critical values c, where c is taken such that the in-control ARL is equal to a pre-specified value. As already mentioned in Section 3.1 the calculation of the critical values turns out to be difficult even for independent variables. If the data show a correlation structure then no explicit formula for the ARL is available. Simulations are necessary. The critical values depend not only on the given value for the in-control ARL ξ, but on all parameters of the target process, too. For that reason tables have to be determined for each process.

What recommendations can be made about the choice of the smoothing parameter λ? As in the case of independent variables there is, of course, no λ which provides the smallest out-of-control ARL for all shifts. However, again the rule of thumb holds that λ should be taken small if a small shift is expected and a large value of λ is better for detecting larger shifts. If the magnitude of the expected shift is known then λ can be chosen such that the corresponding out-of-control ARL takes its minimum. Such tables can be found in Schmid (1997a). An impression about the relationship between the optimal value of λ and the correlation can be obtained by the results in Section 6, too.

A generalization of the modified EWMA chart to the multivariate case is given in Kramer and Schmid (1997).

5.3 Modified CUSUM Charts

In the literature different approaches were used to extend CUSUM charts to time series data. Nikiforov (1983) applied the SPRT to ARMA processes with

Table 1. Critical values of the EWMA chart for an AR(1) process ($\xi = 500$) *(see Kramer (1997))*

λ	α_1								
	-0.8	-0.6	-0.4	-0.2	0.0	0.2	0.4	0.6	0.8
1	2.965	3.055	3.081	3.088	3.090	3.088	3.081	3.055	2.965
0.8	3.004	3.074	3.088	3.090	3.089	3.082	3.064	3.024	2.917
0.6	3.036	3.085	3.090	3.087	3.081	3.064	3.035	2.979	2.859
0.4	3.057	3.084	3.082	3.072	3.054	3.024	2.979	2.910	2.779
0.2	3.048	3.049	3.028	2.997	2.962	2.915	2.857	2.777	2.638
0.1	2.978	2.942	2.899	2.856	2.814	2.764	2.705	2.625	2.487
0.05	2.820	2.753	2.701	2.657	2.615	2.570	2.517	2.445	2.316
0.025	2.574	2.493	2.444	2.405	2.368	2.332	2.288	2.227	2.116
0.01	2.142	2.069	2.030	2.001	1.973	1.949	1.917	1.872	1.788

normal white noise. Yashchin (1993) compared CUSUM schemes based on the original time series data with suitable modified control limits and CUSUM schemes applied to transformed data (similar to the residual chart approach). For the latter transformation Yashchin exploited the same likelihood ratio as later Schmid (1997b). For computational reasons he simplified that ratio. In order to compute ARLs for the first schemes he used an approximation technique, where the original data were replaced by iid data with similar properties. Schmid (1997b) made use of Siegmund's (1985) way of introducing CUSUM. Now, the target process is assumed to be an arbitrary Gaussian process. For independent data all approaches lead to the same CUSUM chart. However, in case of autocorrelated data slightly different schemes are obtained. Note, that Sparks (2000) considered further CUSUM adaptions for AR(1) data, i. e. adjusted reference value k and a modified restarting mechanism.

In the following the approach of Schmid (1997b) is described. For all $t \in \mathbb{N}$ let $\boldsymbol{Y}_t = (Y_1, \ldots, Y_t)'$ be normally distributed with expectation $\boldsymbol{\mu}_t = \mu(1, \ldots, 1)'$ and (positive definite) correlation matrix C_t. Furthermore, we assume that $\gamma_0 = Var(Y_t)$ for all t (homoscedasticity). γ_0, $\boldsymbol{\mu}_t$ and C_t are assumed to be known. With f_0 and $f_{a,m}$ we denote the density of the observed vector $\boldsymbol{X}_t$ in case of no change point at all and a change point at position m, respectively. Then it follows that

$$-2\log\frac{f_0(\boldsymbol{X}_t)}{\max\limits_{1\le m\le t+1} f_{a,m}(\boldsymbol{X}_t)} = 2\max\left\{0, \max_{1\le m\le t} a\,\frac{T_{mt} - a\sqrt{\gamma_0}\,K_{mt}/2}{\sqrt{\gamma_0}}\right\}$$

with $T_{mt} = e'_{mt}\, C_t^{-1}\,(\boldsymbol{X}_t - \boldsymbol{\mu}_t)$ and $K_{mt} = e'_{mt}\, C_t^{-1}\, e_{mt}$. e_{mt} denotes the vector with zeros in the first $m-1$ rows and ones in the remaining $t-m+1$ rows.

The process stops at time t if

$$\max_{1\le m\le t}\left\{T_{mt}-k\sqrt{\gamma_0}\,K_{mt},-(T_{mt}+k\sqrt{\gamma_0}\,K_{mt})\right\}>h\sqrt{\gamma_0}\,.$$

As in the iid case the reference value k is taken equal to $|a|/2$.

The practical computation of this stopping rule is time consuming. However, it is possible to compute the statistic recursively (cf. Schmid (1997b)). E. g., for an AR(1) process with expectation μ we obtain the following run length

$$N_c=\inf\left\{t\in\mathbb{N}: S_t^+>h\sqrt{\gamma_0}\quad\text{or}\quad S_t^-<-h\sqrt{\gamma_0}\right\} \tag{10}$$

with

$$\begin{aligned}
S_t^+ &= \max\left\{0,\varepsilon_t-k\sqrt{\gamma_0},S_{t-1}^+ +(1-\alpha_1)\left(\varepsilon_t-(1-\alpha_1)\,k\sqrt{\gamma_0}\right)\right\}\ ,\ t\ge 2\,,\\
S_1^+ &= \max\left\{0,(1-\alpha_1^2)\,(X_1-\mu-k\sqrt{\gamma_0})\right\},
\end{aligned}$$

$$\begin{aligned}
S_t^- &= \min\left\{0,\varepsilon_t+k\sqrt{\gamma_0},S_{t-1}^- +(1-\alpha_1)\left(\varepsilon_t+(1-\alpha_1)\,k\sqrt{\gamma_0}\right)\right\}\ ,\ t\ge 2\,,\\
S_1^- &= \min\left\{0,(1-\alpha_1^2)\,(X_1-\mu+k\sqrt{\gamma_0})\right\}\,.
\end{aligned}$$

The SPRT approach (cf. Nikiforov (1975 and later)) is based on comparing $f_0(\boldsymbol{X}_t)$ with $f_{a,1}(\boldsymbol{X}_t)$. This leads to

$$T_t\;=\;\log\frac{f_{a,1}(\mathbf{X}_t)}{f_0(\mathbf{X}_t)}\;=\;a\,\frac{T_{1t}-a\sqrt{\gamma_0}\,K_{1t}/2}{\sqrt{\gamma_0}}\,.$$

The related CUSUM chart gives a signal at time t if

$$\max_{1\le m\le t}\left\{T_{1m}-k\sqrt{\gamma_0}\,K_{1m},-(T_{1m}+k\sqrt{\gamma_0}\,K_{1m})\right\}>h\sqrt{\gamma_0}\,. \tag{11}$$

For AR(1) data similar recursions as in (10) are available. The main difference is the vanishing of the middle term $\varepsilon_t\mp k\sqrt{\gamma_0}$ in the recursion rules for S_t^+ and S_t^-, respectively.

The reference value k is selected as in the iid case. For both CUSUM charts (10) and (11) the design parameter h can be determined by the Markov chain approximation (cf. again Brook and Evans (1972)) or for complex processes by simulations. Note that in the next Section (comparison study) the Markov chain approach is used.

Several tables of critical values for the modified EWMA chart and the modified CUSUM chart (10) due to Schmid can be found in Kramer (1997). These values were computed by Monte Carlo simulations for an in-control ARL of $\xi=500$. For a different approach of computing the ARL of CUSUM schemes in case of autoregressive processes we refer to Rowlands (1976). Further, we refer to a recent paper of Lu and Reynolds (2001), who considered modified and residual CUSUM and EWMA schemes for an ARMA(1,1) by computation of the steady-state ARLs.

6 Comparison Study

Comparisons of the behavior of control charts for time series were given by several authors (e. g., Kramer (1997), Schmid (1997b), Kramer and Schmid (1998), Lu and Reynolds (2001)). In this paper we want to compare the control charts described in Section 4 and Section 5. As a measure for the performance the ARL is used which is determined by using Monte Carlo simulations and Markov chain approximations, respectively.

The target process $\{Y_t\}$ is given by an autoregressive process of order 1. Let

$$Y_t = \alpha_1 Y_{t-1} + \varepsilon_t$$

with $\varepsilon_t \sim \mathcal{N}(0,1)$ and $Cov(\varepsilon_s, \varepsilon_t) = 0$ for $s \neq t$. Thus, the target value is equal to 0. The relation between the target process and the observed process is given by the change point model (6) with $m = 1$

$$X_t = Y_t + a\sqrt{\gamma_0}$$

where $a \in \{0, 0.5, 1.0, \ldots, 3.0\}$ and $Var(X_t) = \gamma_0 = 1/(1-\alpha_1^2)$. The following five control charts

(a) modified EWMA chart (cf. 5.2), abbr. EWMAmod,
(b) residual EWMA chart (cf. 4 and 3.1), abbr. EWMAres,
(c) modified CUSUM chart due to Schmid (cf. 5.3), abbr. CUSUMmod,
(d) residual CUSUM chart (cf. 4 and 3.2), abbr. CUSUMres,
(e) CUSUM chart with modified control limits, abbr. CUSUMcla

are adjusted in order to achieve an in-control ARL of $\xi = 500$. The remaining chart parameters (λ for EWMA, k for CUSUM) are taken such that the out-of-control ARL for a given shift a is minimal. Here for EWMAmod λ takes values within $\{0.01, 0.025, 0.05, 0.1, 0.2, 0.4, 0.6, 0.8, 1\}$ (cf. Table 1 or Kramer (1997)) and for EWMAres additionally the values $\{0.3, 0.5, 0.7, 0.9\}$ are permitted. The reference value k for the CUSUM charts is chosen from $\{0, 0.1, 0.2, \ldots, 2.0\}$ for CUSUMmod and CUSUMres, and more roughly from $\{0, 0.25, 0.5, \ldots, 2.0\}$ for CUSUMcla (cf. Kramer (1997)). The ARLs are computed by Monte Carlo simulations for the modified charts EWMAmod and CUSUMclass (Kramer (1997); 100,000 repetitions) and by the Markov chain approach (Brook and Evans (1972) or Lucas and Saccucci (1990)) with matrix dimension 300 for the other charts. Then, the critical values are obtained by iteration algorithms like the regula falsi. After that, the critical values were validated by a Monte Carlo simulation with 1,000,000 repetitions. The ARLs in the out-of-control case for EWMAmod and CUSUMmod are computed by Monte Carlo simulations with 10,000,000 repetitions while for the remaining charts the Markov chain approach was used again.

Now, for all constellations of a and of the control chart parameters λ and k, respectively, the out-of-control ARLs were computed and compared.

Table 2. Critical values of the (standard) CUSUM chart with modified control limits (CUSUMcla) for AR(1) data ($\xi = 500$) *(from Kramer (1997))*

k	α_1								
	-0.8	-0.6	-0.4	-0.2	0.0	0.2	0.4	0.6	0.8
0	32.896	38.994	39.096	35.831	30.481	24.122	17.246	10.494	4.576
0.25	3.846	6.497	8.078	8.737	8.575	7.878	6.588	4.814	2.671
0.5	1.909	3.459	4.489	5.014	5.066	4.826	4.204	3.223	1.904
0.75	1.119	2.258	3.017	3.437	3.538	3.455	3.078	2.414	1.430
1.0	0.854	1.520	2.205	2.565	2.664	2.657	2.405	1.896	1.075
1.25	0.604	1.226	1.638	1.995	2.105	2.130	1.941	1.519	0.808
1.5	0.356	0.972	1.342	1.581	1.707	1.742	1.594	1.231	0.637
1.75	0.111	0.723	1.083	1.293	1.389	1.421	1.315	1.016	0.539
2.0	.[1]	0.473	0.832	1.031	1.109	1.200	1.099	0.859	0.474

[1]: It was not possible to find a value h for $k = 2$ and $\xi = 500$. Therefore, as an approximation $h = 10^{-7}$ was used which led to an in-control ARL of 1120.

The minimal out-of-control ARL for each scheme is given in Table 3. For each possible shift a the corresponding parameter value is shown, too. For a fixed shift the ARL of the fastest chart is bold typed. Furthermore, for autocorrelations $\alpha_1 \in \{-0.6, 0, 0.6\}$ the minimal out-of-control ARLs were plotted in Figure 3.

At first sight, for positive α_1 the modified charts (EWMAmod, CUSUMcla) dominate while for negative α_1 the residual charts provide smaller out-of-control ARLs. With some exceptions ($a = 0.5$ and $\alpha_1 = -0.8$, $a = 1$ and $\alpha_1 = -0.6$, $a = 1$ and $\alpha_1 = -0.4$) we observe the well-known fact that EWMA charts detect small changes somewhat faster than CUSUM charts (remember that we measure by the ARL). The CUSUM charts are faster for shifts a larger than 1 (exceptions: $a = 3$ and $\alpha_1 \in \{-0.4, -0.2, 0.8\}$, $a = 2.5$ and $\alpha_1 = -0.4$).

If we consider the optimal design parameters then it can be observed for the residual charts (EWMAres, CUSUMres) that the values are similar to the optimal design parameters of the standard charts for a shift of size $(1 - \alpha_1)\, a$. For CUSUMres the deviations are smaller than for EWMAres. The chart CUSUMmod which is related to a residual chart shows the most convenient behavior since the optimal k does not depend on α_1. Exceptions can only be observed for strong correlations ($|\alpha_1| = 0.8$). The optimal design parameters of the modified charts behave similarly than those of the residual charts, but are more stable with respect to α_1.

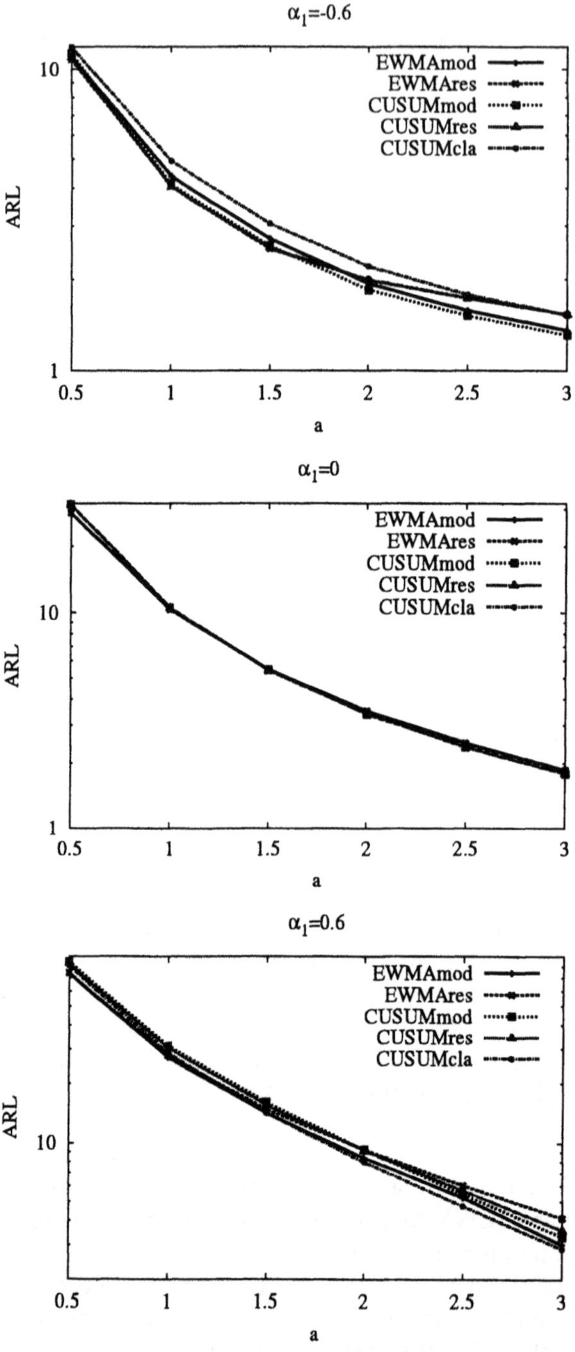

Fig. 3. ARLs in the out-of-control case of the 5 control charts under consideration for AR(1) data; lag-1-autocorrelation α_1 is equal to -0.6, 0, and 0.6.

Table 3. Average Run Length (ARL) in the out-of-control case of the 5 control charts under consideration for AR(1) data (model (6), $\xi = 500$, in parentheses: optimal values for λ and k, respectively)

α_1	chart	shift a 0.5	1.0	1.5	2.0	2.5	3.0
-0.8	EWMAmod	(.1)7.137	(.2)3.193	(.2)2.073	(.4)1.630	(.4)1.408	(.4)1.231
	EWMAres	(.2)6.166	(.6)**2.644**	(.8)2.004	(1)1.864	(1)1.722	(1)1.536
	CUSUMmod	(.3)6.381	(.5)2.681	(1.8)**1.737**	(1.8)**1.486**	(1.8)**1.295**	(1.8)**1.150**
	CUSUMres	(.4)**6.154**	(.8)2.668	(1.1)2.015	(1.8)1.864	(1.8)1.722	(1.8)1.536
	CUSUMcla	(.25)9.382	(.25)3.923	(.25)2.665	(.25)1.960	(.5)1.686	(.75)1.486
-0.6	EWMAmod	(.1)11.102	(.2)4.389	(.4)2.721	(.4)1.938	(.4)1.583	(.4,.6)1.363
	EWMAres	(.1)**10.823**	(.4)4.058	(.6)**2.514**	(.8)1.988	(.9)1.741	(1)1.537
	CUSUMmod	(.3)11.364	(.5)4.128	(.7)2.573	(2.4)**1.851**	(2.4)**1.521**	(2.4)**1.314**
	CUSUMres	(.4)10.964	(.8)**4.053**	(1.1)2.540	(1.4)1.997	(1.8)1.740	(2.4)1.537
	CUSUMcla	(.25)11.962	(.25)4.930	(.5)3.051	(.5)2.206	(.5)1.787	(.75)1.531
-0.4	EWMAmod	(.1)15.947	(.2)5.854	(.4)3.456	(.4)2.339	(.4)**1.824**	(.6)**1.482**
	EWMAres	(.1)**15.738**	(.2)5.716	(.4)3.255	(.6)2.319	(.8)1.863	(.9)1.567
	CUSUMmod	(.3)16.900	(.5)5.817	(.7)3.313	(1.2)2.353	(1.3)1.861	(2.4)1.500
	CUSUMres	(.4)16.571	(.7)**5.706**	(1)**3.251**	(1.2)**2.318**	(1.6)1.855	(1.9)1.564
	CUSUMcla	(.25)16.898	(.5)6.142	(.5)3.538	(.75)2.476	(.75)1.918	(1)1.581
-0.2	EWMAmod	(.05)21.735	(.2)7.782	(.2)4.361	(.4)2.846	(.6)2.120	(.6)1.647
	EWMAres	(.05)**21.711**	(.2)**7.675**	(.3)4.207	(.5)2.820	(.6)2.109	(.8)1.669
	CUSUMmod	(.3)23.410	(.5)7.878	(.7)4.246	(1)2.818	(1.4)2.080	(1.5)**1.645**
	CUSUMres	(.3)22.945	(.6)7.786	(.9)**4.179**	(1.1)**2.770**	(1.4)**2.065**	(1.7)1.646
	CUSUMcla	(.25)23.160	(.5)7.944	(.75)4.314	(.75)2.865	(1)2.105	(1.25)1.664
0.0	EWMAmod	(.05)**28.766**	(.1)**10.333**	(.2)5.501	(.4)3.522	(.5)2.497	(.7)1.865
	EWMAres	(.05)**28.766**	(.1)**10.333**	(.2)5.501	(.4)3.522	(.5)2.497	(.7)1.865
	CUSUMmod	(.3)31.480	(.5)10.519	(.8)**5.458**	(1)**3.413**	(1.3)**2.390**	(1.5)**1.792**
	CUSUMres	(.3)31.480	(.5)10.519	(.8)**5.458**	(1)**3.413**	(1.3)**2.390**	(1.5)**1.792**
	CUSUMcla	(.3)31.480	(.5)10.519	(.8)**5.458**	(1)**3.413**	(1.3)**2.390**	(1.5)**1.792**
0.2	EWMAmod	(.025)38.539	(.1)**13.590**	(.2)7.136	(.4)4.493	(.4)3.073	(.8)2.166
	EWMAres	(.025)**38.520**	(.1)13.677	(.2)7.244	(.3)4.537	(.4)3.104	(.6)2.203
	CUSUMmod	(.3)42.393	(.5)14.353	(.7)7.314	(1)4.409	(1.2)2.925	(1.5)2.068
	CUSUMres	(.2)41.878	(.4)14.325	(.6)7.283	(.8)4.394	(1.1)2.903	(1.3)2.038
	CUSUMcla	(.25)41.473	(.5)14.067	(.75)**7.119**	(1)**4.293**	(1.5)**2.829**	(1.75)**1.988**
0.4	EWMAmod	(.025)**51.486**	(.1)**18.815**	(.2)9.787	(.2)5.977	(.4)3.868	(.8)2.552
	EWMAres	(.025)51.564	(.05)19.123	(.1)10.057	(.2)6.145	(.3)4.116	(.5)2.822
	CUSUMmod	(.3)58.273	(.5)20.300	(.7)10.292	(1)6.040	(1.2)3.807	(1.5)2.516
	CUSUMres	(.2)58.824	(.3)20.120	(.5)10.226	(.6)6.022	(.8)3.819	(1.1)2.503
	CUSUMcla	(.25)56.499	(.5)19.422	(.75)**9.691**	(1.25)**5.642**	(1.5)**3.537**	(2)**2.301**
0.6	EWMAmod	(.025)**74.066**	(.05)**27.286**	(.1)**14.185**	(.2)8.409	(.4)5.238	(1)3.007
	EWMAres	(.01)74.371	(.05)27.800	(.1)14.826	(.1)9.272	(.2)6.057	(.3)4.108
	CUSUMmod	(.2)84.833	(.5)31.238	(.7)16.064	(1)9.242	(1.3)5.468	(1.5)3.303
	CUSUMres	(.1)83.026	(.2)30.347	(.3)15.561	(.4)9.169	(.6)5.708	(.8)3.562
	CUSUMcla	(.25)80.998	(.5)28.636	(.75)14.225	(1.25)**7.985**	(1.75)**4.737**	(2)**2.860**
0.8	EWMAmod	(.01)**119.677**	(.025)**47.398**	(.05)24.982	(.2)14.322	(1)7.494	(1)**3.860**
	EWMAres	(.01)120.784	(.025)48.895	(.05)26.789	(.1)17.027	(.1)11.278	(.2)7.897
	CUSUMmod	(.2)144.146	(.4)60.336	(.7)32.732	(1.3)16.893	(1.4)8.908	(1.5)4.844
	CUSUMres	(.1)149.781	(.1)55.091	(.2)30.236	(.2)17.734	(.3)11.289	(.4)7.189
	CUSUMcla	(.25)133.329	(.5)50.135	(1)**24.878**	(1.5)**13.501**	(2)**7.299**	(2)4.147

7 Summary

The analysis of many industrial datasets has shown that autocorrelation is a widespread pattern of data (Alwan (1989)). If the analyst is unaware of the dependence in the data and he applies standard control charts then more false signals are obtained as supposed and very long delays in detecting the change are frequently observed, respectively. Therefore the classical control charts cannot be applied directly. It is necessary to use the time series structure for constructing new control schemes. Furthermore it has turned out that the detection of structural changes is of great importance in other fields like environmental sciences, finance, medicine, etc. (e. g., Högel (2000), Severin and Schmid (1999)). The extension of the classical charts to time series provides completely new fields of applications and makes these methods more attractive.

In the meantime several control schemes for time series are available. It can be distinguished between two attempts, residual charts and modified control charts. For each type control procedures with memory have been introduced (EWMA, CUSUM). As for independent variables no chart turns out to be overall the best. For positive autocorrelation we recommend (by using the ARL as performance measure) the use of the modified EWMA chart while for negative autocorrelation the classical EWMA chart applied to the residuals or the classical CUSUM chart applied to the residuals must be favored.

References

1. L. C. Alwan (1989) Time Series Modeling for Statistical Process Control, PhD Thesis, Graduate School of Business, Chicago.
2. L. C. Alwan & H. V. Roberts (1988) Time-series modeling for statistical process control, J. Bus. Econ. Stat. **6**, 87-95.
3. M. Bagshaw & R. A. Johnson (1975) The effect of serial correlation on the performance of CUSUM tests II, Technometrics **17**, 73-80.
4. M. Bagshaw & R. A. Johnson (1977) Sequential procedures for detecting parameter changes in a time-series model, J. Am. Stat. Assoc. **72**, 593-597.
5. M. Basseville (1988) Detecting changes in signals and systems – a survey, Automatica, **24**, No. 3, 309-326.
6. P. J. Brockwell & R. A. Davis (1991) Time Series: Theory and Methods, 2nd ed. Springer, Berlin.
7. D. Brook & D. A. Evans (1972) An approach to the probability distribution of CUSUM run length, Biometrika **59**, 539-549.
8. S. V. Crowder(1986) A simple method for studying run-length distributions of exponentially weighted moving average charts, Technometrics **29**, 401-407.
9. A. J. Fox (1972) Outliers in time series, J. Roy. Statist. Soc. B **34**, 350-363.
10. A. L. Goel & S. M. Wu (1971) Determination of A.R.L. and a contour nomogram for CUSUM charts to control normal mean, Technometrics **13**, No. 2, 221-230.
11. P. L. Goldsmith & H. Whitfield (1961) Average run lengths in cumulative chart quality control schemes, Technometrics **3**, 11-20.
12. T. J. Harris & W. H. Ross (1991) Statistical process control procedures for correlated observations, Can. J. Chem. Eng. **69**, 48-57.
13. P. J. Harrison & O. L. Davies (1964) The use of cumulative sum (CUSUM) techniques for the control of routine forecasts of product demand, Oper. Res. **12**, 325-333.
14. D. M. Hawkins & D. H. Olwell (1998) Cumulative Sum Charts and Charting for Quality Improvement, Springer, New York.

15. J. Högel (2000) Applications of statistical process control techniques in medical fields, Allgemeines Statistisches Archiv **84**, 337-359.
16. W. Jiang, K.-L. Tsui & W. H. Woodall (2000) A new SPC monitoring method: the ARMA chart, Technometrics **42**, 399-410.
17. R. A. Johnson & M. Bagshaw (1974) The effect of serial correlation on the performance of CUSUM tests, Technometrics **16**, 103-112.
18. J. Johnston, J. & E. DiNardo (1997) Econometric Methods, MacGraw-Hill, Singapore.
19. H. G. Kramer (1997) On Control Charts for Time Series, Dissertation, Universität Ulm.
20. H. G. Kramer & W. Schmid (1997) EWMA charts for multivariate time series, Sequential Analysis **16**, 131-154.
21. H. G. Kramer & W. Schmid (1998) On the average delay of control schemes, In: E. von Collani, J. Franz, U. Jensen & W. Kahle (Eds.), Advances in Stochastic Models for Reliability, Quality, and Safety, Birkhäuser, Boston, 341-360.
22. H. G. Kramer & W. Schmid (2000) The influence of parameter estimation on the ARL of Shewhart-type charts for time series, Statistical Papers **41**, 173-196.
23. C.-W. Lu & M. R. Reynolds, jr. (1999) EWMA control charts for monitoring the mean of autocorrelated processes, J. Qual. Tech. **31**, 166-188.
24. C.-W. Lu & M. R. Reynolds, jr. (2001) CUSUM charts for monitoring an autocorrelated process, J. Qual. Tech. **33**.
25. J. M. Lucas & M. S. Saccucci (1990) Exponentially weighted moving average control schemes: properties and enhancements, Technometrics **32**, 1-29.
26. D. C. Montgomery (1991) Introduction to Statistical Quality Control, 2nd edition, John Wiley & Sons, New York.
27. D. C. Montgomery & C. M. Mastrangelo (1991) Some statistical process control methods for autocorrelated data, J. Qual. Tech. **23**, 179-204.
28. G. V. Moustakides (1986) Optimal stopping times for detecting changes in distributions. Ann. Stat. **14**, 1379-1387.
29. I. V. Nikiforov (1975) Sequential analysis applied to autoregressive processes, Autom. Remote Control **36**, 1365-1368.
30. I. V. Nikiforov (1979) Cumulative sums for detection of changes in random process characteristics, Autom. Remote Control **40**, 192-200.
31. I. V. Nikiforov (1983) Posledovatel'noe obnaruzhenie izmeneniya svojstv vremennykh ryadov, Nauka, Moskva (in Russian).
32. I. V. Nikiforov (1984) Sequential detection of change of properties on the basis of a modified algorithm of cumulative sums, Stat. Prob. Upr. **65**, 146-154 (in Russian).
33. E. S. Page (1954) Continuous inspection schemes, Biometrika **41**, 100-115.
34. E. S. Page (1961) Cumulative sum charts, Technometrics **3**, No. 1, 1-9.
35. Y. Ritov (1990) Decision theoretic optimality of the CUSUM procedure, Ann. Stat. **18**, 1464-1469.
36. S. W. Roberts (1959) Control chart tests based on geometric moving averages, Technometrics **1**, 239-250.
37. S. W. Roberts (1966) A comparison of some control chart procedures, Technometrics **8**, 411-430.
38. R. J. Rowlands & G. B. Wetherill (1991) Quality Control, In: B. K. Ghosh & P. K. Sen (Eds.) Handbook of Sequential Analysis, New York etc.: Marcel Dekker.

39. R. J. Rowlands (1976) Formulae for performance characteristics of CuSum schemes when the observations are autocorrelated, Ph.D. thesis, University of Wales, Cardiff.
40. W. Schmid & A. Schöne (1997) Some properties of the EWMA control chart in the presence of autocorrelation, Ann. Stat. **25**, 1277-1283.
41. W. Schmid (1995) On the run length of a Shewhart chart for correlated data, Statistical Papers **36**, 111-130.
42. W. Schmid (1997a) On EWMA charts for time series, In: H.-J. Lenz & P.-Th. Wilrich (Eds.), Frontiers in Statistical Quality Control, Physica, Heidelberg, Germany, 115-137.
43. W. Schmid (1997b) CUSUM control schemes for Gaussian processes, Statistical Papers **38**, 191-217.
44. A. Schöne, W. Schmid & S. Knoth (1999) On the run length of the EWMA scheme: A monotonicity result for normal variables, J. Stat. Plann. Inf. **79**, 289-297.
45. T. Severin & W. Schmid (1999) Monitoring changes in GARCH processes, Allgemeines Statistisches Archiv **83**, 281-307.
46. D. Siegmund (1985) Sequential Analysis, Springer, New York.
47. W. A. Shewhart (1926) Quality control charts, Bell System Technical Journal, 593-603.
48. W. A. Shewhart (1931) Economic Control of Quality of Manufactured Product, D. van Nostrand Company, Inc., Toronto, Canada.
49. R. H. Shumway & D. S. Stoffer (2000) Time Series Analysis and its Applications, Springer, New York.
50. R. S. Sparks (2000) CUSUM charts for AR1 data: Are they worth the effort? Austral. & New Zealand J. Statist. **42**, 25-42.
51. M. S. Srivastava & Y. Wu (1997) Evaluation of optimum weights and average run lengths in EWMA control schemes, Commun. Stat., Theory Methods **26**, No. 5, 1253-1267.
52. M. S. Srivastava (1994) Comparison of CUSUM and EWMA procedures for detecting a shift in the mean or an increase in the variance, J. Appl. Stat. Sci. **1**, No. 4, 445-468.
53. L. Vance (1986) ARL of cumulative sum control chart for controlling normal mean, J. Qual. Tech. **18**, 189-193.
54. A. V. Vasilopoulos & A. P. Stamboulis (1978) Modification of control chart limits in the presence of data correlation, J. Qual. Tech. **10**, 20-30.
55. K.-H. Waldmann (1986a) Bounds for the distribution of the run length of geometric moving average charts, J. Roy. Stat. Soc. C **35**, 151-158
56. K.-H. Waldmann (1986b) Bounds for the distribution of the run length of one-sided and two-sided CUSUM quality control schemes, Technometrics **28**, No. 1, 61-67.
57. D. G. Wardell, H. Moskowitz & R. D. Plante (1994a) Run length distributions of residual control charts for autocorrelated processes, J. Qual. Tech. **26**, 308-317.
58. D. G. Wardell, H. Moskowitz & R. D. Plante (1994b) Run-length distributions of special-cause control charts for correlated processes. (With discussion). Technometrics **36**, 3-27.
59. G. B. Wetherill (1977) Sampling Inspection and Quality Control, Chapman and Hall, London.
60. E. Yashchin (1993) Performance of CUSUM control schemes for serially correlated observations, Technometrics **35**, 37-52.

A Note on Stochastic Ordering Criteria in SPC for the Case of Correlated Output

Manuel Cabral Morais and António Pacheco

Department of Mathematics and Centre for Mathematics and its Applications
Instituto Superior Técnico, Technical University of Lisbon
Av. Rovisco Pais, 1049-001 Lisboa, Portugal

Abstract. The assumption that the output can be modelled by a sequence of independent random variables is standard in statistical process control. However, if the random variables are not independent but correlated the performance of the control schemes may be affected significantly. A typical effect of correlation is for instance a decrease in the in-control average run length (ARL) of a standard scheme — yielding to a higher false alarm rate, as reported by several numerical studies in the SPC literature. Moreover, W. Schmid and collaborators proved in a series of papers that, under mild conditions, the presence of correlation leads to a decrease of the survival function of the in-control RL (and therefore of the in-control ARL) of several modified schemes for correlated output, if we falsely assume that the underlying process is a sequence of independent random variables.

Bearing in mind that the use of the ARL to measure the ability to detect a process shift gives an incomplete picture of how a control scheme performs, establishing stochastic ordering results in the line of work pioneered by W. Schmid — as opposed to numerical results, organized in tables and graphs — provides a qualitative basis for a more objective assessment of the impact of correlation in the performance of quality control schemes.

In this paper we establish stochastic ordering results concerning the (in-control and out-of-control) RL of residual schemes for the mean of stationary autoregressive processes of order 1 or 2.

Keywords: SPC; Correlated Output; Run Length; Stochastic Ordering.

1 Background

The effects and implications of correlation have been frequently addressed in the SPC literature. For a detailed review refer to Knoth and Schmid (2001). These issues are usually tackled numerically: take for instance the investigations by Johnson and Bagshaw (1974), Alwan (1992), Maragah and Woodall (1992), Wardell, Moskowitz and Plante (1994), Runger, Willemain and Prabhu (1995), VanBrackle III and Reynolds Jr. (1997) and Lu and Reynolds Jr. (1999). These and several other papers have tables and graphs, usually referring to the ARL, to provide evidence that the performance of the appealing traditional control schemes is severely compromised by the presence of serial correlation.

Analytical investigations on the effects and implications of correlation have been so far presented in a few papers, e.g., Schmid (1995, 1997a, 1997b), Schmid and Schöne (1997), Schöne, Schmid and Knoth (1999) and Kramer and Schmid (2000). These papers provide in general a comparison between the survival functions of the *RLs* of two modified schemes for monitoring the mean of two (weakly) stationary Gaussian processes with different autocorrelation functions. Most of these stochastic order relations are established by using the fact that the joint distribution of any finite collection of X_t's from a stationary Gaussian process $\{X_t\}$ is a multivariate normal distribution.

In these latter investigations, emphasis is given to ARL based interpretations of the results because the ARL is by far the most popular of the RL related performance measures and has been extensively used to describe the likely performance of a control scheme. However, we should have in mind that confronting two *ARLs* essentially means comparing unidimensional and possibly misleading snapshots of the performances of the two schemes and ignores detailed information about the probabilistic behaviour of the *RLs*. On the other hand, confronting the survival functions, the hazard rate functions (Shaked and Shanthikumar (1994, p. 13)) or the equilibrium rate functions (Shaked and Shanthikumar (1994, p. 435)) of two *RLs*, thus establishing stochastic order relations, provides a more informative comparison of the performance of the control schemes we are dealing with than the comparisons based entirely on *ARLs*.

Now we describe and provide interpretations of three stochastic order relations between the performance measures of (two) control schemes 1 and 2, say RL_1 and RL_2, respectively.

Stochastically smaller in the usual stochastic sense ($\leq_{st}$)

$$RL_1 \leq_{st} RL_2 \Leftrightarrow P(RL_1 > t) \leq P(RL_2 > t), \ t = 1, 2, \ldots \tag{1}$$

$RL_1 \leq_{st} RL_2$ means that scheme 1 signals within the first t samples more frequently than scheme 2, for any value of t. Thus, the probability of obtaining a signal at the first sample is greater in scheme 1 than in scheme 2.

Since

$$RL_1 \leq_{st} RL_2 \ \Rightarrow \ P(RL_1 \leq RL_2) \geq 1/2,$$

it follows that if $RL_1 \leq_{st} RL_2$ then scheme 1 signals before scheme 2 more often. In addition,

$$RL_1 \leq_{st} RL_2 \ \Leftrightarrow \ E[f(RL_1)] \leq E[f(RL_2)]$$

for all increasing (cost) functions for which the expectations exist (see Shaked and Shanthikumar (1994, p.4)). Therefore,

$$RL_1 \leq_{st} RL_2 \ \Rightarrow \ ARL_1 \leq ARL_2.$$

Stochastically smaller in the hazard rate sense ($\leq_{hr}$)

$$RL_1 \leq_{hr} RL_2 \Leftrightarrow \lambda_{RL_1}(t) \geq \lambda_{RL_2}(t), \; t = 1, 2, \ldots \tag{2}$$

where

$$\lambda_{RL}(t) = P(RL = t)/P(RL \geq t)$$

represents the hazard rate function of the RL and was proposed by Margavio *et al.* (1995) as the *alarm rate* function of the control scheme.

If $RL_1 \leq_{hr} RL_2$, we can state that, for any value of t, scheme 1 is more likely to signal at sample t than scheme 2, given that the previous $t-1$ samples were not responsible for triggering a signal in any of the two schemes; or, following Margavio *et al.* (1995), we can add that scheme 1 has a larger alarm rate than scheme 2, regardless of the sàmple number value t.

The stochastic comparison based on hazard rate functions is stronger than the one based on survival functions,

$$RL_1 \leq_{hr} RL_2 \;\Rightarrow\; RL_1 \leq_{st} RL_2,$$

and the hazard rate comparison provides a performance confrontation in a specific conditional setting.

Stochastically smaller in the likelihood ratio sense ($\leq_{lr}$)

$$RL_1 \leq_{lr} RL_2 \Leftrightarrow P(RL_1 = t)/P(RL_2 = t) \downarrow_t \text{ over the set } \{1, 2, \ldots\}. \tag{3}$$

Thus, the odds of scheme 1 signalling at the sample t against scheme 2 triggering a signal at the same sample decreases as we progress with the sampling procedure.

Note that

$$RL_1 \leq_{lr} RL_2 \Rightarrow RL_1 \leq_{hr} RL_2$$

and

$$RL_1 \leq_{lr} RL_2 \Leftrightarrow (RL_1 | a \leq RL_1 \leq b) \leq_{st} (RL_2 | a \leq RL_2 \leq b)$$

whenever $a \leq b$ (Shaked and Shanthikumar (1994, p. 29)). Therefore, establishing the order relation $RL_1 \leq_{lr} RL_2$ provides a comparison between the performances of schemes 1 and 2 in a stricter conditional setting than the one imposed by $RL_1 \leq_{hr} RL_2$. Furthermore, it is worth noticing that

$$RL_1 \leq_{lr} RL_2 \Leftrightarrow r_{RL_1}(t) \geq r_{RL_2}(t), \; t = 1, 2, \ldots \tag{4}$$

(see Shaked and Shanthikumar (1994, p. 436)) where r_{RL} denotes the equilibrium rate of RL, i.e. $r_{RL}(1) = 0$ and

$$r_{RL}(t) = P(RL = t-1)/P(RL = t), t = 2, 3, \ldots.$$

Stochastic ordering has been largely excluded from the SPC literature despite the fact that it provides insight into how schemes work in practice, and, in the case of correlated output, enables us to assess the impact of serial correlation in the performance of schemes.

This paper focus in this latter aspect and special attention is given to *Shewhart* (*CUSUM* and *EWMA*) residual schemes for the mean of stationary autoregressive models of order 1 and 2 (of order 1), described in the following sections.

Two final remarks: the proofs of all the results are in the Appendix and all the numerical results were produced by programs for the package *Mathematica* (Wolfram (1996)).

2 Shewhart residual schemes

One of the two following approaches is usually adopted to build control schemes for the mean of correlated output. In the first approach, the original data is plotted in a traditional scheme (*Shewhart*, *CUSUM*, *EWMA*, etc.), however, with readjusted control limits to account for the correlation; the resulting monitoring tool is the so-called *modified scheme* (Vasilopoulos and Stamboulis (1978), Schmid (1995, 1997a) and Zhang (1998)).

The second approach also makes use of a traditional scheme, but the residuals of a time-series model are plotted instead of the original data; this sort of scheme is termed a *special-cause control scheme* (Alwan and Roberts (1988) and Wardell, Moskowitz and Plante (1994)) or, more commonly, a *residual scheme* (Runger, Willemain and Prahbu (1995) and Zhang (1997)).[1] The rationale behind residual schemes is that the residuals are independent in case the time-series model is valid, thus, they meet one of the standard assumptions of traditional schemes, which facilitates the evaluation of *RL* related measures.

There are a few points in favour of residual schemes. According to the comparison studies in Schmid (1995) and Schmid (1997b), and, as mentioned by Kramer and Schmid (2000), residual schemes tend to be better than modified schemes (in the *ARL* sense) in the detection of shifts in the mean of a stationary Gaussian autoregressive of order 1 model, when the autoregressive parameter is negative. In addition, the critical values required to implement the residuals schemes do not depend on the underlying in-control process, as those needed by modified schemes. Finally, the *ARL* of modified schemes are usually obtained using simulations (see Schmid (1995, 1997a, 1997b)) and can only have closed expressions for very special cases like exchangeable normal variables (Schmid (1995)).

[1] According to Knoth and Schmid (2001), Harrison and Davies (1964) were the first authors to use control charts for residuals.

Let $\{X_t\}$ be a Gaussian autoregressive model of order 2 ($AR(2)$) satisfying the equation

$$(1 - \phi_1 B - \phi_2 B^2)(X_t - \mu) = a_t \tag{5}$$

where: μ represents the process mean; $\{a_t\}$ corresponds to Gaussian white noise with variance σ_a^2; and B represents the backshift operator defined as $BX_t = X_{t-1}$ and $B^j X_t = X_{t-j}$. Also, recall that $\{X_t\}$ is a stationary process if $\phi_1 + \phi_2 < 1$, $\phi_2 - \phi_1 < 1$ and $-1 < \phi_2 < 1$ (Box, Jenkins and Reinsel (1994, p. 60)) and we are dealing with a stationary autoregressive model of order 1 ($AR(1)$) in case $\phi_2 = 0$ and $-1 < \phi_1 = \phi < 1$.

The residual scheme proposed by Zhang (1997) for $AR(2)$ output makes use of the residuals

$$e_t = (1 - \phi_1 B - \phi_2 B^2)(X_t - \mu_0) \tag{6}$$

where μ_0 is the in-control value of the process mean, $E(X_t)$. It is worth noticing that these residuals are summary statistics since they only depend on the in-control value of the process mean. Thus, they do not vary with the true value of the process mean as the residuals used in Wardell, Moskowitz and Plante (1994).

The main purpose of the *Shewhart* residual scheme we consider is to detect a single step change in the process mean from the in-control value

$$E(X_t) = \mu_0, \; t = \ldots, -1, 0,$$

to

$$E(X_t) = \mu_0 + \delta\sigma_x, \; t = 1, 2, \ldots,$$

where $\delta \neq 0$ and $\sigma_x^2 = V(X_t)$ remains constant.

In the absence of an assignable cause, the residuals given by (6) verify $e_t \overset{iid}{\sim} N(0, \sigma_a^2)$, where

$$\sigma_a^2 = (1 + \phi_2)\{(1 - \phi_2)^2 - \phi_1^2\}\, \sigma_x^2/(1 - \phi_2)$$

(Box, Jenkins and Reinsel (1994, p. 62)). As a consequence, the control scheme ought to trigger a signal at time t if

$$e_t < -\xi\,\sigma_a \quad \text{or} \quad e_t > \xi\,\sigma_a \tag{7}$$

where ξ is a positive constant usually selected by fixing the ARL in two situations: one being when $\delta = 0$ and one when δ is equal to some fixed nonzero value. Moreover, the *RLs* of the *Shewhart* residual schemes sharing the same ξ are all matched in-control and have *Geometric*$(2[1 - \Phi(\xi)])$ in-control distribution. Thus, these *RLs* are identically distributed to the one of a standard *Shewhart* X-scheme for *i.i.d.* output, with control limits $\mu_0 \pm \xi\sigma_x$, as mentioned by Wardell, Moskowitz and Plante (1994).

In the out-of-control situation, we still get independent residuals with variance σ_a^2. However:

$$E(e_1) = \delta\sigma_x \text{ and } E(e_t) = \delta(1-\phi)\sigma_x, t = 2, 3, \ldots,$$

for the $AR(1)$ model; and

$$E(e_1) = \delta\sigma_x,\ E(e_2) = \delta(1-\phi_1)\sigma_x,\ E(e_t) = \delta(1-\phi_1-\phi_2)\sigma_x, t = 3, 4, \ldots,$$

for the $AR(2)$ model. All these properties enable us to independently determine the probability that the residual e_t is beyond the control limits and, therefore, to assess the performance of the *Shewhart* residual scheme in a straighforward manner, as we shall see below.

2.1 AR(1) model

Let $RL(\phi, \delta)$ be the run length of the *Shewhart* residual scheme for an $AR(1)$ process, conditioned on the fact that the autoregressive parameter is equal to ϕ and the mean μ equals μ_0 for $t = \ldots, -1, 0$, and $\mu = \mu_0 + \delta\sigma_x$, for $t = 1, 2, \ldots$ and $-\infty < \delta < +\infty$. Then, the survival function, the hazard rate (or alarm rate) function and the equilibrium rate function of $RL(\phi, \delta)$ are given, respectively, by

$$P[RL(\phi,\delta) > t] = \beta(1;\phi,\delta)\,[\beta(2;\phi,\delta)]^{t-1},\ t = 1, 2, \ldots \tag{8}$$

$$\lambda_{RL(\phi,\delta)}(t) = \begin{cases} 1-\beta(1;\phi,\delta),\ t=1 \\ 1-\beta(2;\phi,\delta),\ t = 2, 3, \ldots \end{cases} \tag{9}$$

$$r_{RL(\phi,\delta)}(t) = \begin{cases} 0, & t=1 \\ \frac{1-\beta(1;\phi,\delta)}{\beta(1;\phi,\delta)[1-\beta(2;\phi,\delta)]}, & t=2 \\ \frac{1}{\beta(2;\phi,\delta)}, & t = 3, 4, \ldots \end{cases} \tag{10}$$

where

$$1-\beta(t;\phi,\delta) = 1 - \{\Phi\,[\xi - \delta\, f(t;\phi)] - \Phi\,[-\xi - \delta\, f(t;\phi)]\} \tag{11}$$

with

$$f(t;\phi) = \frac{|E(e_t)|}{\delta\sigma_a} = \begin{cases} \frac{1}{\sqrt{1-\phi^2}},\ t=1 \\ \sqrt{\frac{1-\phi}{1+\phi}},\ t = 2, 3, \ldots. \end{cases} \tag{12}$$

being what Zhang (1997) called the *detection capability index*.

Given that no signal has been triggered before time t, $[1-\beta(\min\{t,2\};\phi,\delta)]$ represents the probability that sample t is responsible for triggering a signal, and according to Equation (9) it also denotes the alarm rate at sample t.

According to Equation (9), the alarm rate takes at most two distinct values: $1-\beta(1;\phi,\delta)$ at sample 1, and $1-\beta(2;\phi,\delta)$ at the following samples. As a consequence, $RL(\phi,\delta)$ has either increasing, constant or decreasing hazard rate function depending on whether the alarm rate at sample 1 is smaller, equal or greater than the alarm rate at the subsequent samples. Therefore the monotone character of $\lambda_{RL(\phi,\delta)}(t)$ allows the comparison of the scheme ability to trigger a signal at sample 1 and at the following samples.

Theorem 1 *$RL(\phi, \delta)$ has an increasing (constant, decreasing) hazard rate function, if and only if $-1 < \phi < 0$ ($\phi = 0, 0 < \phi < 1$).*

The increasing behaviour of the alarm rate function for $-1 < \phi < 0$ is consistent with the fact that when the model has a negative autoregressive parameter "the one step-ahead forecast moves in the opposite direction of the shift" (Wardell, Moskowitz and Plante (1994)). This yields increasingly large residuals and hence increasing hazard rates. This fact is illustrated by the numerical results in Table 1. Conversely, for positively correlated output, the alarm rate decreases because residuals tend to get smaller; thus, early detection is more likely to happen.

The next theorem concerns the increasing stochastic behaviour of $RL(\phi, \delta)$ in terms of the autoregressive parameter. Let $RL_{iid}(\delta) = RL(0, \delta)$, so that $RL_{iid}(\delta)$ represents the RL of a standard *Shewhart* X-scheme whose summary statistic and control limits are given by Equations (6) and (7) with $\phi_1 = \phi_2 = 0$.

Theorem 2 *If $-1 < \phi \leq 0$ then*

$$RL(\phi, \delta) \uparrow_{hr} \text{ with } \phi. \tag{13}$$

As a consequence $RL(\phi, \delta) \leq_{hr} RL_{iid}(\delta)$, for $-1 < \phi < 0$.

Theorem 2 can be phrased more clearly by noting that the detection ability (or the alarm rate) of the *Shewhart* residual scheme for any stationary Gaussian $AR(1)$ model decreases with a nonpositive autoregressive parameter ϕ. For the numerical illustration of this property please refer to Table 1.

Remark 1 a) $\beta(1; \phi, \delta)$ decreases with ϕ whereas $\beta(2; \phi, \delta)$ increases with ϕ, for positively autocorrelated $AR(1)$ output. Therefore, it comes as no surprise that we are unable to establish the stochastic monotone behaviour of $RL(\phi, \delta)$, in the usual, the hazard rate or likelihood ratio senses, for $0 < \phi < 1$.

b) It is worth mentioning in passing that

$$RL(\phi, \delta) \not\leq_{lr} RL_{iid}(\delta), \text{ for } -1 < \phi < 0,$$

because the equilibrium rate can be a nonmonotonous function of ϕ (for fixed t). Furthermore, when $r_{RL(\phi,\delta)}(2)$ and $r_{RL(\phi,\delta)}(3)$ are monotone functions of ϕ, for $-1 < \phi < 0$, they usually have distinct behaviour and we cannot assert that the odds of a signal at sample t has a monotone behaviour with regard

Table 1. *ARLs*, alarm rates and equilibrium rates of the *Shewhart* residual scheme for $AR(1)$ output ($\xi = 3$).

δ	ϕ	ARL	Alarm rate $t=1$	$t \geq 2$	Equilibrium rate $t=2$	$t \geq 3$
0	$\phi \in$ (-1,1)	370.4	0.002700	0.002700	1.002707	1.002707
0.10	-0.75	273.2	0.003007	0.003663	0.823405	1.003677
	-0.50	322.2	0.002878	0.003105	0.929754	1.003114
	-0.25	342.1	0.002842	0.002923	0.975109	1.002932
	0	352.9	0.002833	0.002833	1.002842	1.002842
0.50	-0.75	22.1	0.012500	0.046767	0.270673	1.049061
	-0.50	61.2	0.007877	0.016478	0.481861	1.016754
	-0.25	106.6	0.006722	0.009407	0.719371	1.009497
	0	155.2	0.006442	0.006442	1.006484	1.006484
0.75	-0.75	7.3	0.031031	0.154890	0.206759	1.183277
	-0.50	23.1	0.016478	0.044484	0.376628	1.046555
	-0.25	47.7	0.013107	0.021126	0.628686	1.021581
	0	81.2	0.012313	0.012313	1.012466	1.012466
1.00	-0.75	3.6	0.068360	0.361576	0.202933	1.566358
	-0.50	10.5	0.032513	0.102409	0.328153	1.114093
	-0.25	23.3	0.024607	0.043734	0.576857	1.045734
	0	43.9	0.022782	0.022782	1.023313	1.023313
2.00	-0.75	1.5	0.509460	0.989033	1.050088	91.181141
	-0.50	2.1	0.244909	0.678713	0.477880	3.112477
	-0.25	3.4	0.175047	0.337970	0.627837	1.510504
	0	6.3	0.158656	0.158656	1.188574	1.188574
3.00	-0.75	1.1	0.937679	1.000000	15.045836	2.524×10^6
	-0.50	1.3	0.678713	0.985959	0.985959	71.222445
	-0.25	1.6	0.539187	0.808664	0.808664	5.226406
	0	2.0	0.500000	0.500000	2.000000	2.000000

to ϕ. The mentioned properties are apparent in Table 1.

c) Theorem 2 strenghtens Conclusion 2(a) of Zhang (1997, p. 484):

$$ARL(\phi, \delta) \leq ARL_{iid}(\delta), \text{ for } -1 < \phi < 0. \tag{14}$$

d) Theorem 1 in Schmid and Schöne (1997) refers to modified *EWMA* schemes and reads as follows: $RL_{iid}(0) \leq_{st} RL_G(0)$, where $RL_{iid}(0)$ and $RL_G(0)$ refer here to the in-control *RLs* of *EWMA* modified schemes for *i.i.d.* output and a stationary Gaussian model with nonnegative autocovariance function. $RL(\phi,\delta) \leq_{hr} RL_{iid}(\delta)$, for $-1 < \phi < 0$, is a consequence of result (13) and resembles Schmid and Schöne's result; however, the stochastic order relation we obtained is not only reversed but also stronger, and refers to both in-control and out-of-control situations. These differences probably stem from the fact that *Shewhart* residual schemes have simpler RL characteristics than the modified $EWMA$ schemes considered by Schmid and Schöne (1997), and that we are dealing with a stricter class of stationary processes (autoregressive models) with different correlation behaviour than the class considered by those authors.

e) Theorem 2 is also in agreement with the notes in the last paragraph of page 182 of Kramer and Schmid (2000). ■

2.2 AR(2) model

The survival, hazard rate (or alarm rate) and equilibrium rate functions of $RL(\phi_1,\phi_2,\delta)$, the run length of the *Shewhart* residual scheme for stationary $AR(2)$ output conditioned on ϕ_1, ϕ_2 and δ, are equal, respectively, to

$$P[RL(\phi_1,\phi_2,\delta) > t] = \begin{cases} \beta(1;\phi_1,\phi_2,\delta), & t=1 \\ \beta(1;\phi_1,\phi_2,\delta)\beta(2;\phi_1,\phi_2,\delta) \\ \qquad \times[\beta(3;\phi_1,\phi_2,\delta)]^{t-2}, & t=2,3,\ldots \end{cases} \tag{15}$$

$$\lambda_{RL(\phi_1,\phi_2,\delta)}(t) = \begin{cases} 1-\beta(1;\phi_1,\phi_2,\delta),\ t=1 \\ 1-\beta(2;\phi_1,\phi_2,\delta),\ t=2 \\ 1-\beta(3;\phi_1,\phi_2,\delta),\ t=3,4,\ldots \end{cases} \tag{16}$$

$$r_{RL(\phi_1,\phi_2,\delta)}(t) = \begin{cases} 0, & t=1 \\ \frac{1-\beta(t-1;\phi_1,\phi_2,\delta)}{\beta(t-1;\phi_1,\phi_2,\delta)[1-\beta(t;\phi_1,\phi_2,\delta)]}, & t=2,3 \\ \frac{1}{\beta(3;\phi_1,\phi_2,\delta)}, & t=4,5,\ldots \end{cases} \tag{17}$$

where

$$\beta(t;\phi_1,\phi_2,\delta) = \Phi\left[k-\delta f(t;\phi_1,\phi_2)\right] - \Phi\left[-k-\delta f(t;\phi_1,\phi_2)\right],\ t=1,2,\ldots \tag{18}$$

with the detection capability index given by

$$f(t;\phi_1,\phi_2) = \begin{cases} \sqrt{\frac{1-\phi_2}{1+\phi_2}\frac{1}{(1-\phi_2)^2-\phi_1^2}}, & t=1 \\ |1-\phi_1|\, f(1;\phi_1,\phi_2), & t=2 \\ (1-\phi_1-\phi_2)\, f(1;\phi_1,\phi_2), & t=3,4,\ldots, \end{cases} \tag{19}$$

following the expressions (6)–(8) and (A1)–(A2) of Zhang (1997, p. 478 and p. 489).

Let us define the following sets in $\mathbb{R}^2$:

$$
\begin{aligned}
A &= \{(\phi_1,\phi_2) : \phi_1+\phi_2<1, \phi_2-\phi_1<1, -1<\phi_2<1\}\\
B &= \{(\phi_1,\phi_2) : \phi_1<0, \phi_2<0\}\\
C &= \{(\phi_1,\phi_2) : \phi_1>0, (\phi_2>0 \text{ or } 2\phi_1+\phi_2>2)\}\\
D &= \left\{(\phi_1,\phi_2) : \phi_1 < \left[(1-\phi_2)-\sqrt{(1-\phi_2)(1-\phi_2-2\phi_2^2)}\right]/2 \quad \text{or}\right.\\
&\qquad \left.\phi_1 > \left[(1-\phi_2)+\sqrt{(1-\phi_2)(1-\phi_2-2\phi_2^2)}\right]/2\right\};\\
E &= \left\{(\phi_1,\phi_2) : \phi_1+\phi_2-\phi_2^2<0\right\}.
\end{aligned}
$$

Recall that A corresponds to the stationarity region for the $AR(2)$ model. The meaning of the remaining sets will be discussed in the next theorem and lemma.

The following result corresponds to the analogue of Theorem 1 for $AR(2)$ models.

Theorem 3 *$RL(\phi_1,\phi_2,\delta)$ has increasing (constant, decreasing) hazard rate if $(\phi_1,\phi_2)\in A\cap B$, ($(\phi_1,\phi_2)=(0,0)$, $(\phi_1,\phi_2)\in A\cap C$).*

The increasing behaviour of the alarm rate for $\phi_1,\phi_2<0$ is illustrated in Table 2, for $\delta=0.10,1$.

Before we proceed, we would like to remind the reader that the detection capability index of a standard *Shewhart* X-scheme is unitary, as observed by Zhang (1997); also, the alarm rate function is constant. Moreover, note that, as in the $AR(1)$ model, the alarm rate, $1-\beta(t;\phi_1,\phi_2,\delta)$, is also an increasing function of the detection capability index. Summing up these facts, we can assert that the alarm rate of the *Shewhart* residual scheme for $AR(2)$ output is larger than the one of the standard *Shewhart* X-scheme in case $f(t;\phi_1,\phi_2)>1, t=1,2,3,\ldots$. Hence, it is very important to identify the sets where these inequalities hold, as we do in the next lemma that corresponds to Theorem 1 of Zhang (1997).

Lemma 1 *For fixed t, the detection capability index satisfies*

$$
f(t;\phi_1,\phi_2,\delta)>1 \Leftrightarrow (\phi_1,\phi_2)\in \begin{cases} A, & t=1\\ A\cap D, & t=2\\ A\cap E, & t=3,4,\ldots \end{cases} \tag{20}
$$

Let $RL_{iid}(\delta)=RL(0,0,\delta)$. Then an analogue of Theorem 2 holds for the stationary $AR(2)$ model.

Theorem 4 *For $(\phi_1,\phi_2)\in A\cap D\cap E$,*

$$
RL(\phi_1,\phi_2,\delta) \leq_{hr} RL_{iid}(\delta). \tag{21}
$$

Table 2. *ARLs*, alarm rates and equilibrium rates of the *Shewhart* residual scheme for $AR(2)$ output ($\xi = 3$).

δ	ϕ_1	ϕ_2	ARL	Alarm rate $t=1$	$t=2$	$t \geq 3$	Equilibrium rate $t=2$	$t=3$	$t \geq 4$
0	$(\phi_1, \phi_2) \in A$		370.4	0.002700	0.002700	0.002700	1.002707	1.002707	1.002707
0.10	-0.50	-0.50	284.3	0.002901	0.003156	0.003522	0.921759	0.899073	1.003534
		-0.25	310.1	0.002870	0.003085	0.003227	0.932858	0.959012	1.003237
		0	322.2	0.002878	0.003105	0.003105	0.929754	1.003114	1.003114
	-0.25	-0.50	306.0	0.002883	0.002988	0.003270	0.967858	0.916438	1.003281
		-0.25	329.4	0.002848	0.002933	0.003036	0.974035	0.968682	1.003046
		0	342.1	0.002842	0.002923	0.002923	0.975109	1.002932	1.002932
	0	-0.50	322.2	0.002878	0.002878	0.003105	1.002887	0.929754	1.003114
		-0.25	342.1	0.002842	0.002842	0.002923	1.002850	0.975109	1.002932
		0	352.9	0.002833	0.002833	0.002833	1.002841	1.002841	1.002841
1.00	-0.50	-0.40	5.9	0.033501	0.106052	0.217532	0.326839	0.545359	1.278007
	-0.25		9.6	0.029326	0.053285	0.120942	0.566996	0.465379	1.137582
	0		15.4	0.028158	0.028158	0.070449	1.028974	0.411276	1.075789
	-0.50	0	10.5	0.032513	0.102409	0.102409	0.328153	1.114093	1.114093
	-0.25		23.3	0.024607	0.043734	0.043734	0.576857	1.045734	1.045734
	0		43.9	0.022782	0.022782	0.022782	1.023313	1.023313	1.023313
	-0.50	0.40	4.0	0.152412	0.484359	0.203620	0.371251	4.613153	1.255683
	-0.25		39.6	0.035963	0.066850	0.023892	0.558032	2.998391	1.024477
	0		100.0	0.028158	0.028158	0.009634	1.028974	3.007654	1.009727

Furthermore, for fixed ϕ_2 $(-1 < \phi_2 < 1)$,

$$RL(\phi_1, \phi_2, \delta) \uparrow_{hr} \text{ with } \phi_1 \text{ over the interval } (\phi_2 - 1, 0]. \tag{22}$$

The stochastic increasing behaviour in the hazard rate sense of $RL(\phi_1, \phi_2, \delta)$ with respect to ϕ_1 is apparent in Table 2, for $\delta = 1$. Although the ARL increases with ϕ_2, Table 2 illustrates the nonmonotonous behaviour of the alarm rate in terms of ϕ_2 (for $\delta = 0.10$) — thus the ARL gives partial information about the scheme performance.

Remark 2 a) Corollary 2 of Zhang (1997, p. 481) can be read as follows:

$$ARL(\phi_1, \phi_2, \delta) \leq ARL_{iid}(\delta) \text{ when } -1 < \phi_1 < -0.25.$$

Equation (21) clearly strengthens this corollary.

b) Once again, we cannot compare the *RL* of the *Shewhart* residual scheme and the *RL* of the standard *X*-scheme for uncorrelated output in the likelihood ratio sense. Actually, for the same reasons pointed out for the $AR(1)$ model, $RL(\phi_1, \phi_2, \delta) \not\leq_{lr} RL_{iid}(\delta)$, as we can see from the numerical results regarding the equilibrium rate function in Table 2. ■

3 CUSUM and EWMA residual schemes

The improvement of the sensitivity of Shewhart residual schemes to small and moderate shifts through the adoption of *CUSUM* and *EWMA* residuals schemes has been also addressed in the SPC literature. For example, Harris and Ross (1991), Runger, Willemain and Prabhu (1995) and Lu and Reynolds Jr. (1999) discuss the application of the *CUSUM* and the *EWMA* techniques and conclude that the resulting residual schemes have superior performance than their Shewhart counterparts, in *ARL* terms.

In this section we assess the impact of the correlation in the performance of upper one-sided *CUSUM* and *EWMA* schemes for residuals of a stationary $AR(1)$ model, by means of stochastic ordering. The main purpose of the upper one-sided *CUSUM* and *EWMA* residual schemes, described in the next two subsections, is the detection of upward shifts in the process mean — from μ_0 to $\mu_0 + \delta\sigma_x$, where $\delta > 0$.

3.1 CUSUM residual schemes

The detection of upward shifts can be done by the upper one-sided *CUSUM* scheme which uses the following summary statistic

$$V_t = \begin{cases} v, & t = 0 \\ \max\{0,\ V_{t-1} + (e_t - k\,\sigma_a)\}, & t = 1, 2, \ldots \end{cases} \tag{23}$$

and lower and upper control limits

$$LCL_C = 0 \text{ and } UCL_C = h\,\sigma_a. \tag{24}$$

v denotes the initial value given to the upper one-sided *CUSUM* statistic. Let

$$v = LCL_C + \beta(UCL_C - LCL_C)$$

for some $\beta \in [0, 1)$. If $\beta \in (0, 1)$ (or $\beta = 0$) a $\beta \times 100\%$ head start (or no head start) has been given to the chart.

By virtue of the fact that the reference value and the upper control limit are both multiples of σ_a, the *RLs* of the *CUSUM* residual schemes are identically distributed in the absence of an assignable cause ($\delta = 0$) for all values of the parameter ϕ. Thus, all these *CUSUM* residual schemes have matched in-control performance, as well as the Shewhart residual schemes considered in the previous section.

The Markov chain approach, introduced by Brook and Evans (1972), was used by Runger, Willemain and Prabhu (1995) to obtain an approximation to the ARL of an upper one-sided $CUSUM$ residual scheme for an $AR(p)$ model. This approach comprises the discretization of the summary statistic V_t, and can be also applied to obtain approximations to other RL related measures like its survival function, percentage points and alarm rate function.

Let $RL_C^{\lfloor\beta(x+1)\rfloor}(\phi, \delta; x)$, where $\lfloor y \rfloor$ represents the integer part of the real number y, be the Markov approximation for the run length of the upper one-sided $CUSUM$ residual scheme for stationary $AR(1)$ output with an $\beta \times 100\%$ head start, based on an absorbing Markov chain with discrete state space $\{0, 1, \ldots, x+1\}$ and absorbing state $(x+1)$. Note that a $\beta \times 100\%$ head start corresponds to the initial state $\lfloor\beta(x+1)\rfloor$ in the Markov approximation.

Since, in the presence of a shift in the process mean, the residuals e_t have their expected value equal to $\delta\sigma_x$ (for $t = 1$) and $\delta(1-\phi)\sigma_x$ (for $t = 2, 3, \ldots$), the first transition and the remaining transitions between the transient states of this Markov chain are ruled by two distinct substochastic matrices $\mathbf{Q}(\phi, \delta; x)$ and $\mathbf{Q}(\phi, \delta(1-\phi); x)$ (respectively), where $\mathbf{Q}(.,.;x)$ is given in the appendix.

On account of the first transition, $RL_C^{\lfloor\beta(x+1)\rfloor}(\phi, \delta; x)$ is associated to a time-non-homogeneous Markov chain. In this setting, the approximating RL is related to a first passage time: $T_C^{\underline{\alpha}(\phi,\delta;x)}(\phi, \delta(1-\phi); x)$, that takes values in the set $\{0, 1, \ldots\}$ and concerns a time-homogeneous absorbing Markov chain whose random initial state equals the discretized version of V_1 and whose transitions between transient states are governed by the substochastic matrix $\mathbf{Q}(\phi, \delta(1-\phi); x)$. Thus,

$$P[RL_C^{\lfloor\beta(x+1)\rfloor}(\phi, \delta; x) = t] = P[T_C^{\underline{\alpha}(\phi,\delta;x)}(\phi, \delta(1-\phi); x) = t - 1] \tag{25}$$

for $t = 1, 2, \ldots$.

$T_C^{\underline{\alpha}(\phi,\delta;x)}(\phi, \delta(1-\phi); x)$ has a discrete phase-type distribution (see Neuts (1981, Chap.2) for a discussion on continuous and discrete phase-type distributions) with parameters

$$(\underline{\alpha}(\phi, \delta; x), \mathbf{Q}(\phi, \delta(1-\phi); x)) \tag{26}$$

where

$$\underline{\alpha}^\top(\phi, \delta; x) = \underline{\mathbf{e}}_{\lfloor\beta(x+1)\rfloor}^\top \mathbf{Q}(\phi, \delta; x). \tag{27}$$

Here $\underline{\mathbf{e}}_u$ represents the $(u+1)^{th}$ unit-vector of dimension $x+1$ and $\underline{\mathbf{a}}^\top$ denotes the transpose of $\underline{\mathbf{a}}$. Considering $\underline{\mathbf{1}}$ a $(x+1)$-dimensional vector of ones,

$$(\underline{\alpha}^\top(\phi, \delta; x), 1 - \underline{\alpha}^\top(\phi, \delta; x)\underline{\mathbf{1}}) \tag{28}$$

corresponds to the probability vector of the discretized version of V_1, i.e., the initial state of the absorbing Markov chain concerning $T_C^{\underline{\alpha}(\phi,\delta;x)}(\phi, \delta(1-\phi); x)$.

As far as its survival function is concerned, $RL_C^{\lfloor\beta(x+1)\rfloor}(\phi,\delta;x)$ is defined as follows:

$$P[RL_C^{\lfloor\beta(x+1)\rfloor}(\phi,\delta;x)>t]=\underline{\mathbf{e}}_{\lfloor\beta(x+1)\rfloor}^{\top}\,[\mathbf{Q}(0,\delta;x)]^t\,\underline{\mathbf{1}},\ t=1,2,\ldots$$

for *i.i.d.* output; and

$$P[RL_C^{\lfloor\beta(x+1)\rfloor}(\phi,\delta;x)>t]= \\ =\begin{cases}\underline{\alpha}^{\top}(\phi,\delta;x)\,\underline{\mathbf{1}}, & t=1\\ \underline{\alpha}^{\top}(\phi,\delta;x)\,[\mathbf{Q}(\phi,\delta(1-\phi);x)]^{t-1}\,\underline{\mathbf{1}}, & t=2,\ldots.\end{cases} \tag{29}$$

for $AR(1)$ output. Moreover, the Markov approximation to ARL,

$$ARL_C^{\lfloor\beta(x+1)\rfloor}(\phi,\delta;x),$$

equals:

$$\underline{\mathbf{e}}_{\lfloor\beta(x+1)\rfloor}^{\top}\,[\mathbf{I}-\mathbf{Q}(0,\delta;x)]^{-1}\,\underline{\mathbf{1}} \tag{30}$$

for *i.i.d.* output; and

$$1+\underline{\alpha}^{\top}(\phi,\delta;x)\,[\mathbf{I}-\mathbf{Q}(\phi,\delta(1-\phi);x)]^{-1}\,\underline{\mathbf{1}} \tag{31}$$

for $AR(1)$ output. Also the alarm rate function is obviously given by

$$\lambda_{RL_C^{\lfloor\beta(x+1)\rfloor}(\phi,\delta;x)}(t)=1-\frac{P[RL_C^{\lfloor\beta(x+1)\rfloor}(\phi,\delta;x)>t]}{P[RL_C^{\lfloor\beta(x+1)\rfloor}(\phi,\delta;x)>t-1]} \tag{32}$$

for $t=1,2,\ldots$.

In Subsection 2.1 we were able to obtain simple expressions for the *RL* related measures of the *Shewhart* residual schemes and — through the behaviour of the detection capability index — give straight answers for questions concerning the assessment of the impact of a change in the autogressive parameter ϕ on the performance of a control scheme.

However, Markov-type control schemes lead to formulae that, although easy to handle numerically, require some mathematical work on the special features of the matrix $\mathbf{Q}(.,.;x)$ in order to obtain the sort of results derived for the Shewhart residual schemes. So far we have been able to assess the stochastic behaviour of $RL_C^{\lfloor\beta(x+1)\rfloor}$, with regard to ϕ, in the usual sense.

Furthermore, since $RL_C^{\lfloor\beta(x+1)\rfloor}(\phi,\delta;x)$ converges in law to the exact RL of this scheme, $RL_C^{\beta}(\phi,\delta)$, as $x\to\infty$, the next theorem still holds for $RL_C^{\beta}(\phi,\delta)$.

Theorem 5 *If* $-1<\phi\leq 0$ *then, for fixed* k *and* h,

$$RL_C^{\lfloor\beta(x+1)\rfloor}(\phi,\delta;x)\uparrow_{st}\quad with\quad \phi. \tag{33}$$

Namely,

$$RL_C^{\lfloor\beta(x+1)\rfloor}(\phi,\delta;x)\leq_{st}RL_C^{\lfloor\beta(x+1)\rfloor}(0,\delta;x),\ for\ -1<\phi<0.$$

Theorem 5 allows us to assert that, for any stationary Gaussian $AR(1)$ model with nonpositive parameter, the *RL* of the *CUSUM* residual scheme increases with the autoregressive parameter ϕ.

3.2 EWMA residual schemes

Now we provide a brief description of the upper one-sided $EWMA$ residual scheme for the stationary $AR(1)$ model.

This residual scheme uses the summary statistic

$$W_t = \begin{cases} w, & t = 0 \\ \max\{0,\ (1-\lambda)\,W_{t-1} + \lambda\, e_t\}, & t = 1, 2, \ldots \end{cases} \tag{34}$$

and the lower and upper control limits

$$LCL_E = 0 \text{ and } UCL_E = \gamma\sqrt{\lambda(2-\lambda)^{-1}}\,\sigma_a, \tag{35}$$

where $\lambda \in (0,1]$ corresponds to the weight given to the most recent sample residual and w is the initial value of the summary statistic.

To avoid repetition, we merely mention that the Markov approximation to the RL of this residual scheme, $RL_E^{\lfloor\beta(x+1)\rfloor}(\phi, \delta; x)$, has a similar distribution to the one of the upper one-sided $CUSUM$ residual scheme previously described — the two substochastic matrices, $\mathbf{Q}(\phi, \delta; x)$ and $\mathbf{Q}(\phi, \delta(1-\phi); x)$ have to be conveniently replaced. (For the underlying details of these two matrices please refer to the appendix.)

Furthermore, $RL_E^{\lfloor\beta(x+1)\rfloor}(\phi, \delta; x)$ and the exact RL of this $EWMA$ residual scheme, $RL_E^{\beta}(\phi, \delta)$, also have a stochastic increasing behaviour in terms of ϕ $(-1 < \phi \leq 0)$.

Theorem 6 *If $-1 < \phi \leq 0$ then, for fixed λ and γ,*

$$RL_E^{\lfloor\beta(x+1)\rfloor}(\phi, \delta; x) \uparrow_{st} \text{ with } \phi. \tag{36}$$

Thus,

$$RL_E^{\lfloor\beta(x+1)\rfloor}(\phi, \delta; x) \leq_{st} RL_E^{\lfloor\beta(x+1)\rfloor}(0, \delta; x), \text{ for } -1 < \phi < 0.$$

3.3 Numerical illustrations

Now we proceed into the numerical illustration of the properties stated in Theorems 5 and 6 and other properties of the upper one-sided $CUSUM$ and $EWMA$ residual schemes.

The numerical results in Table 3 refer to the ARL and alarm rate function of an upper one-sided CUSUM residual scheme with no head start, $k = 0.125$, $h = 9.9609$ and $x + 1 = 30$ transient states, yielding an in-control ARL of approximately 370.409 samples.

Similarly, Table 4 comprises the same performance measures for an upper one-sided $EWMA$ residual scheme with $\lambda = 0.05$, $\gamma = 2.4432$ and, once again, with 30 transient states; with this constelation of parameters we got an in-control ARL of 370.382 samples.

Table 3. *ARLs* and alarm rates of the *CUSUM* residual scheme for *AR*(1) output ($k = 0.125, h = 9.9609; x = 29$).*

δ	ϕ	*ARL*	Alarm rate $t=2$	$t=3$	$t=4$	$t=5$	$t=10$
0	$\phi \in$ (-1,1)	370.4				0.000002	0.000194
0.10	-0.75	54.2			0.000002	0.000020	0.002472
	-0.50	86.8			0.000001	0.000009	0.001110
	-0.25	116.1			0.000001	0.000006	0.000737
	0	144.5				0.000005	0.000560
0.50	-0.75	9.3		0.000049	0.002737	0.023931	0.341026
	-0.50	14.2		0.000003	0.000193	0.002132	0.096229
	-0.25	19.4		0.000001	0.000046	0.000527	0.035666
	0	25.7			0.000018	0.000199	0.015614
0.75	-0.75	6.3	0.000001	0.001769	0.053872	0.238243	0.736166
	-0.50	9.4		0.000057	0.002889	0.023974	0.328881
	-0.25	12.5		0.000009	0.000484	0.004711	0.142409
	0	16.3		0.000003	0.000137	0.001382	0.062660
1.00	-0.75	4.9	0.000016	0.026329	0.321621	0.666664	0.934758
	-0.50	7.1		0.000699	0.024186	0.130327	0.605919
	-0.25	9.3		0.000081	0.003563	0.027027	0.328620
	0	12.0		0.000018	0.000840	0.007180	0.163284
2.00	-0.75	2.9	0.111184	0.973210	0.999664	0.999964	0.999997
	-0.50	3.8	0.001292	0.294935	0.837998	0.956693	0.993862
	-0.25	4.8	0.000070	0.045220	0.377360	0.684913	0.925994
	0	5.9	0.000010	0.008185	0.119386	0.344764	0.755253
	0.75	13.1	0.000005	0.000587	0.005833	0.020255	0.122137
	0.90	18.8	0.000211	0.003474	0.012649	0.024945	0.065966
3.00	-0.75	2.0	0.956633	1.000000	1.000000	1.000000	1.000000
	-0.50	2.9	0.164324	0.979785	0.999641	0.999951	0.999995
	-0.25	3.3	0.015067	0.645757	0.960742	0.990412	0.998560
	0	4.0	0.002161	0.248735	0.736428	0.896450	0.976054

*The empty cells correspond to alarm rate values that up to six decimal places are equal to 0.000000.

We note that our *ARL* values for the upper one-sided *CUSUM* residual scheme differ from those obtained by Runger, Willemain and Prabhu (1995) not only because of the constelation of parameters but also for other reasons: these authors considered the discretization proposed by Brook and Evans (1972), whereas we followed the one in Morais and Pacheco (2001); Runger, Willemain and Prabhu (1995) considered the magnitude of the upward shift equal to δ instead of $\delta\sigma_x$; and, in addition, σ_a was taken equal to 1 by these authors.

Table 4. *ARLs* and alarm rates of the *EWMA* residual scheme for *AR*(1) output ($\lambda = 0.05, h = 2.4432; x = 29$).*

			Alarm rate				
δ	ϕ	*ARL*	$t=2$	$t=3$	$t=4$	$t=5$	$t=10$
0	$\phi \in$ (-1,1)	370.4		0.000002	0.000017	0.000070	0.000944
0.10	-0.75	57.2		0.000010	0.000119	0.000530	0.007756
	-0.50	95.1		0.000005	0.000064	0.000277	0.004012
	-0.25	127.7		0.000004	0.000047	0.000202	0.002855
	0	158.1		0.000003	0.000039	0.000164	0.002266
0.50	-0.75	7.8	0.000021	0.003512	0.036154	0.112973	0.391210
	-0.50	12.4	0.000003	0.000430	0.005117	0.020052	0.141798
	-0.25	17.6	0.000001	0.000143	0.001685	0.006865	0.064591
	0	24.3	0.000001	0.000070	0.000784	0.003161	0.033304
0.75	-0.75	5.3	0.000385	0.043672	0.252497	0.472803	0.751732
	-0.50	7.9	0.000026	0.003856	0.036830	0.111089	0.378395
	-0.25	10.8	0.000007	0.000951	0.009941	0.034961	0.191760
	0	14.5	0.000003	0.000367	0.003775	0.013827	0.099256
1.00	-0.75	4.1	0.004340	0.233316	0.650779	0.823683	0.935195
	-0.50	5.9	0.000207	0.022930	0.151943	0.327040	0.632614
	-0.25	7.9	0.000042	0.004887	0.041574	0.117537	0.376109
	0	10.3	0.000015	0.001613	0.014442	0.045754	0.211409
2.00	-0.75	2.4	0.629697	0.998391	0.999941	0.999985	0.999996
	-0.50	3.2	0.068988	0.734780	0.949054	0.979868	0.993286
	-0.25	3.9	0.011018	0.303690	0.679317	0.823076	0.925811
	0	4.8	0.002924	0.109529	0.368147	0.552753	0.763822
	0.75	11.4	0.001573	0.015601	0.042530	0.070282	0.139912
	0.90	17.8	0.017660	0.040531	0.054688	0.062002	0.067155
3.00	-0.75	2.0	0.999493	1.000000	1.000000	1.000000	1.000000
	-0.50	2.3	0.713931	0.998649	0.999926	0.999979	0.999994
	-0.25	2.8	0.259578	0.925537	0.988458	0.995323	0.998320
	0	3.3	0.084119	0.665812	0.891416	0.942356	0.974353

*The empty cells correspond to alarm rate values that up to six decimal places are equal to 0.000000.

The first apparent feature from both tables is the magnitude of the alarm rate values for $t = 1, 2$: most of the time, both schemes lead to values that up to six decimal places are equal to 0.000000, which are then ommited. This fact happens with stronger emphasis in the *CUSUM* case and leads, in particular, to the omission of all the alarm rate values referring to $t = 1$ from Tables 3 and 4. This is essentially due to the well know initial inertia of *CUSUM* and *EWMA* summary statistics which causes the skewness to the right of the *RL* of these schemes.

These two tables also suggest that the alarm rate of both upper one-sided *CUSUM* and *EWMA* residual schemes increases as we collect more samples for fixed ϕ ($\phi \in (-1, 0]$) and δ and it decreases with ϕ ($\phi \in (-1, 0]$), as in the *Shewhart* case. Thus, the *RLs* of both upper one-sided *CUSUM* and *EWMA* residual schemes seem to also increase with ϕ ($\phi \in (-1, 0]$) in the hazard rate sense.

In addition, the numerical results concerning *ARL*, in both tables, misleadingly suggest that the *RL* stochastically increases with ϕ in the interval $[0, 1)$. However, additional investigations made us realize that the same does not occur to some *RL* percentage points concerning the upper one-sided *EWMA* residual scheme, leading to the conclusion that the increasing behaviour does not hold in the usual sense.

The upper one-sided *EWMA* residual scheme leads in general to larger alarm rate values than the matched in-control upper one-sided *CUSUM* residual scheme, thus suggesting a greater ability to detect upward shifts.

Replacing the *Shewhart* residual scheme described in the previous section for any of the two upper one-sided *CUSUM* and *EWMA* residual schemes leads to smaller *ARL* values for small and moderate values of δ. Besides this well know feature of the *CUSUM* and *EWMA* schemes, we got smaller alarm rate values for $t = 1, 2$ and significantly larger values for $t \geq 3$ with small or moderate values of δ ($\delta = 0.10, 0.50, 0.75, 1.00$). These results would suggest an occasionally early detection of upward shifts by the *Shewhart* residual schemes and a more persistent early detection by the upper one-sided *CUSUM* and *EWMA* residual schemes, for shifts with small and moderate magnitude.

Additional numerical investigations with an upper one-sided *Shewhart* residual scheme with an in-control *ARL* of 370.4 samples lead to the same conclusions.

4 Concluding notes

Special attention has been given to the impact of correlation in the alarm rate function of a few residual schemes for $AR(1)$ and $AR(2)$ output. Sufficient conditions have been established to guarantee that the run length stochastically increases (in the hazard rate sense, in the Shewhart case; and in the usual sense, in the *CUSUM* and *EWMA* cases) with the first autoregressive parameter.

As a final remark, we would like to add that stochastic ordering is a vital and powerful tool in the study of the statistical performance of control schemes. As we have seen, it provides insight into how schemes work in practice, and we strongly believe that it leads to an effective assessment of the impact of correlation in an objective and informative manner.

Appendix

Shewhart residual schemes

In the first part of the appendix we provide proofs of Theorems 1, 2, 3, 4, 5 and 6. Lemma 1 is proved in Zhang (1997, pp. 489-490).

Proof (Theorem 1) The detection capability index verifies:

- $f(t;0) = 1$;
- $f(t;\phi) > 1$, if $-1 < \phi < 0$ (Corollary 1, Zhang (1997, p. 479)).

Moreover, since the sign of the derivative $\frac{d\beta(t;\phi,\delta)}{d\,f(t;\phi)}$ equals the sign of $-2\delta \sinh[\xi\delta f(t;\phi)]$, we can assert that $\beta(t;\phi,\delta)$ is a decreasing function of the detection capability index, for any $-\infty < \delta < +\infty$. Then, by noting that

$$f(t;\phi) \geq f(1;\phi), \text{ for } -1 < \phi < 0 \text{ and } t = 2,3,\ldots,$$

we immediately conclude that

$$\lambda_{RL(\phi,\delta)}(t) = 1 - \beta(t;\phi,\delta) \geq 1 - \beta(1;\phi,\delta) = \lambda_{RL(\phi,\delta)}(1).$$

In view of (9), we conclude that $\lambda_{RL(\phi,\delta)}$ is an increasing hazard rate function.

Analogously, we get, for $0 < \phi < 1$ and $t = 2,3,\ldots$, $f(t;\phi) \leq f(1;\phi)$. Thus,

$$\lambda_{RL(\phi,\delta)}(t) = 1 - \beta(t;\phi,\delta) \leq 1 - \beta(1;\phi,\delta) = \lambda_{RL(\phi,\delta)}(1),$$

and $\lambda_{RL(\phi,\delta)}(t)$ becomes a decreasing hazard rate function for $0 < \phi < 1$.

For $\phi = 0$ we have a constant alarm rate:

$$\lambda_{RL(0,\delta)}(t) = 1 - \beta(1;0,\delta), t = 1,2,\ldots. \quad \blacksquare$$

Proof (Theorem 2) Recall that $\beta(t;\phi,\delta)$ is a decreasing function of $f(t;\phi)$, for $-1 < \phi < 0$, in which case $f(1;\phi)$ and $f(2;\phi)$ are both decreasing functions of ϕ. The result follows immediately since $1 - \beta(t;\phi,\delta)$ turns out to be a decreasing function of ϕ in the interval $(-1,0]$. $\blacksquare$

Proof (Theorem 3) The alarm rate function $\lambda_{RL(\phi_1,\phi_2,\delta)}(t)$ increases with t if

$$f(1;\phi_1,\phi_2) < |1-\phi_1|\, f(1;\phi_1,\phi_2) < (1-\phi_1-\phi_2)\, f(1;\phi_1,\phi_2).$$

Taking into account that (ϕ_1, ϕ_2) must belong to set A to guarantee the process stationarity, the double inequality $1 < |1-\phi_1| < 1-\phi_1-\phi_2$ holds if

$$(\phi_1,\phi_2) \in \{(\phi_1',\phi_2') \in A : (\phi_1' < 0 \text{ or } \phi_1' > 2), (\phi_2' < 0 \text{ and } 2\phi_1' + \phi_2' < 2)\}$$
$$= \{(\phi_1',\phi_2') \in A : \phi_1' < 0, \phi_2' < 0\} = A \cap B$$

The inequalities are reversed for the decreasing behaviour and the result also follows by adding the stationarity condition:

$$(\phi_1,\phi_2) \in \{(\phi_1',\phi_2') \in A : (\phi_1' > 0 \text{ and } \phi_1' < 2), (\phi_2' > 0 \text{ or } 2\phi_1' + \phi_2' > 2)\}$$
$$= \{(\phi_1',\phi_2') \in A : \phi_1' > 0, (\phi_2' > 0 \text{ or } 2\phi_1' + \phi_2' > 2)\} = A \cap C. \quad \blacksquare$$

Proof (Theorem 4) First, note that in the stationarity region A the two following conditions hold: $1-\phi_1-\phi_2 > 0$ and $1-\phi_2+\phi_1 > 0$. As the sign of $\frac{\partial f(1;\phi_1,\phi_2)}{\partial \phi_1}$ equals the one of $\phi_1[(1-\phi_2)^2 - \phi_1^2]$, $\frac{\partial f(1;\phi_1,\phi_2)}{\partial \phi_1}$ is nonpositive if $\phi_1 \le 0$. Moreover, since

$$\frac{\partial f(2;\phi_1,\phi_2)}{\partial \phi_1} = -f(1;\phi_1,\phi_2) + (1-\phi_1)\frac{\partial f(1;\phi_1,\phi_2)}{\partial \phi_1}, \tag{37}$$

and

$$\frac{\partial f(3;\phi_1,\phi_2)}{\partial \phi_1} = -f(1;\phi_1,\phi_2) + (1-\phi_1-\phi_2)\frac{\partial f(1;\phi_1,\phi_2)}{\partial \phi_1}, \tag{38}$$

a sufficient condition for $\frac{\partial f(t;\phi_1,\phi_2)}{\partial \phi_1} \le 0, t = 1,2,3,\ldots$ is to have the pair of parameters (ϕ_1,ϕ_2) belonging to the set

$$\{(\phi_1,\phi_2) \in A : \phi_1 \le 0\} = \{(\phi_1,\phi_2) : -1 < \phi_2 < 1, \phi_2 - 1 < \phi_1 \le 0\}.$$

In this last set, the alarm rate function decreases with ϕ_1 since $\lambda_{RL(\phi_1,\phi_2,\delta)}(t)$ is an increasing function of $f(t;\phi_1,\phi_2)$. $\blacksquare$

CUSUM and EWMA residual schemes

This second part of the appendix includes the definition of the substochastic matrices the Markov approximation for the *RLs* depends upon for upper one-sided *CUSUM* and *EWMA* schemes. It also comprises the proofs of Theorems 5 and 6 concerning stochastic properties of the *RLs* of these schemes in terms of the autoregressive parameter ϕ.

The Markov approach is used to compute approximations to *RL* related measures. This approach starts with the division of the decision interval $[LCL, UCL)$ in $(x+1)$ sub-intervals, $[e_i, e_{i+1})$, with equal length $\Delta = (UCL - LCL)/(x+1)$. These sub-intervals are then associated with the $(x+1)$ transient states of an absorbing Markov chain, say $\{S_t(\phi,\delta;x), t = 0,1,\ldots\}$, with discrete state space $\{0,1,\ldots,x+1\}$ and absorbing state $(x+1)$.

The transitions between the transient states of $\{S_t(\delta, \phi; x), t = 0, 1, \dots\}$ are governed by the substochastic matrices $\mathbf{Q}(\phi, \delta; x)$, for the first transition, and $\mathbf{Q}(\phi, \delta(1-\phi); x)$, for the subsequent transitions.

The matrices $\mathbf{Q}$ have the following generic form

$$\mathbf{Q}(\phi, \delta'; x) = [q_{i\ j}(\phi, \delta'; x)]_{i,j=0}^{x} = [a_{i\ j}(\phi, \delta'; x) - a_{i\ j-1}(\phi, \delta'; x)]_{i,j=0}^{x} \quad (39)$$

where $a_{i\ j}(\phi, \delta'; x)$ denotes the left partial sum

$$\sum_{l=0}^{j} q_{i\ l}(\phi, \delta'; x), \text{ for } i, j = 0, \dots, x$$

and $a_{i\ -1}(\phi, \delta'; x) = 0$ for $i = 0, \dots, x$.

The particular form of the substochastic matrix $\mathbf{Q}(\phi, \delta')$ for the *CUSUM* and *EWMA* schemes is defined in terms of

$$a_{i\ j}(\phi, \delta'; x) = \Phi\left(k + \frac{h\,[(j+1) - (i+1/2)]}{x+1} - \frac{\delta'}{\sqrt{1-\phi^2}}\right) \quad (40)$$

and

$$a_{i\ j}(\phi, \delta'; x) = \Phi\left(\frac{\gamma\,[(j+1) - (1-\lambda)(i+1/2)]}{(x+1)\sqrt{\lambda(2-\lambda)}} - \frac{\delta'}{\sqrt{1-\phi^2}}\right) \quad (41)$$

for $i, j = 0, \dots, x$, respectively.

Proof (Theorems 5-6) Let $\{S_t(\phi, \delta; x), t = 0, 1, \dots\}$ be the absorbing Markov chain used to obtain the approximation of the run length $RL^{\lfloor \beta(x+1) \rfloor}(\phi, \delta; x)$. Analogously,

$$\{U_t(\phi, \delta(1-\phi); x), t = 1, 2, \dots\}$$

denotes the absorbing Markov chain whose absorption time is represented by $T^{\underline{\alpha}(\phi,\delta;x)}(\phi, \delta(1-\phi); x)$.

The initial state of this last Markov chain, $U_1(\phi, \delta; x)$, has probability vector

$$(\underline{\alpha}^{\top}(\phi, \delta; x), 1 - \underline{\alpha}^{\top}(\phi, \delta)\, \underline{1})$$

where, as mentioned earlier,

$$\underline{\alpha}^{\top}(\phi, \delta; x) = \underline{\mathbf{e}}_{\lfloor \beta(x+1) \rfloor}^{\top}\, \mathbf{Q}(\phi, \delta; x).$$

In addition, the state transitions are governed by the stochastic matrix

$$\begin{aligned} \mathbf{P}(\phi, \delta(1-\phi); x) &= [p_{i\ j}(\phi, \delta(1-\phi); x)]_{i,j=0}^{x+1} \\ &= \begin{bmatrix} \mathbf{Q}(\phi, \delta(1-\phi); x) & [\mathbf{I} - \mathbf{Q}(\phi, \delta(1-\phi); x)]\, \underline{1} \\ \underline{\mathbf{0}}^{\top} & 1 \end{bmatrix} \end{aligned} \quad (42)$$

where $\underline{0}$ and $\mathbf{I}$ represent a $(x+1)$ vector of zeroes and the identity matrix with rank $x+1$.

Given that $a_{i\,j}$ increases with $\phi \in (-1,0]$, the initial state $U_1(\phi,\delta(1-\phi);x)$ stochastically decreases with ϕ:

$$U_1(\phi,\delta(1-\phi);x) \geq_{st} U_1(\phi',\delta(1-\phi');x), \text{ for } -1<\phi\leq\phi'\leq 0. \tag{43}$$

Another consequence of the increasing behaviour of $a_{i\,j}(\phi,\delta)$ in terms of ϕ (see Equations (40) and (41)) is:

$$\sum_{l=0}^{j} p_{i\,l}(\phi,\delta(1-\phi);x) \leq \sum_{l=0}^{j} p_{i\,l}(\phi',\delta(1-\phi');x), \tag{44}$$

for $-1<\phi\leq\phi'\leq 0$; i.e., the probability transition matrices $\mathbf{P}(\phi,\delta(1-\phi);x)$ and $\mathbf{P}(\phi',\delta(1-\phi');x)$ can be ordered in the usual sense (Kulkarni (1995, pp.148-149):

$$\mathbf{P}(\phi,\delta(1-\phi);x) \geq_{st} \mathbf{P}(\phi',\delta(1-\phi');x), -1<\phi\leq\phi'\leq 0. \tag{45}$$

Therefore, according to Theorem 3.31 of Kulkarni (1995, p. 149), the Markov chains $\{U_t(\phi,\delta(1-\phi);x), t=1,2,\ldots\}$ and $\{U_t(\phi',\delta(1-\phi');x), t=1,2,\ldots\}$ can be also ordered in the usual sense:

$$\{U_t(\phi,\delta(1-\phi);x), t=1,2,\ldots\} \geq_{st} \{U_t(\phi',\delta(1-\phi');x), t=1,2,\ldots\}, \tag{46}$$

for $-1<\phi\leq\phi'\leq 0$.

The stochastic order relation in the usual sense in (46) is equivalent to

$$E\left[g(\{U_t(\phi,\delta(1-\phi);x), t=1,2,\ldots\})\right] \geq \\ \geq E\left[g(\{U_t(\phi',\delta(1-\phi');x), t=1,2,\ldots\})\right] \tag{47}$$

for every increasing functional g for which the expectations in (47) exist. Recall that a functional g is called increasing if

$$g(\{b_t, t=1,2,\ldots\}) \leq g(\{b'_t, t=1,2,\ldots\})$$

whenever $b_t \leq b'_t, t=1,2,\ldots$ (see Shaked and Shanthikumar (1994, p. 124)).

Finally, note that

$$P[T^{\underline{\alpha}(\phi,\delta;x)}(\phi,\delta(1-\phi);x) > t-1] = \\ = 1 - E\left(\min\left\{1, \sum_{m=1}^{t} I_{\{x+1\}}[U_m(\phi,\delta(1-\phi');x)]\right\}\right) \tag{48}$$

where $I_{\{x+1\}}(a)$ is the indicator function of the set $\{x+1\}$; i.e., $I_{\{x+1\}}(a)$ equals one if $a=x+1$ and is equal to zero otherwise. Since

$$g(b_1,\ldots,b_t) = \min\left\{1, \sum_{m=1}^{t} I_{\{x+1\}}(b_m)\right\}$$

is an increasing functional in $\{0, 1, \ldots, x+1\}$, applying (47), we get

$$P[T^{\underline{\alpha}(\phi,\delta;x)}(\phi, \delta(1-\phi); x) > t-1] \leq P[T^{\underline{\alpha}(\phi',\delta;x)}(\phi', \delta(1-\phi'); x) > t-1], \quad (49)$$

for all $t = 1, 2, \ldots$ and $-1 < \phi \leq \phi' \leq 0$. That is,

$$RL^{\lfloor\beta(x+1)\rfloor}(\phi, \delta; x) \leq_{st} RL^{\lfloor\beta(x+1)\rfloor}(\phi', \delta; x), -1 < \phi \leq \phi' \leq 0. \quad (50)$$

This concludes the proof. ■

Acknowledgements: This research was supported in part by FCT (Fundação para a Ciência e a Tecnologia), grant SFRH/BSAB/251/01 and the projects POSI/34826/CPS/2000 and POSI/42069/CPS/2001.

References

1. Alwan, L.C. (1992). Effects of autocorrelation on control chart performance. *Communications in Statistics — Theory and Methods* **21**, 1025–1049.
2. Alwan, L.C. and Roberts, H.V. (1988). Time-series modeling for statistical process control. *Journal of Business and Economic Statistics* **6**, 87–95.
3. Box, G.E.P., Jenkins, G.M. and Reinsel, G.C. (1994). *Time Series Analysis: Forecasting and Control.* Prentice-Hall, Englewood Cliffs, New Jersey.
4. Brook, D. and Evans, D.A. (1972). An approach to the probability distribution of CUSUM run length. *Biometrika* **59**, 539–549.
5. Harris, T.J. and Ross, W.H. (1991). Statistical process control procedures for correlated observations. *The Canadian Journal of Chemical Engineering* **69**, 48–57.
6. Harrison, P. J. and Davies, O.L. (1964). The use of cumulative (CUSUM) techniques for the control of routine forecasts of product demand. *Operations Research* **12**, 325–333.
7. Johnson, R.A. and Bagshaw, M. (1974). The effect of serial correlation on the performance of CUSUM test. *Technometrics* **16**, 103–112.
8. Knoth, S. and Schmid, W. (2001). Control charts for time series: a review. In *Proceedings of the VIIth International Workshop on Intelligent Statistical Quality Control* (Session: Statistical Product & Process Control II), University of Waterloo, Canada, September 5–7, 2001.
9. Kramer, H. and Schmid, W. (2000). The influence of parameter estimation on the ARL of Shewhart type charts for time series. *Statistical Papers* **41**, 173–196.
10. Kulkarni, V.G. (1995). *Modeling and Analysis of Stochastic Systems.* Chapman and Hall, London.
11. Lu, C.-W. and Reynolds, M.R., Jr. (1999). Control charts for monitoring the mean and variance of autocorrelated processes. *Journal of Quality Technology* **31**, 259–274.
12. Maragah, H.D. and Woodall, W.H. (1992). The effect of autocorrelation on the retrospective $\bar{X}$-chart. *Journal of Statistical Simulation and Computation* **40**, 29-42.
13. Margavio, T.M., Conerly, M.D., Woodall, W.H. and Drake, L.G. (1995). Alarm rates for quality control charts. *Statistics & Probability Letters* **24**, 219-224.

14. Morais, M.C. and Pacheco, A. (2001). Some stochastic properties of upper one-sided $\bar{X}$ and EWMA charts for μ in the presence of shifts in σ. *Sequential Analysis* **20**(1/2), 1–12.
15. Neuts, M.F. (1981). *Matrix-Geometric Solutions in Stochastic Models.* Johns Hopkins University Press, Baltimore, MD.
16. Runger, G.C., Willemain, T.R. and Prabhu, S. (1995). Average run lengths for CUSUM control charts applied to residuals. *Communications in Statistics — Theory and Methods* **24**, 273–282.
17. Schmid, W. (1995). On the run length of a Shewhart chart for correlated data. *Statistical Papers* **36**, 111–130.
18. Schmid, W. (1997a). On EWMA charts for time series. In *Frontiers of Statistical Quality Control*, Vol. 5, (Lenz, H.J., Wilrich, P.-Th., Eds.), 115-137. Physica-Verlag, Heidelberg.
19. Schmid, W. (1997b). CUSUM control schemes for Gaussian processes. *Statistical Papers* **38**, 191–217.
20. Schmid, W. and Schöne, A. (1997). Some properties of the EWMA control chart in the presence of autocorrelation. *The Annals of Statistics* **25**, 1277–1283.
21. Schöne, A., Schmid, W. and Knoth, S. (1999). On the run length of the EWMA scheme: a monotonicity result for normal variables. *Journal of Statistical Planning and Inference* **79**, 289–297.
22. Shaked, M. and Shanthikumar, J.G. (1994). *Stochastic Orders and Their Applications.* Academic Press, Inc., San Diego.
23. VanBrackle III, L.N. and Reynolds, Jr., M.R. (1997). EWMA and CUSUM control charts in the presence of correlation. *Communications in Statistics – Simulation and Computation* **26**, 979–1008.
24. Vasilopoulos, A.V. and Stamboulis, A.P. (1978). Modification of control chart limits in the presence of data correlation. *Journal of Quality Technology* **10**, 20–30.
25. Wardell, D.G., Moskowitz, H. and Plante, R.D. (1994). Run-length distributions of special-cause control charts for correlated processes. *Technometrics* **36**, 3–27.
26. Wolfram, S. (1996). *The Mathematica Book* — 3rd edition. Wolfram Media, Inc.
27. Zhang, N.F. (1997). Detection capability of residual control chart for stationary process data. *Journal of Applied Statistics* **24**, 475–492.
28. Zhang, N.F. (1998). A statistical control chart for stationary process data. *Technometrics*, **40**, 24–38.

Part 3

Off-line Control

A New Statistical Tool for Finding Causal Conditions for Small Data Sets and Its Software

Kazunori Yamaguchi[1], Yasunari Kono[2] and Chooichiro Asano[3]

[1] Rikkyo University, Department of Industrial Relations, Tokyo, Japan
[2] Rikkyo University, Graduate School of Social Relations, Tokyo, Japan
[3] Soka University, Faculty of Engineering, Tokyo, Japan

Abstract. We propose a method that finds target groups such that the response probability of a dichotomous dependent variable is larger than a pre-specified target value. This is enabled by a selection of independent variables with high order interactions. The method is effective to extract interesting patterns from datasets. We developed software for this method. The software incorporates various types of tools like data management, model building, visualization, and so on.

1 Introduction

The computer technology progress allows collecting large amounts of data easily. Although various methods have been introduced for analyzing such data, it is not easy to construct the method to detect useful information from huge databases. Data mining is the search for valuable information in large volumes of data, which is an interdisciplinary field bringing together techniques from machine learning, pattern recognition, statistics, database theory, and visualization (Cavena et al. 1997). Data mining is mainly used for 'explanatory' analysis (Iwasaki 1999). Several methods of data mining exist such as classification, regression, clustering, dependency modelling, change and deviation detection, decision tree, discovery of association rule and neural networks (Fayyad et al. 1996). Weiss and Indurkhya (1998) have divided some generic types of data mining problems into two general categories:

(a) prediction (classification, regression, time series) and

(b) knowledge discovery (deviation detection, database segmentation, clustering, association rules, summarization, visualization, text mining).

Confirmatory techniques construct a model and explain a specific part of the data by other items (Shiota 1997). Data mining by classification is mainly analyzed by decision tree such as C4.5 (Quinlan 1998) or CART (Breiman et al. 1984). Explanatory techniques like clustering,, on the other hand, aim at the simple representation of the contents or the structure of the data to understand them when there is no dependent variable.

In this paper, we propose a model to detect subgroups in a large data set according to the response probabilities. The method finds target groups such that response probabilities of a dichotomous dependent variable are larger than a pre-specified target value. This means that such subgroups give us a hint to causal condition for the dependent variable. This algorithm selects independent variables with a high order interaction. The method is quite effective to extract some interesting patterns or groups from huge datasets, and it can be used for extracting sufficient conditions such that the dependent variable has positive response with high probability. The method also can be considered as a regression-type version of association rule which has been proposed by Agrawal and Srikant (1994). We also introduce software which offers many facilities for the proposed method.

2 Proposed Method

2.1 Basic idea

A dichotomous multivariate data matrix is assumed to be given. Let $X_1, X_2, \ldots, X_p$ be p independent variables, and Y be a dependent binary variable. The goal of the analysis is to explore equations on the condition $D(X_1, X_2, \ldots, X_p)$ that satisfied probability of a positive response $P(Y = 1|D) > c$, where c is a positive value. The method, in other words, finds the groups such that response probability of a dichotomous dependent variable is larger than the target value c.

The algorithm of our method consists of the following five steps;

Step 0: Set the target value c.

Step 1: Set the percent cutoff value.

Step 2: Get the minimized equation by the Quine-McClusky algorithm.

Step 3: Set the confidence intervals of response probabilities for each group.

Step 4: Check the results (Back to Step 1 if the result leaves still room for improvement).

In Step 0, the percentage of the target group is set for the purpose of the research. It is possible to change the target value as the need arises. In Step 1, the analysts make a contingency table from the data. The first percent cutoff value is generally set to the same percentage as the target value. In Step 2, the target groups are nominated according to minimized equations detected by the Quine-McClusky algorithm (see the next section). In Step 3, analysts decide the level of confidence intervals, and find the lower limit of the confidence intervals for the nominated groups. In Step 4, it is checked whether the target vale is less than all of lower limit values of confidence intervals or not. Even if it is less than all of the values, it is desirable to repeat the above steps with smaller cutoff values, because lager groups are desirable.

2.2 Quine-McClusky algorithm

The Quine-McClusky algorithm (Quine 1952) is a tool for reducing a logical equation to its simplest reduced form. It has the advantage of being easily implemented on the computer and it is possible to work on many variables.

This section will describe the Quine-McClusky algorithm to minimize Boolean expressions used for the proposed method. This algorithm consists of the following four steps:

1. Given binary representations (which is called as "term") of conditions for the positive response of the dependent variable, classify each term into a group where each term in a group contains the same number of one's. Arrange these groups in ascending order with respect to the number of one's in each group.
2. Compare each term of a group with every term of the next group. If two terms are logically adjacent combine the two terms to make a new term. The new term has the same bit values as the original two terms except the one bit.
3. Mark each term that is used in a combination.
4. Repeat the process of combining and marking with the new combined terms until no more combinations are possible. The remaining unmarked terms cannot be reduced further and are prime implicants.

The original Quine-McClusky algorithm includes the essential prime implicants selecting process, but we do not use such process here, because another selecting process is designed for our method (see step 3 in the above section). For the details of this algorithm, see Miller (1965).

2.3 Example

Let X_1, X_2, X_3 be three independent binary variables. The data set is summarized in Table 1. The sample size is 3633. We try to find conditions that the probability of a positive response of Y should be larger than 70 percents.

Table 1. Example Data With Three Independent Variables

Condition			Positive Response	Cases
X_1	X_2	X_3		
0	0	0	195	1965
0	0	1	120	150
0	1	0	225	872
0	1	1	69	101
1	0	0	0	34
1	0	1	43	65
1	1	0	387	501
1	1	1	4	5
				3693

Minimized equations of the 60, 70 and 80 percent of the response probability are represented as follows (a capital letter represents positive response and a small letter shows negative):

$$60\%: Y = X_1\, X_2 + X_3 \quad (1)$$

$$70\%: Y = x_1 x_2\, X_3 + X_1\, X_2 \quad (2)$$

$$80\%: Y = x_1 x_2\, X_3 + X_1\, X_2\, X_3 \quad (3)$$

The patterns of the groups of dependent variables are found according to a response probability of dependent variables. When it is set to 60 percent, Equation (1) shows the selecting of individuals who respond positively to X_3 or to both X_1 and X_2. Quine-McClusky algorithm generally selects larger groups of positive respondent probability than the setting percentages. It represents 77 (74) percent in case of setting 60 percent that the individuals who respond positively to $X_1\, X_2$ (X_3) number 391 of 506 (236 of 321).

The following example shows four steps to find groups with the response probability more than 70 percent by the proposed method.

Step 0: Set the target value to 70 percent.

Step 1: Set the percent cutoff to 70 percent.

Step 2: Get the minimized equation and nominate the groups.

Step 3: Set the confidence intervals for response probabilities of the above two groups.

Step 4: Check the results

The minimized equation for 70 percent cutoff is $x_1\, x_2\, X_3 + X_1 X_2$ as shown Equation (2). This equation nominates two groups, $x_1\, x_2\, X_3$ and $X_1 X_2$, for target groups. In this example, the 80 percent confidence intervals for response probabilities are used. The result is as follows:

	Estimated response probability	Lower limit
$x_1\, x_2\, X_3$	80.0%	75.8%
$X_1\, X_2$	77.3%	74.9%

The both lower limit of the 80 percent confidence intervals for $x_1\, x_2\, X_3$ and $X_1\, X_2$ are larger than 70 percent. The results accept these two groups of Equation (2). Repeat the above steps with the 60 percent cutoff value to check whether larger groups than these two groups that meet the requirements exist or not.

Step 1': Set the percent cutoff to 60 percent.

Step 2': Get the minimized equation.

Step 3': Set the confidence intervals for response probabilities of the above two groups.

Step 4': Check the results

The minimized equation for 60 percent cutoff is $X_1\, X_2 + X_3$ as shown Equation (1). This equation nominates two groups, $X_1\, X_2$ and X_3, for target groups. In this example, the 80 percent confidence intervals for response probabilities are used. The result is as follows:

	Estimated response probability	Lower limit
$X_1\, X_2$	77.3%	74.9%
X_3	73.5%	70.4%

The both lower limits of confidence intervals with the 80 percent confidence for the groups $X_1\, X_2$ and X_3 are larger than 70 percent. The results accept these two groups of Equation (1). These two groups of Equation (1), $X_1\, X_2$ and X_3, also accept the target value of the 70 percent. The iteration is stopped at this point because the output does not change until 26 percent cutoff.

As the result stated above, this method finds $X_1\, X_2$ and X_3 as the groups that meet the requirements.

3 Software

We developed software for the analysis introduced in the Section 2. This software provides the following facilities:

Data management : The proposed method requires a binary data matrix. The user can transform to a dichotomous variables from several types of variables, continuous, ordered-multilevel, or nominal.

Modelling and estimation : Here, the user can specify the dependent and independent variables, target values, levels of confidence and so on.

Visualization : Several visualization tools are also provided, for original data, transformed variables, groups and so on.

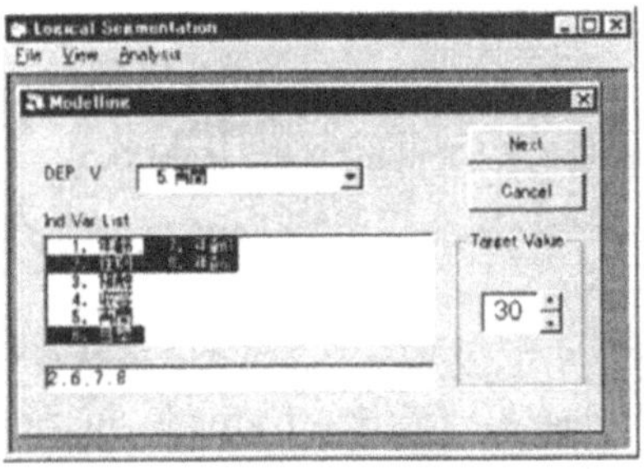

Fig. 1. A dailog box for modelling

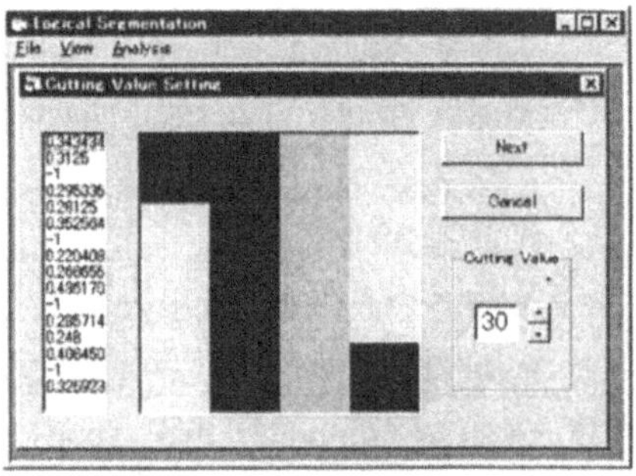

Fig. 2. An output for response variable pattern

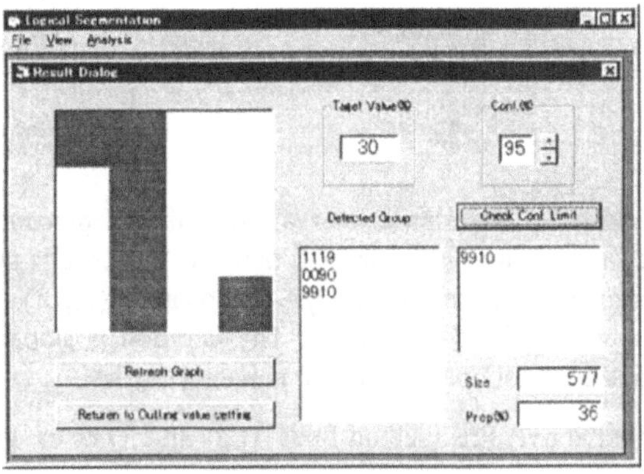

Fig. 3. An output of the result

This software runs under the Microsoft Windows. The programming languages used are C++, Visual Basic and Fortran. Figures 1-3 show examples of user interfaces and outputs of a prototype version of the software.

4 Concluding Remarks

We presented an algorithm for discovering subgroups in a dataset according to the response probabilities. This method can be considered as a regression-type version of association rule which has been proposed by Agrawal and Srikant (1994). Agrawal and Srikant (1994)'s algorithm, Apriori, is useful for a huge database. The proposed method uses the Quine-McClusky algorithm, which is inefficient for huge datasets. When the method is applied to such dataset, we have to do some modifications for our method to reduce calculation tasks. One of the effective methods is to make a restriction of the number of variables which are used for constructing subgroups. If we specify the maximum number of variables, we apply the proposed method to the contingency tables with the specified number of variables. However there are so many sub-tables which have to be analyzed. We have to find an effective criterion to get useful sub-tables for our purpose.

Acknowledgements

The authors would like to thank the Editor and referee for their helpful comments. This research was supported, in part, by Grant-in-Aid for Scientific Research, Contract Number 12558022.

References

1. Agrawal, R. and Srikant, R. (1994) Fast Algorithms for Mining Association Rule, Proc. the 20th VLDB Conference Santiago, Chile.
2. Breiman, L., Friedman, J., Olshen, R. S., Stone, C. (1984) Classification and Regression Trees. Wadsworth International Group. Belmont, CA
3. Cavena, P., Hadjinian P., Stadler, R., Verhees, J., Zanasi (1997) Discovering Data Mining, From Concept to Implementation. Prentice-Hall, Inc. NJ
4. Fayyad U., Piatestsky-Shapiro, G., Smyth, P. (1996) Advances in Knowledge Discovery and Data Mining. AAAI/MIT Press. Menlo Park, CA
5. Iwasaki, M. (1999) Data Mining and Knowledge Discovery, From a Statistical Viewpoint. The Japanese Journal of Behaviormetrics, 26, 46-58
6. Miller, R.E. (1965) Switching Theory, Wiley, New York
7. Quine, W. V. (1952) The Problem of Simplifying Truth Functions. American Mathematics Monthly, October, 59, 521-531
8. Quinlan, J. (1998) C4.5: Programs for Machine Learning. Morgan Kaufmann Publishers, Inc. San Francisco, CA
9. Shiota, C. (1997) Data Mining - Techniques and Applications. Bulletin of The Computational Statistics of Japan, 10, 127-144
10. Weiss, S. M. , Indurkhya, N. (1998) Predictive DataMining, A Practical Guide. Morgan Kaufmann Publishers, Inc. San Francisco, CA

What Has Data Mining Got to Do with Statistical Process Monitoring? A Case Study

Ross Sparks

CSIRO Mathematical and Information Sciences,
Locked Bag 17, North Ryde, NSW 2113, Australia

Abstract. A simple usable approach to monitoring complex multivariate systems is proposed. Simple charting and data mining principles are used to easily discover the necessary information for targeted decisions. It also limits the number of charts that deserve attention for efficient process management.

1 Introduction

Process monitoring applications are inherently multivariate and typically complex. Unfortunately in service industry settings, the literature is not helpful in that it focuses on traditional statistical methods, such as Hotellings T^2 statistic, PCA, and PLS, which may be useful in a manufacturing setting, but are difficult to apply in other situations. Managers in industry do not generally understand such methods, and therefore their use as a source of information for decision-making is compromised.

This paper deals with an application in a company involved with the construction and maintenance of railway networks. The recent safety record of rail passengers carriers is poor (as is seen from the increase in tragic rail incidents in Britain, Australia and elsewhere). The general public is demanding greater safety accountability from passenger transport companies. Monitoring of both minor and serious rail safety incidents has typically been poor.

This paper will combine traditional data mining, such as slicing and dicing of the data, with traditional monitoring methods to help signal four W's:

- **What** incidents are increasing with time?
- **Who** was mostly responsible for the increases in incidents (e.g., the Civil Group, Signals Group, etc)?
- **Where** are these incidents increasing? (e.g., Which network, Sector within a network, Location within a Sector?)
- **When** did the increase start and finish.

The technology was kept simple, allowing managers to immediately access information for timely decision-making. The product design integrated usability with the power of traditional monitoring techniques. The reports differentiated incident counts from the risk score of these incidents, thus helping management control both the frequency of incidents and the associated risk of these incidents.

2 The Approach

Summary of the separate steps used during this work is now outlined.

Data familiarisation: Meetings were held with each custodian of four different data sources in the organisation. Each of these data sources were discussed in terms of accuracy, reliability and level of current usage.

Assessment of Management Needs: Meetings were held with the important decision-makers (such as CEO and key General Managers of the company) who use such data to make policy decisions. Key needs that came out of these meetings were:

- The presentation of performance history, eg, improving monthly reports.
- Performance forecasts.
- The systematic assessment of contractual performance for each of their current 12 large maintenance contracts.
- Performance as a good corporate citizen, mindful of the company's public image.
- Linked performance information from different databases.
- Close monitoring of important operational measurements that link to the Corporate Key Performance Indicators (KPI). Quick and easy method of identifying roots causes of any movement in KPI scores.
- Reporting system should be simple (almost everyone in the organisation should be able to use it) and statistically powerful (provide early warning of problems without too many false alarms).
- Flagging when process performance has changed significantly, i.e., flagging when the process requires management attention.
- Developing a Pareto mentality amongst managers.
- Encouraging a culture of management using data and information.
- The ability to facilitate root cause discovery.
- Data management systems and reporting systems that are designed to fulfil the accreditation body standards.

Development of Prototype Reporting System: Initially it was decided to focus on a single database, and the safety incident database was selected. This step designed several different approaches of presenting the data to managers. The charts selected were via a process of constructive feedback and positive criticism. The charts finally selected were Pareto charts and control charts. These charts were run over three separate months as a trial. Manager feedback was recorded and appropriate adjustments to the reporting process was made.

Development of the Data Mining System: A conceptual framework for mining was developed to flag areas that deserve management's attention. A prototype WWW based reporting system was developed using Stat Server that interfaces with the Splus software. This prototype software allowed users to drill down on the four W's (*what, when, where* and *who*).

Final Ownership Building: Decide on a modified reporting process. Although the reporting process has been decided on after a trial of three months, the prototype WWW based reporting system is still at the decision stage. Therefore the WWW reporting system has not been fully developed yet.

3 The Monitoring Methodology

The safety incident data was selected as the starting point. Two control charts are used:

- time between consecutive incidents for less frequent safety incidents, and
- frequency of incidents per week/month for frequent incidents.

Control limits were constructed for these measurements assuming the exponential distribution or Poisson distribution, respectively. The justification for this now follows.

The time between consecutive incidents reported no significant auto-correlation for most contracts in the company for the past three years of data. Two of the twelve contracts reported significant auto-correlation, albeit not very high (lag 1 auto-correlation less than 0.35). The *qq*-plot (Wilk and Gnanadesikan, 1968) for the exponential distribution and the time between consecutive safety incidents within a contract indicated that this distribution fitted the data rather well for all 12 contracts (see Fig. 1 for a contract that that had significant autocorrelated inter-arrival times).

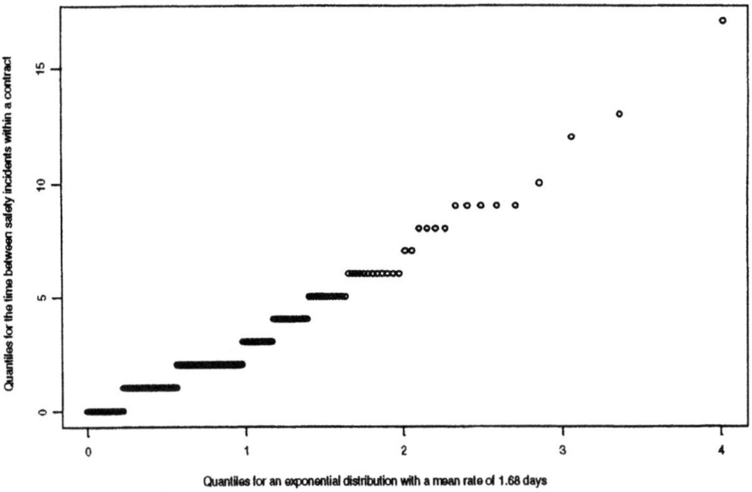

Figure 1: QQ-plot of the time between safety incidents for an exponential distribution

The exponential distribution for the time between consecutive safety incidents, and the Poisson distribution for the frequency of incidents per week or months were assumed good operating distributions for flagging unusual process behaviour. However the mean rate of the distribution was assumed to be able to wander to accommodate the influence of autocorrelation. This wander was estimated by the ewma statistic very similar to Mastrangelo and Montgomery (1991).

3.1 Estimating the Moving Mean Rate and One-Step-Ahead Forecasts

The exponentially weighted moving average is used to give a one-step ahead forecast of the mean rate for the time intervals. For example, let N_t be the frequency of incidents in a random month, and M_t be the number of days in month t, then

$$x_t = N_t / M_t \quad \text{for } t = 1,2,\ldots,24$$

be the rate of incident counts for over the most recent 24 months. Start the exponentially weighted moving average statistic (E) taking its initial value (E_0) equal to

$$E_0 = (x_{-1} + x_{-2} + x_{-3})/3$$

where the minus sign indicates the number of months prior to the start of monitoring. Thereafter, for months $t = 1, 2, 3, \ldots, 24$

$$E_t = \alpha \times x_t + (1-\alpha) \times E_{t-1}$$

where: $0 < \alpha < 1$. The value used is $\alpha = 0.2$.

E_t is an estimate of the moving mean rate, and in the application it is used as the one-step-ahead forecast of the value we expect at time $t+1$, i.e., the one-step-ahead forecast for x_{t+1} is E_t.

3.2 Forecasts for More Structured Safety Incidents

Incidents such as a broken rail are influenced by temperature, that is, a rail track is more likely to break in colder weather than hot weather. Temperature values at all locations were unavailable, therefore the Cosine and Sin functions were used to model monthly frequency of incidents. The Poisson regression models were used to model the mean daily rate μ_t as follows

$$\log_e(\mu_t) = \alpha_0 + \alpha_1 \times \cos(2 \times \prod \times \mathrm{T}/12) + \alpha_2 \times \sin(2 \times \prod \times T/12),$$

where T is 1, 2,...,12 depending on whether the month is January, February, ..., December, respectively. The response being fitted is $x_t = N_t / M_t$.

This model is fitted over the longest set of in-control data available. The fitted mean daily rates for each month are the in-control mean rates. These are then used to establish the control limits assumming a Poisson distribution. Lag variable $x_{t-1} = N_{t-1} / M_{t-1}$ could be included as an explanatory variable to improve one-step-ahead forecasts when temperature variations are locally similar but vary significantly for year to year within a season.

These models provide forecasts for the frequency of monthly broken rails for the year ahead. These forecasts together with their standard errors are used to construct control charts (e.g., see Fig. 2).

3.3 Control Chart for Processes with a Moving Mean

If a monthly incident count differs significantly from its one-step-ahead forecast, then it is assumed to be due to special causes and not due to the natural variation in the process.

To flag when differences are significant, the 95% confidence interval for the one-step-ahead forecast is calculated as follows

Upper limit (U) is found as follows: $\sum_{x=0}^{U} e^{E_{t-1}} (E_{t-1})^x / x! = 0.975$, i.e., the 97.5 percentile for a Poisson distribution with a mean rate of E_{t-1}.

Lower limit (L) is found as follows: $\sum_{x=0}^{L} e^{E_{t-1}} (E_{t-1})^x / x! = 0.025$, i.e., the 2.5 percentile for a Poisson distribution with a mean rate of E_{t-1}.

If either $x_t > U$ or $x_t < U$, then the count x_t is flagged as unusual and should be investigated. This approach of using one-step-ahead forecast errors was advocated by Mastrangelo and Montgomery (1991), but for normally distributed data. This approach can be less efficient at signalling process drifts (see Sparks, 2000), but this is partially compensated by using narrower control limits than usual in the application considered. However, this compensation is likely to increase false alarm signals.

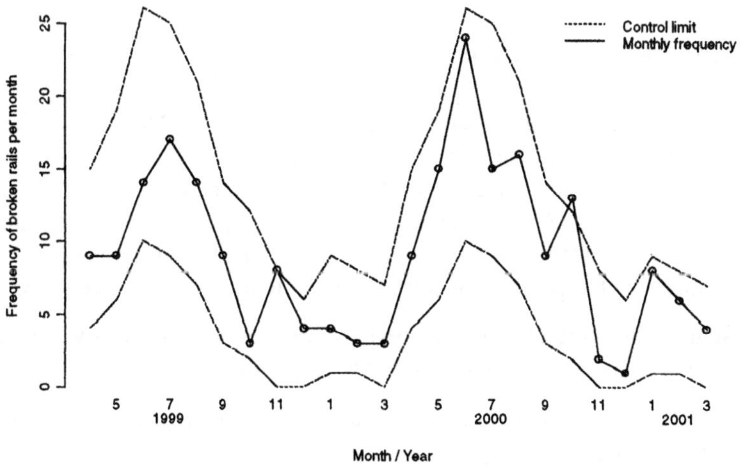

Figure 2: Control chart for the monthly frequency of broken rails

Fig. 2 presents an example of the one-step-ahead forecast of broken rail monthly frequencies with control limits for these forecasts. One signal in the past 24 months flags a significant increase in broken rail incidents in October 2000.
Fig. 3 presents an example of a control chart for the one-step-ahead forecast for the total frequency of incidents of a contract. Note that, although all points are within the control limits, the trend for monthly counts is slightly upwards for the past 17 months. This generally indicates that the trend is not caused by an unusual event, but appears to be part of the process.

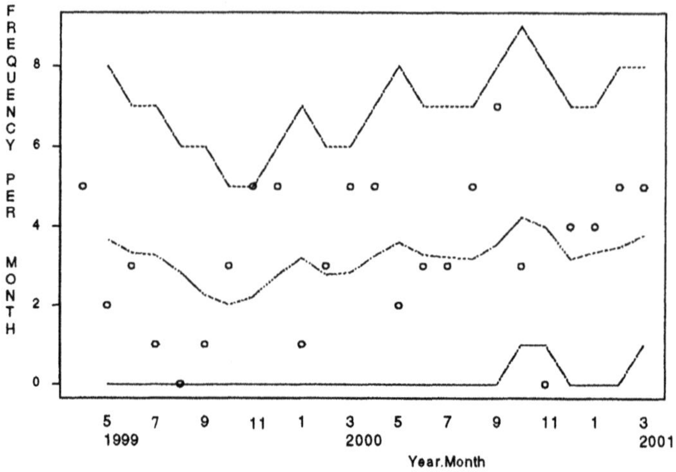

Figure 3: Control chart comparing monthly incident counts to their one-step-ahead forecast for a particular contract

3.4 Pareto Principle and Pareto Charts

The aim is to focus the organisation's attention on the 20% of the Modes, Causes, Reasons or Hazard Codes that cause 80% of the unsafe incidents, and also to identify which contracts, sectors, and locations contribute to the frequency of unsafe incidents. Therefore, Pareto charts (eg, see Fig. 4) are used to indicate major sources of incidents. Frequency bars on the Pareto chart are also coloured to flag significant changes, i.e., the respective bar in the barplot are coloured:

- Orange if the current temporal period indicates no statistically significant change in frequency over the past month or 24 months(see Fig. 4);
- Red if the current temporal period indicates a statistically significant increase in unsafe incidents (see Fig. 4);
- Green if the current temporal period indicates a statistically significant reduction in the frequency of unsafe incidents.

The expected mean rate for colouring the bars in the Pareto plot is taken as the long-term average mean rate, and not the one-step-ahead forecasts as used for control charts.

In this way, decision-makers can identify the few issues that cause the majority of safety incidents, as well as flagging which issues have improved or deteriorated significantly. For example, in Fig. 4, the most frequent mode of incidents over the last 24 months is mode G (Potential Strike).

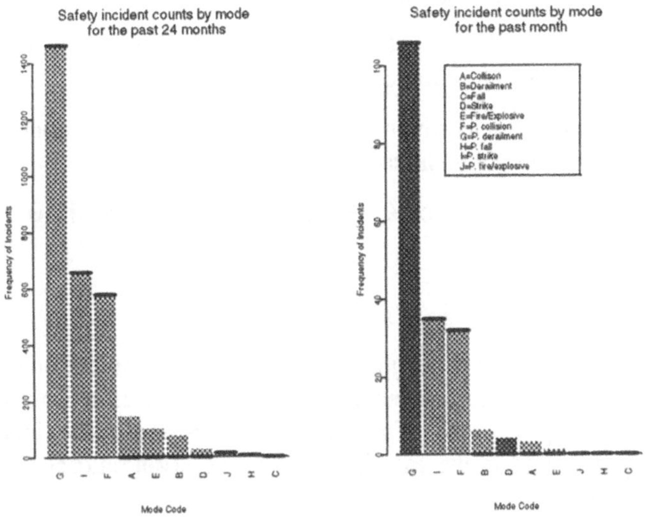

Figure 4: Pareto chart indicating the most frequent mode for safety incidents.

Past 24 months (left-hand side barplot): Potential strike accounts for 47% of incidents in the past 24 months and since this bar is coloured red, the mean rate of these incidents has increased significantly. The second most frequent incident is potential derailment and this accounts for 22% of the total incidents. The third is potential collisions, which account for 12% of incidents.

Past month (right-hand side barplot): There were 2 potential strikes, 2 potential derailment and 1 fire/explosion incident(s). None of these are significantly higher or lower than expected from their mean rates (estimated from past data).

Causes of incidents, reasons for incidents occurring, and hazard code frequencies are displayed identically to Fig. 4 (e.g., see Fig. 5) to indicate which causes, reasons or hazards most contribute to unsafe incidents, and also which of these have increased significantly.

These all answer the question of ***what*** breached the safety codes. For example, if in addition to Fig. 4 we find that the most frequent reason for incidents is unlawfully thrown, then it is likely that potential strikes are caused by something being unlawfully thrown.

A series of Pareto charts similar to Fig. 4 can be used to display, which:

- contracts within the organisation contribute to significant breaches of safety codes,

- sectors within a contract contribute to significant breaches of safety codes within the contract,
- line codes within a sector contribute to significant breaches of safety codes within the sector, or
- locations (station) within a sector contributes to significant breaches of safety codes within the sector.

This helps identify ***where*** breaches in safety codes have increased significantly.

3.5 Data Mining

The data mining principle is designed to make it easy to discover major causes of incidents or the significant contributors to significant changes in incidents reported at the higher aggregated levels. For example, we notice that safety incidents have significantly increased in the past 12months at the organisational level, and we now wish to know

- ***what*** contributed to this increase,
- ***where*** in the system did these significant changes occur,
- ***when*** in the past 12 months did this change occur, and
- ***who*** was responsible for these changes?

We will now demonstrate how to establish this desirable knowledge. The reporting system offers many drill down options (there are well over 5000 control charts that are possible for all levels of management). We certainly do not want to look at all of them. The hierarchical nature of the charts helpfully divert managers' attention to areas of greater importance, and thus reducing the number of false alarm signals they need to devote attention to. We, therefore, wish to adopt an approach that targets those relatively few groups of charts that deserve management's attention. We start by producing a set of Pareto charts and a control chart of monthly frequency of incidents at the organisational level. Rough examples of some of these charts are reported in the first row of Fig. 5.

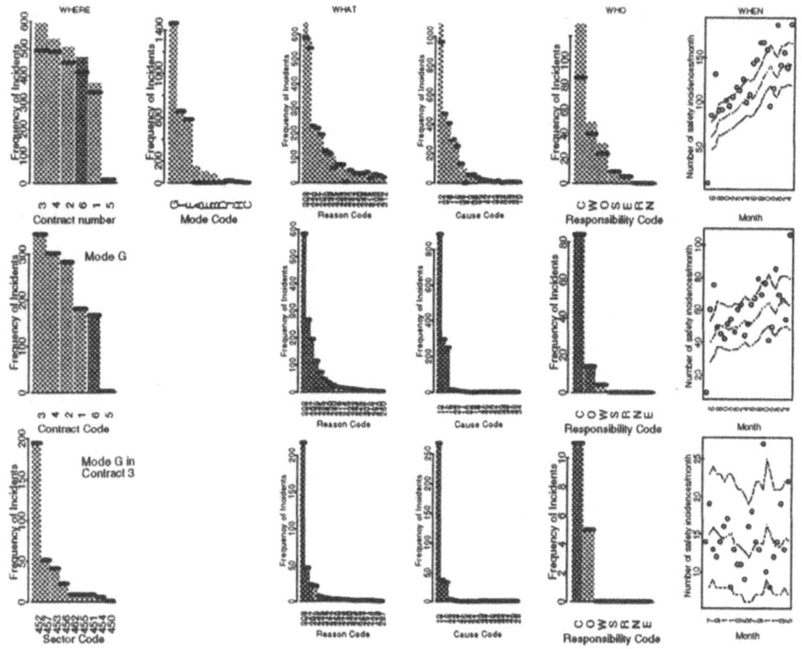

Figure 5: Illustraing the drill-down on Mode G safety incidents

The first in this row is a Pareto chart indicating how the organisation's frequency of unsafe incidents is broken down by different contracts over the past two years, i.e., each bar indicates a contract's contribution to the organisation's performance. The second, third and fourth Pareto charts indicates how the organisation's frequency of unsafe incidents over the past two years is split into the different modes, reasons, and causes, respectively. The fifth chart indicates how the organisation's frequency of major unsafe incidents over the past two year is assigned to the responsibility of various groups inside and outside the organisation. The sixth chart is the control chart for the monthly frequency of incidents for recent months. Drilling down to the next level of reports is achieved by clicking on the bar in one on these charts thus performing a conditional analysis of reports at the next level. Fig. 5's second row illustrates the result of clicking on Mode I in the second Pareto Chart. The charts in row 2 illustrate:

- First, the contract annual frequency for mode I incidents.
- Second, the breakdown of reasons for mode I incidents.
- Third, the breakdown of causes for mode I incidents.
- Fourth, the breakdown of who is responsibly for major mode I incidents.
- Lastly, the control chart for the monthly frequency of mode I incidents.

By clicking on any of the other charts we can produce a set of charts for the next level of reporting by further conditioning. Fig. 5's third row illustrates a further conditioning on Contract 3. Therefore, the charts in the third row present information:

- First, the frequency for mode I incidents broken down by sectors within contract 3.
- Second, the breakdown of reasons for mode I incidents within contract 3.
- Third, the breakdown of causes for mode I incidents within contract 3.
- Fourth, the breakdown of who is responsible for major mode I incidents within contract 3.
- Lastly, the control chart for the monthly frequency of mode I incidents within contract 3.

The user can click on any bars in the third row to drill down to next level of reporting with further conditioning.

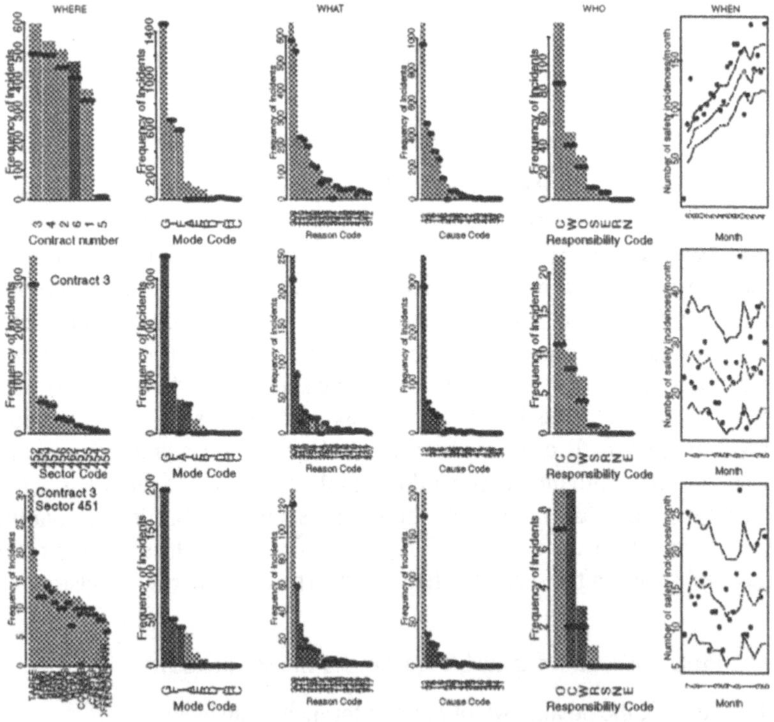

Figure 6: Drilling down on WHERE incidents occur in the organisation

Fig. 6 illustrates a simple drilling down example. The first row is the organisational report of total incidents, where the second row conditions on Contract 3. The third row further conditions on Sector 452 within Contract 3. All these are designed to establish ***where*** significant incidents occur, and therefore where to focus improvement efforts.

It is also possible to drill down from the control charts to produce monthly reports, but this will not be covered in this paper. The data mining methodology represents a rich and simple way for managers to interact with the software, and thus find targeted and relevant information quickly.

4 Conclusion

Senior managers are advised to start by looking first at the organisation performance report (i.e., the charts in the first row of Fig. 5 or 6) first before deciding what, where, when and who to explore further. The slicing and dicing of the data according to statistical significance changes, helps focus on significant issues and avoids looking at many lower level control charts. By interacting with graphs via the WWW, it's easy for managers to find the information they need.

We recognised that Sector managers may not want to start exploring reports at the organisational level, because sectors are nested within one contract. They may choose to start the report by conditioning on their contract. Therefore, the software allows users to enter at the organisational, contract, sector or line code level. This means that users can select where they are going to start mining the data.

This approach takes the pain out of finding relevant information for managing complex systems. It also treats the process as a multivariate system without adding to the complexity of the visualisation process. All the statistical complexity is hidden from the user and they only need to respond to trends in the control charts or changes in colours on the Pareto charts.

The monitoring process is useable and simple to understand, but rich in information.

References

1. Montgomery, D.M. and Mastrangelo, C. (1991). Some statistical process control methods for autocorrelated data, *Journal of Quality Technology*, 23, 179-193.
2. Ross, S.M. (1970). *Applied Probability Models with Optimization Applications.* Holden-Day, San Francisco.
3. Sparks, R.S. (2000). CUSUM charts for AR1 data: are they worth the effort? Australian & New Zealand Journal of Statistics, 42, 25-42.
4. Wilk, M.B. and Gnanadesikan, R. (1968). Probability Plotting Methods for the Analysis of Data. *Biometrika*, 55, 1-17.

Profile Analysis Based on Repeated Measurements – A New Approach to the Analysis of Interaction

Chihiro Hirotsu and Eri Ohta

Faculty of Science and Technology, Meisei University
2-1-1 Hodokubo, Hino-City, Tokyo 191-8506, Japan

Abstract. An overview of the analytical methods for interaction is given with particular interest in the row- and/or column-wise multiple comparison procedures for the case where there is a natural ordering in the rows and/or columns. The methods are useful not only for the analysis of variance model, but also for two-way contingency tables. In particular, a new approach is introduced to classifying rows according to their convex, flat or concave response patterns along the columns. A real example of classifying 147 subjects based on their 24 hours blood pressure measurements is also given.

1 Introduction

The analysis of interaction is a key of many statistical inferences. It is not narrowly restricted to the ANOVA model but more widely interpreted including the contingency table and the nonparametric regression model for analyzing association among several factors (Hirotsu, 1998, 2000). There are a variety of analytical methods for interaction including the case of ordered rows and/or columns as well as the nominal case without such ordering. Among them the row- and/or column-wise multiple comparison procedures are most useful for practical problems in industry and biometry. This suggests an interaction model with only a few parameters whereas the usual overall χ^2 test cannot suggest such a model (Hirotsu, 1983a, b, 1993, 1997, 1998, 2000). The purpose of the present paper is therefore to overview the procedure for the interaction analysis and add a new approach to classifying rows based on their concave, flat or convex response patterns along the columns.

2 Basic idea

Suppose we have data from a two-way layout without replication,

$$y_{ij} = \mu_{ij} + \varepsilon_{ij}, \quad i = 1, \ldots, a; \; j = 1, \ldots, b, \tag{1}$$

where we assume the ε_{ij} to be independently distributed according to a normal distribution $N(0, \sigma^2)$. The null hypothesis for interaction

$$H_0 : \mu_{ij} = \bar{\mu}_{i.} + \bar{\mu}_{.j} - \bar{\bar{\mu}}_{..}$$

can be written in the matrix notation as

$$H_0 : (P_a' \otimes P_b')\,\boldsymbol{\mu} = \mathbf{0},$$

where $\boldsymbol{\mu}$ is the vector of μ_{ij} arranged in alphabetical order, $\otimes$ a Kronecker product, P_k a $k \times k-1$ orthogonal matrix satisfying $P_k'P_k = I_{k-1}$ and $P_kP_k' = I_k - k^{-1}\boldsymbol{j}\boldsymbol{j}'$ with P' transpose of a matrix P, I an identity matrix, $\boldsymbol{j}$ vector of $1's$, and we follow the usual dot bar notation for the sum or the average with respect to a suffix. Then the usual sum of squares for interaction

$$S_{A\times B} = \sum_i \sum_j (y_{ij} - \bar{y}_{i\cdot} - \bar{y}_{\cdot j} + \bar{\bar{y}}_{..})^2$$

can be written in the matrix form as

$$S_{A\times B} = \| (P_a' \otimes P_b')\boldsymbol{y} \|^2,$$

where $\boldsymbol{y}$ is the vector of y_{ij} arranged in alphabetical order.

The row- and/or column-wise multiple comparisons are performed based on the maximal contrast type statistics

$$S_1 = \max_{\boldsymbol{a}'\boldsymbol{j}=0,\boldsymbol{a}'\boldsymbol{a}=1} \| (\boldsymbol{a}' \otimes P_b')\boldsymbol{y} \|^2$$

and/or

$$S_2 = \max_{\boldsymbol{b}'\boldsymbol{j}=0,\boldsymbol{b}'\boldsymbol{b}=1} \| (P_a' \otimes \boldsymbol{b}')\boldsymbol{y} \|^2,$$

where the maximum is taken with respect to $\boldsymbol{a}$ or $\boldsymbol{b}$ subject to the specified conditions. Both S_1 and S_2 follow the same distribution of the largest root of a Wishart matrix and therefore the procedures for the rows and columns are symmetrical (Hirotsu, 1983a).

Now, $\boldsymbol{a}$ is a vector of contrast comparing rows and a practical procedure will be first to compare every two rows by specializing $\boldsymbol{a}'$ to such a type as

$$(0\cdots 0, -1/\sqrt{2}, 0\cdots 0, 1/\sqrt{2}, 0\cdots 0) \tag{2}$$

and then to combine two rows which give the smallest S_1 value into one group. Repeat the same procedure considering the resulting group as a single row with two times weight. Some adjustment of the procedure is given in Hirotsu(1991).

3 Procedure for ordered columns

If only the columns have natural ordering in the two-way layout the procedure cannot be symmetrical for the rows and columns. We therefore mention the method of the row-wise multiple comparisons in §3.1 and that for columns in §3.2.

3.1 Row-wise multiple comparisons based on the maximal cumulative chi-squared statistic

If one is interested in classifying rows according to their up- or downward tendencies along the columns a procedure based on

$$S_r^* = \max_{\boldsymbol{a}'\boldsymbol{j}=0,\boldsymbol{a}'\boldsymbol{a}=1} \| (\boldsymbol{a}' \otimes P_b^{*\prime})\boldsymbol{y} \|^2 \tag{3}$$

has been proposed in Hirotsu (1991), where

$$P_b^{*\prime} = D(D_b' D_b)^{-1} D_b' \tag{4}$$

with a differential matrix

$$D_b' = \begin{bmatrix} -1 & 1\,0 \cdots & 0\,0 \\ 0 & -1\,1 \cdots & 0\,0 \\ \cdots & \cdots\cdots & \cdots \\ 0 & 0\,0 \cdots & -1\,1 \end{bmatrix}_{b-1\times b}$$

and a diagonal matrix D for standardization. The matrix $P_b^{*\prime}$ has been specialized for testing the null hypothesis

$$H_0' : \mu_{i1} - \mu_{i'1} = \mu_{i2} - \mu_{i'2} = \cdots = \mu_{ib} - \mu_{i'b} \tag{5}$$

against the ordered alternative

$$H_1 : \mu_{i1} - \mu_{i'1} \leq \mu_{i2} - \mu_{i'2} \leq \cdots \leq \mu_{ib} - \mu_{i'b}, \quad i \neq i', \tag{6}$$

with at least one inequality strong. The null hypothesis asserts the parallelism of the responses of the rows i and i', and the alternative hypothesis stresses the departure from it in the direction of up- and/or downward tendencies when the second suffix j increases, see Hirotsu(1978) for a detailed explanation. The null distribution of S_r^* is given by the distribution of the largest latent root of the Wishart matrix $W(\sigma^2 P_b^{*\prime} P_b^*, a-1)$. Although the distribution is not easy to handle there is a very good chi-squared approximation for it since the largest root of $P_b^{*\prime} P_b^*$ is dominating the other roots (Hirotsu, 1991). A statistic for cancelling out the variance σ^2 is also proposed in the paper, which has inverse characteristic to S_r^*, namely, less affected by the ordered effects.

We applied the procedure for cholesterol measurements measured once every month for 6 months at 23 subjects. The data are therefore obtained as the 23×6 two-way table given in Table 1. The procedure nicely classified those 23 subjects into three classes(I_1, I_2, I_3) according to their downward(I_1), flat(I_2) and upward(I_3) tendencies of the time profiles of the cholesterol measurements, which are shown in Fig. 1.

Table 1. Total cholesterol amounts

		Period					
Treatment	Subject	1	2	3	4	5	6
	1	317	280	275	270	274	266
	2	186	189	190	135	197	205
	3	377	395	368	334	338	334
	4	229	258	282	272	264	265
	5	276	310	306	309	300	264
Drug	6	272	250	250	255	228	250
	7	219	210	236	239	242	221
	8	260	245	264	268	317	314
	9	284	256	241	242	243	241
	10	365	304	294	287	311	302
	11	298	321	341	342	357	335
	12	274	245	262	263	235	246
	13	232	205	244	197	218	233
	14	367	354	358	333	338	355
	15	253	256	247	228	237	235
	16	230	218	245	215	230	207
	17	190	188	212	201	169	179
Placebo	18	290	263	291	312	299	279
	19	337	337	383	318	361	341
	20	283	279	277	264	269	271
	21	325	257	288	326	293	275
	22	266	258	253	284	245	263
	23	338	343	307	274	262	309

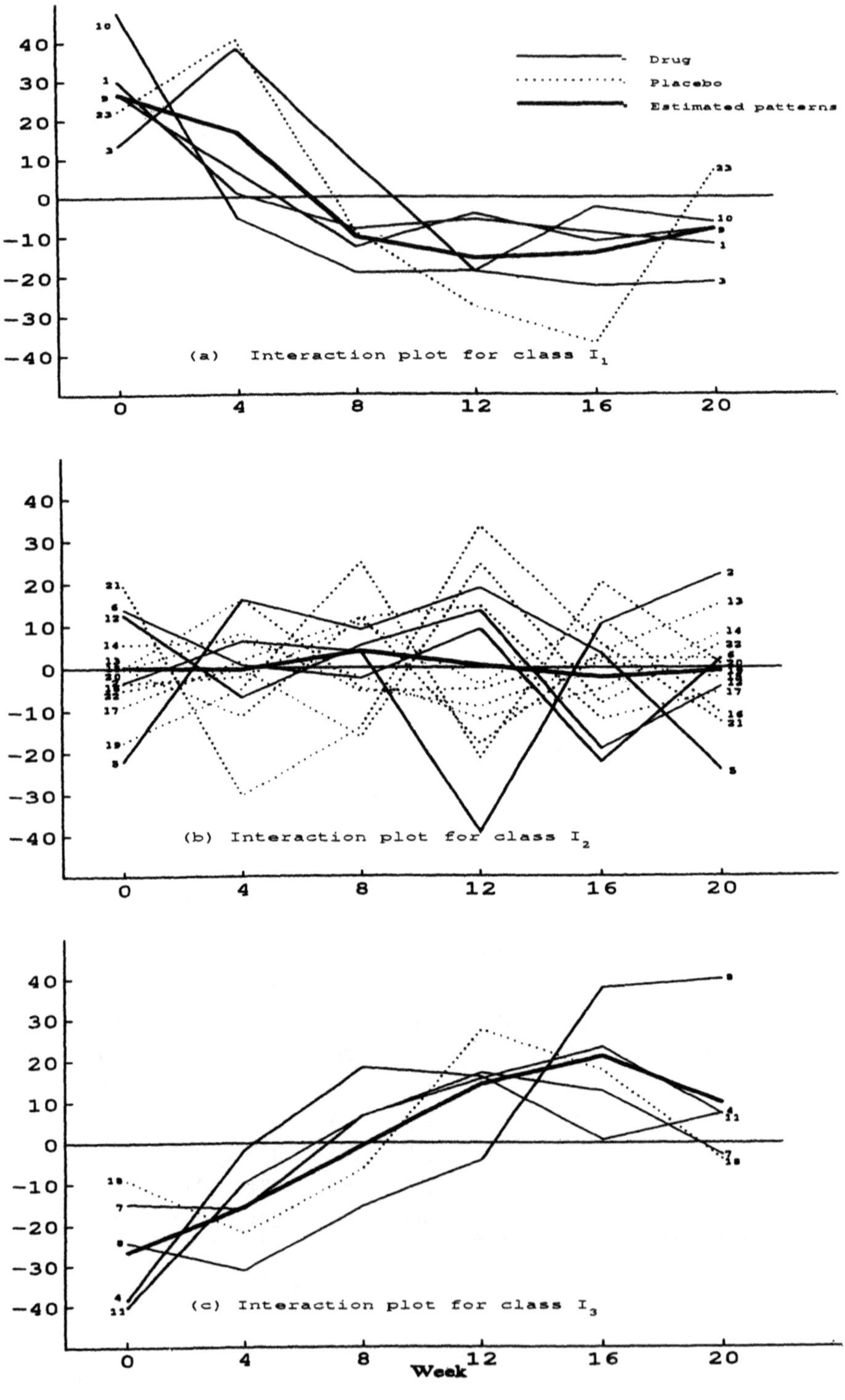

Fig. 1. Profile patterns of each of three subgroups

3.2 Multiple comparisons of ordered columns based on maximal chi-squared statistic

The multiple comparisons of ordered columns are carried out based on the maximal chi-squared statistic

$$S_c^* = \max_{j=1,\cdots,b-1} \| (P_a' \otimes \boldsymbol{p}_{j(b)}')\boldsymbol{y} \|^2,$$

where

$$\boldsymbol{p}_{j(b)}' = \{bj(b-j)\}^{-1/2}\,(j-b \cdots j-b\ j \cdots j) \tag{7}$$

is the jth row of $P_b^{*\prime}$ detecting a step type change in columns.

The successive components of S_c^* are correlated chi-squared variables with $(a-1)$ degrees of freedom following the first order Markov process, which gives an exact and very efficient algorithm for the probability calculation, see Hirotsu et al.(1992) for details. Note here that S_c^* is making maximal contrast along columns while forming independent sum of squares along rows whereas S_r^* is making maximal contrast along rows while forming correlated sum of squares along columns. A quadratic form of $\boldsymbol{y}$ based on the matrix $P_b^* P_b^{*\prime}$ has been called a cumulative chi-squared statistic by the cumulative nature of the rows of $P_b^{*\prime}$ suggested by equation (7).

All these ideas can be extended to ordered categorical data and we give in §4 an example of a two-way contingency table with natural ordering in both rows and columns.

4 Methods for ordered rows and columns

First we give an example from a dose response experiment in Table 2.

Table 2. Usefulness in a dose response experiment

Drug	1 Un- desirable	2 Slightly undesirable	3 Not useful	4 Slightly useful	5 Useful	6 Excellent	Total
Placebo	3	6	37	9	15	1	71
AF 3mg/kg	7	4	33	21	10	1	76
AF 6mg/kg	5	6	21	16	23	6	77

This is an example of two-way ordered categorical data and we are interested in testing the null hypothesis

$$H_0'' : \frac{p_{i+11}}{p_{i1}} = \frac{p_{i+12}}{p_{i2}} = \cdots = \frac{p_{i+16}}{p_{i6}}, \quad i = 1, 2,$$

against an ordered alternative

$$H_1' : \frac{p_{i+11}}{p_{i1}} \leq \frac{p_{i+12}}{p_{i2}} \leq \cdots \leq \frac{p_{i+16}}{p_{i6}}, \quad i = 1, 2$$

with at least one inequality strong, which implies that the higher dose is more useful than the lower dose. In this case we have natural ordering in both rows and columns and can form the cumulative chi-squared type, maximal contrast type and linear rank statistics in each of the rows and columns. However, in considering the purpose of this experiment to find out an optimal dose we prefer the multiple comparison procedure for comparing rows rather than the overall chi-squared type test and the most appropriate will be the maximal contrast type statistic. Then we can apply any of the three types of statistics to columns to obtain the result shown in Table 3. The idea of forming these statistics is very similar to that of ANOVA case and an interested reader should refer to Hirotsu(1993).

Table 3. Comparing three types of tests

Procedure for columns	Statistics	Two-sided p-value
Linear rank	7.63	0.011
Cumulative chi-squared	24.07	0.005
Maximal contrast	10.033	0.014

In this case all the three procedures give a significant result suggesting the superiority of the highest dose against the lower doses. The p-values vary among the procedures among which the procedure based on the cumulative chi-squared gives the most significant result. However, it depends on the data and the maximal contrast type has some advantage in possessing an exact algorithm for the probability calculation (Hirotsu, 1997).

5 An extension to the new type of data

The data in Fig. 2 are the 24 observations of blood pressure measured every hour for 24 hours for 147 subjects. We give the summary of data in Table 4. Those measurements are known normally to decrease slightly in the night and too large a decline is abnormal as well as a flat or an upward tendency. Since the measurements return approximately to the same level with the initial point, the procedure introduced in previous sections for detecting up- or downward trend does not work well in this case. We therefore propose a modification of the previous method to suit our present interest.

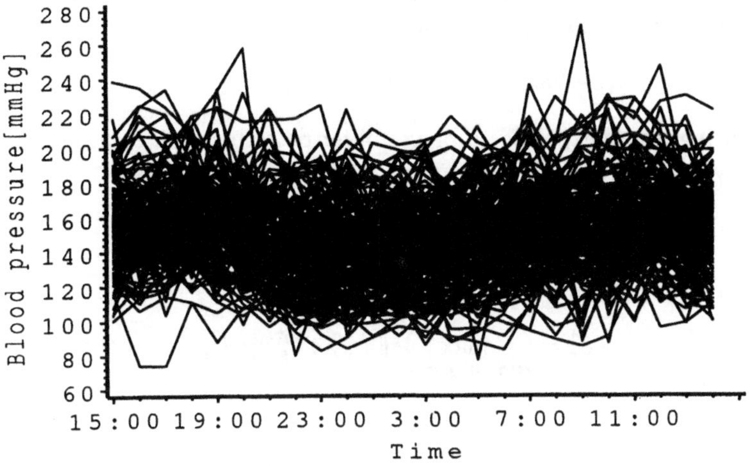

Fig. 2. Plots of original data

Table 4. Summary of data of 24 hours blood pressure measurements(147 subjects)

Time	15	16	17	18	19	20	21	22	23	24	1	2
Average	155	157	160	160	157	152	149	143	140	141	140	140
max.	239	235	234	219	234	258	223	217	224	222	211	201
min.	101	74	74	111	88	98	97	80	83	84	91	91
Time	3	4	5	6	7	8	9	10	11	12	13	14
Average	142	143	147	148	154	155	154	155	159	159	152	153
max.	204	218	212	205	236	229	271	230	229	247	230	222
min.	86	88	77	83	94	89	85	87	99	96	99	100

We denote by y_{ij} the jth observation of the ith subject and assume a two-way layout model (1). Now the basic idea is to assume a systematic departure from the null model

$$H_0''' : \mu_{ij} - \mu_{ij+1} - (\mu_{i'j} - \mu_{i'j+1}) = \mu_{ij+1} - \mu_{ij+2} - (\mu_{i'j+1} - \mu_{i'j+2}), \quad (8)$$
$$i, i' = 1, \cdots, a(i \neq i'); \; j = 1, \cdots, b-2$$

in the direction

$$H_2 : \mu_{ij} - \mu_{ij+1} - (\mu_{i'j} - \mu_{i'j+1}) \leq \mu_{ij+1} - \mu_{ij+2} - (\mu_{i'j+1} - \mu_{i'j+2}), \quad (9)$$
$$i, i' = 1, \cdots, a(i \neq i'); \; j = 1, \cdots, b-2$$

with at least one inequality strong. Note here the differences between the equations (5) and (8), and (6) and (9), respectively. The null model H_0'''

asserts the parallelism of slope changes (not slope itself) between the profiles of subjects i and i'. Rewriting H_2 as

$$H_2 : \mu_{ij} - 2\mu_{ij+1} + \mu_{ij+2} \leq \mu_{i'j} - 2\mu_{i'j+1} + \mu_{i'j+2},$$

the resulting test is expected to be useful for detecting a difference in convex or concave pattern. We therefore replace $P_b^{*\prime}$ in (3) by

$$P_b^{\dagger\prime} = diag(\eta_i^{1/2}\delta_i)(L_b' L_b)^{-1} L_b', \tag{10}$$

to obtain

$$S_r^{\dagger} = \max_{a'j=0, a'a=1} \| (a' \otimes P_b^{\dagger\prime}) y \|^2,$$

where

$$L_b' = \begin{bmatrix} 1 & -2 & 1 & 0 & \cdots & 0 & 0 & 0 \\ 0 & 1 & -2 & 1 & \cdots & 0 & 0 & 0 \\ \cdots & & & & & & & \cdots \\ 0 & 0 & 0 & 0 & \cdots & 1 & -2 & 1 \end{bmatrix}_{b-2 \times b}$$

is a second order differential matrix and

$$\eta_i = 1 + 2(i-1)(b-2-i)/\{3(b-1)\},$$

and

$$\delta_i = -\left[\frac{6b(b^2-1)}{i(i+1)(b-i)(b-i-1)\{(b-1)(2i+1) - 2(i^2-1)\}}\right]^{1/2}$$

are the constants for standardization. Note the similarity between the equations (4) and (10). Let $S^{\dagger}(i;i')$ denote the statistic $\| (a' \otimes P_b^{\dagger\prime}) y \|^2$ when a' is specialized to (2) with $-1/\sqrt{2}$ and $1/\sqrt{2}$ as its i and i'th elements, respectively, then $S^{\dagger}(i;i')$ can be expanded in the form

$$S^{\dagger}(i;i') = \left\{\gamma_2 \chi_{(2)}^2 + \gamma_3 \chi_{(3)}^2 + \cdots \gamma_{b-1} \chi_{(b-1)}^2\right\} \sigma^2$$

with

$$\gamma_k = \frac{2b(b+1)}{(k-1)k(k+1)(k+2)}, \tag{11}$$

whereas $S^*(i;i')$ similarly defined to $S^{\dagger}(i;i')$ based on (3) is expanded like

$$S^*(i;i') = \left\{\frac{b}{1 \cdot 2}\chi_{(1)}^2 + \frac{b}{2 \cdot 3}\chi_{(2)}^2 + \cdots + \frac{b}{(b-1)b}\chi_{(b-1)}^2\right\} \sigma^2,$$

where $\chi_{(l)}^2$ is the chi-squared variable with $a-1$ degrees of freedom detecting the departure from a null model in the direction of the Chebyshev's lth order orthogonal polynomial, see Hirotsu(1986) for details. This expansion characterizes well the statistic $S^{\dagger}$ in contrast to S^* suggesting that $S^{\dagger}$ will work well for detecting the convex or concave patterns whereas S^* is appropriate for detecting up- or downward trend.

When $a \geq b$ the null distribution of $S_r^{\dagger}$ is the distribution of the largest root of a Wishart matrix $W(\sigma^2 P_b^{\dagger\prime} P_b^{\dagger}, a-1)$ which is very well approximated by $\sigma^2 \gamma_2 \chi^2_{(2)}$, where γ_2 obtained from (11) by substituting $k = 2$ is the largest root of a matrix $P_b^{\dagger\prime} P_b^{\dagger}$. The approximation is indeed good since γ_2 is dominating the other $b-3$ roots of $P_b^{\dagger\prime} P_b^{\dagger}$, see Hirotsu(1991) for the detail of the chi-squared approximation.

6 Cancelling out the nuisance parameter σ^2

Since the null distribution of $S_r^{\dagger}$ involves a nuisance parameter σ^2 we need to cancel it out. A possible approach will be to cancel σ^2 by dividing $S_r^{\dagger}$ by an appropriate quadratic form less affected by the systematic change in μ_{ij}. For this purpose we propose

$$S^{\dagger -} = y'\{P_a P_a' \otimes (P_b^{\dagger\prime} P_b^{\dagger})^{-1}\}y$$

by inverting a quadratic form $S_r^{\dagger}$. Under the null hypothesis $S^{\dagger -}$ is obviously expanded as

$$S^{\dagger -} = \{\gamma_2^{-1}\chi^2_{(2)} + \cdots + \gamma_{b-1}^{-1}\chi^2_{(b-1)}\}\sigma^2.$$

Then the higher order terms have larger weight and this is what we intended since a chi-squared statistic for a higher order polynomial is less systematic and should be more appropriate for evaluating the noise variation.

Now our test statistic is the ratio $S_r^{\dagger}/S^{\dagger -}$ of the two statistics. Under the assumption of the independence among errors, we can give a very good approximation for the p value,

$$\begin{aligned} p &= \Pr\{S_r^{\dagger}/S^{\dagger -} \geq s_0\} \\ &= \Pr\left\{F_{a-1,f} \geq \frac{s_0(\gamma_3^{-1} + \cdots + \gamma_{b-1}^{-1})}{\gamma_2 - s_0\gamma_2^{-1}}\right\}, \end{aligned} \tag{12}$$

where $F_{a-1,f}$ follows an F distribution with the degrees of freedom $(a-1, f)$.

Actually, however, we have a very high correlation between the subsequent observations and we cannot apply the formula (12) as it is. In this regards the application given in the next section is a tentative approximate result.

7 Application

We apply the approximate procedure proposed in previous sections to real data of Fig. 2. As stated in the previous section we expect a high correlation between the subsequent observations. To reduce it to some extent we therefore try to form two new data sets,

A: employing only 6 data at every 4 hours,
B: averaging subsequent 4 points,

both forming 147×6 two-way data. Now following exactly the same classification procedure of subjects in Hirotsu(1991) excepting the difference of the basic statistics we obtained the result in Table 5, where the percent contribution of the classification in the total sum of squares $S_T^{\dagger} = \| (P_a' \otimes P_b^{\dagger\prime})y \|^2$ is also given.

Table 5. p-value and percent contribution of the classification

Dataset	Number of groups	p-value	Percent contribution
A	2	0.867	46.7
	3	0.051	64.5
	4	0.021	69.7
	5	0.007	74.0
B	2	6.85E-6	55.0
	3	7.94E-10	70.1
	4	8.64E-13	78.9
	5	5.81E-13	82.0

The results given in Table 5 show extremely high significance levels due to the large number of degrees of freedom for the denominator statistics. Although the data set B might be suffering from a considerable false positive due to possible correlations among observations we expect data set A is escaping from it. Due to the smoothing effect of the averaging process the result for data set B is clearer than that of A and we give only the result of classification based on data set B. Now by moving from 4 to 5 groups the gains in percent contribution and p-value are very small whereas there is a jump in those values from 3 to 4 groups. We therefore employ the classification into 4 groups and show the resulting profiles in Fig. 3. Obviously group 1 is concave, group 2 is slightly concave, group 3 is flat and group 4 is slightly convex. It is very interesting to compare Fig. 1 and Fig. 3, which characterize very well the procedures based on S_r^* and $S^{\dagger}$, respectively.

There is a classification of the data by a clinical doctor into 3 groups of Declined, Flat and Elevated. We therefore give in Table 6 a cross classification of subjects by his and our methods.

It is seen from Table 6 that clinical doctor's classification Decline, Flat and Elevated correspond mainly to our (G1 and G2), G3 and G4, respectively,

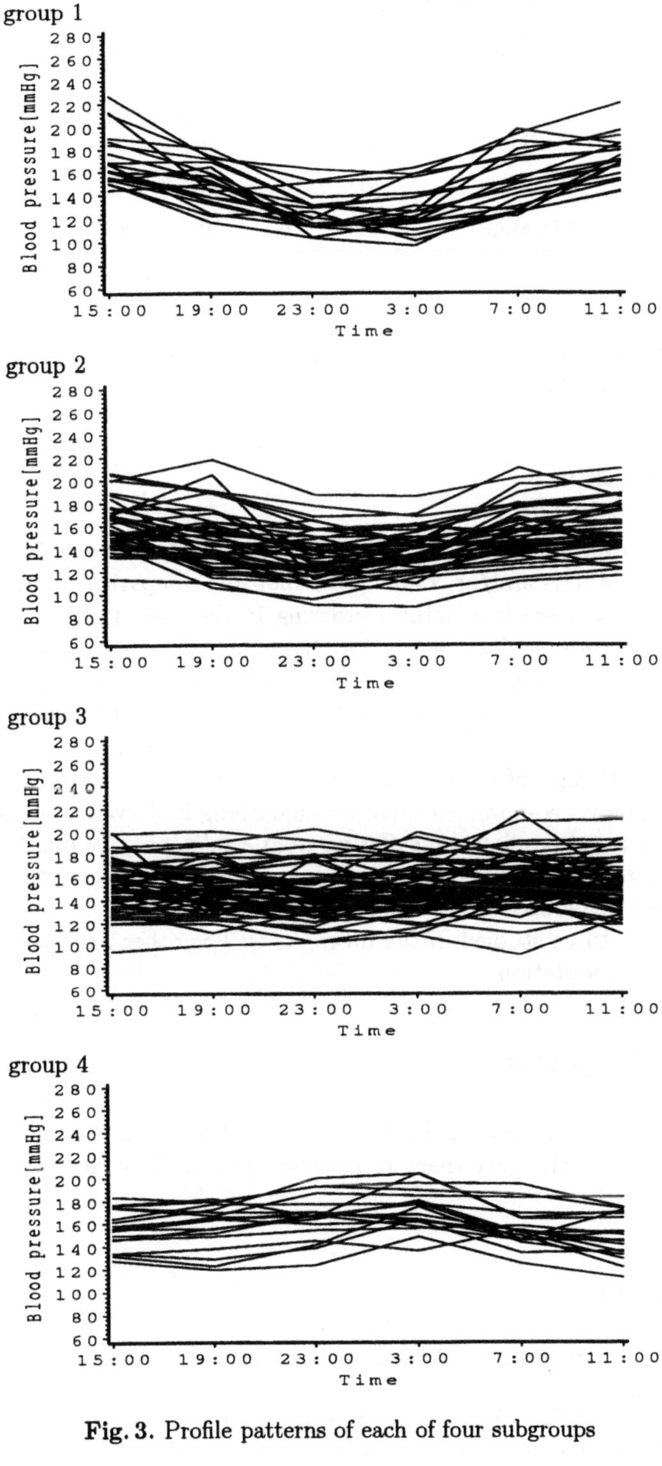

Fig. 3. Profile patterns of each of four subgroups

Table 6. Cross-classification of subjects

Present method / Clinical doctors	G1	G2	G3	G4
Decline	20	40	7	0
Flat	0	5	56	11
Elevated	0	0	0	8

and the latter gives a refinement of the former. One advantage of our method is that it is computerized, efficient and not subjective.

8 Discussion

We gave an overview of the analytical methods for interaction with particular interest in the row- and/or column-wise multiple comparison procedures for the case where there is a natural ordering in the rows and/or columns. In particular we proposed a method for classifying subjects according to their convex, flat or concave patterns of 24 hours blood pressure measurements. An application of the method to real data is given in §8 where it nicely classified 147 subjects into 4 groups. Especially the result for data set B is clear as compared with that of A partly because the original data set is so noisy and the averaging process was effective for smoothing it. However, as far as the test is concerned it might be suffering from a false positive due to possible correlations between subsequent measurements and the observed significance level should be reduced to some extent. The appropriate amount of reduction is a problem to be studied in the future as well as other methods to reduce the effect of correlation.

Acknowledgement

The authors are greatly indebted to Dr.'s N. Hirose and K. Shimizu of the Keio University who gave them an interesting example of convex or concave profile patterns.

References

1. Hirotsu, C. (1978). Ordered alternatives for interaction effects. Biometrika **65**, 561–570.
2. Hirotsu, C. (1983a). An approach to defining the pattern of interaction effects in a two-way layout. Annals Inst. Statist. Math. **A 35**, 77–90.

3. Hirotsu, C. (1983b). Defining the pattern of association in two-way contingency tables. Biometrika **70**, 579–590.
4. Hirotsu, C. (1986). Cumulative chi-squared statistic as a tool for testing goodness of fit. Biometrika **73**, 165–173.
5. Hirotsu, C. (1991). An approach to comparing treatments based on repeated measures. Biometrika **78**, 583–594.
6. Hirotsu, C. (1993). Beyond analysis of variance techniques : Some applications in clinical trials. International Statistical Review **61**, 183–201.
7. Hirotsu, C. (1997). Two-way change point model and its application. Austral. J. Statistics **39**, 205–218.
8. Hirotsu, C. (1998). Technological aspect of TQM. Proc. Int. Sympos. Quality Improvement Through Statistical Methods. Boston:Birkhauser, 275–287.
9. Hirotsu, C. (2000). Modeling and analyzing the generalized interaction. In statistical monitoring and Optimization for process control Ed. S. H. Park and G. G. Vining, New York:Marcel Dekker, 395–409.
10. Hirotsu, Kuriki and Hayter (1992). Multiple comparison procedures based on the maximal component of the cumulative chi-squared statistic. Biometrika **79**, 381–392.

Trouble-Shooting by Graphical Models

Hans-J. Lenz

Institute of Statistics and Econometrics, Freie Universität Berlin, Garystr. 21,
D-14195 Berlin, Germany

Abstract. Warranty claims, call backs, accidents and troubles during production can give raise to massive error backtracking or diagnosis. One way to support diagnosis is to use Bayesian belief network models (BBNs) as a kind of graphical models. They allow to backtrack error symptoms to their causes or to predict effects of interactive engineering settings. Trouble shooting with such models is based on a two-step procedure. Firstly, the causes-effects structure is designed as a directed, acyclic graph. Secondly, (subjective) probabilities are allocated to each node. According to the inner degree of a node the probabilities are either of the marginal or conditional type. Due to the singularity of call back events no data-driven procedure can be applied. Instead, domain experts have to allocate subjective probabilities to the set of variables. Various sensitivity studies are used in order to highlight extreme scenarios. These scenarios give the management of Company X a fair chance to identify individual households who use risky ovens. HUGIN Light is used as an appropriate software tool.

1 Introduction

We start with an illustrating example from a case study to introduce the kind of diagnosis problems we are interested in. We consider "Company X" which is a manufacture of gas ovens. In the year 1998 a call back action was started. The campaign involved about 120000 gas stoves of series 8XY, which were installed all over the country. They were produced in a limited number of lots with an underlying identical design. The kick-off of the affair started when the stove of a private household exploded. This accident was caused by rust corrosion and a leakage. Fortunately enough, the damage was negligible. Solid investigations by engineers of "Company X" showed that the design of the insulation around the three or four fire rings on the top of the gas ovens was not effective enough under factors like location of the stove or individual behaviour of a house wife like cooking and baking habits and frequencies. In order to prevent the company from further troubles, industrial liabilities and loss of good-will, the question arose how to predict corrosion and leakage rates of ovens from the given production lot at the earliest convenience. Note, that in such cases there is typically no data at hand to start mining or to begin with a sound statistical analysis based on general linear models like regression, log-linear models etc. The same is mostly true for case-based reasoning, cf. Breese and Heckerman (1996), due to a lack of similar cases corresponding to influential variations across production lots. Even process trouble shooting is different as can be learned from the classical paper of Girshik and Rubin (1952).

An alternative technique is known as failure analysis, which uses FMEA diagrams. However, these diagrams are descriptive only, and don't allow any proper quantitative risk assessment. However, a FMEA diagram can be considered as a directed, acyclic graph (DAG). Such diagrams can be easily re-designed and extended by local probability distributions. Local means, that under reasonable conditions it is sufficient to know the marginal and conditional distributions, i.e. partial knowledge of the system under study, in order to compute the joint probability distribution. When the joint distribution is at hand, any query of interest can be answered in a straightforward way. Therefore we end up with Bayesian Belief Networks (BBNs) for trouble shooting, which were under world wide research since the middle of the eighties, cf. Pearl (1988), Lauritzen and Spiegelhalter (1988), Shafer (1996) and Lauritzen (1996). They can be considered a useful sub-class of graphical models, see Jordan (1998) for more details.

2 Trouble Shooting Model

BBNs heavily relate to causality, i.e. the relationship between causal factors and effects is modelled. In our case the first main effect is (rust) corrosion on the pipes in the body of a gas stove. If corrosion starts there, then it will cause almost always a leakage of the pipe earlier or later as the second main effect of interest. The corrosion itself is dominated by two factors. One is the location of the gas stove. It may be installed in the kitchen of a solid house with central heating or, in the worst case, in a cottage with no heating at the countryside. Another factor is the usage mode. This factor is a composition of at least three sub-factors: Cooking style, cooking frequency and baking frequency of the housewife. The cooking style of the housewife is important in our case. If the responsible person is sloppy, there is a higher risk of spoiling water, soups or other fluids, which will penetrate through the insufficient insulation of the fire rings and contaminate tubes and pipes in the body of the gas stove. This will happen with a higher possibility if the usage mode 'cooking style' interacts in the same direction with the baking frequency. If an oven is used only from time to time (spotty baking), then the stove will never be dried totally, especially when the environment is 'wet'. This might be the case if the stove is placed in an unheated cottage. The usage mode as well as the location can have a very different effect. They can cause sediments asdust, which is more a question of life style, mud, chalk stone or grease. All these effects do contribute neither to (rust) corrosion nor to a leakage. Therefore this causal chain will be ignored later. Summarising we introduce the following variables and list them together with their corresponding ranges. This is the first part of the modelling process called "Qualitative Analysis – Step 1". The step 2 follows by defining the causal relationships and taxanomies.

I. Factors (Causes):

a) Usage Mode:
   Cooking Style: fussy/ sloppy
   Cooking frequency: spotty / weekly or more frequent
   Baking frequency: spotty / weekly or more frequent

b) Location: cottage (humid) / solid house (dry)

II. Effects:

a) (Rust) Corrosion: no / yes

b) Leakage: no / yes

c) Sediments: no / Mud or Chalk Stone or Grease.

Next we present the DAG, which reflects the causal relationships between the above variables. Note the intermediate role of the variable "Corrosion".

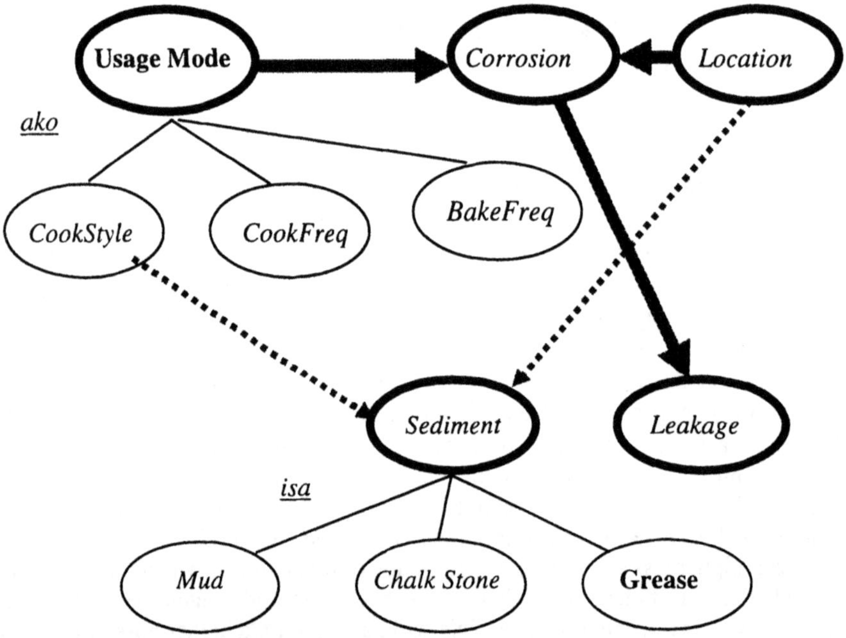

Figure 1: Causal relationships of gas stove example (fat lines represent the main model, dotted lines gives a hint to a sediment model; ako (isa) is a predicate expressing "a_kind_of" ("is_a")

The second part of the modelling process concerns "quantitative analysis" in so far as (discrete) probability distributions are allocated to the variables (nodes of the DAG). As such problems are isolated events with no experience from the past, and the incoming data are collected too slowly relative to the time pressure of decision making concerning a call back action, say, only subjective probabilities can be used. A task force of experts of "Company X" determined the subjective probabilities during a round table discussion called brainstorming. A good starting point was to fix the causal graph in Fig.1, which was accepted as a common view. We present below the marginal (MPTs) and conditional (CPTs) probability tables corresponding to the nodes of the DAG. We note that the DAG as well the probability tables are somehow distorted in order to protect the anonymity of "Company X".

Table 1: Marginal Probability Tables (MPTs)

Variable	**values**	**probabilities in %**
Cooking Style	fussy	55
	sloppy	45
Cooking Frequency	spotty	30
	weekly_or_more	70
Baking Frequency	spotty	50
	weekly_or_more	50
Location	cottage (humid)	30
	solid house (dry)	70

Note, that the compound variable 'user mode' is resolved.

Table 2: Conditional Probability Table (CPT) "Leakage by Corrosion" in %

		Corrosion	
		no	yes
Leakage	no	99,0	90,5
	yes	1,0	9,5

Table 3: Conditional Probability Table (CPT) "Sediments by Cook Style and Location"

CookStyle	fussy		sloppy	
Location	humid	dry	humid	dry
Mud_Chalk_Greese	10	1	20	6,25
no sediment	90	99	80	93,75

Note, that Table 3 reflects the fact, that there exists strong interaction between the factors "CookStyle" and "Location" on the effect "Sediment". As all four conditional distributions P(Sediment|CookStyle, Location) are different, we say that there does not exist local independency on factor levels, i.e. 'local independency'.

Table 4: Conditional Probability Tables (CPTs) "Corrosion by Location, Baking Frequency, Cooking Frequency and Cooking Style" in %, CPT No.I-IV

Location	cottage (humid)			
BakeFreq	spotty			
CookFreq	spotty		weekly or more	
CookStyle	fussy	sloppy	fussy	sloppy
Corrosion no	99	90	99,9	95
yes	1	10	0,1	5

Location	cottage (humid)			
BakeFreq	weekly or more			
CookFreq	spotty		weekly or more	
CookStyle	fussy	sloppy	fussy	sloppy
Corrosion no	99,5	95	99,9	97,5
yes	0,5	5	0,1	2,5

Location	solid house (dry)			
BakeFreq	spotty			
CookFreq	spotty		weekly or more	
CookStyle	fussy	sloppy	fussy	sloppy
Corrosion no	99,5	92	99,95	97,5
yes	0,5	8	0,05	2,5

Location	solid house (dry)			
BakeFreq	weekly or more			
CookFreq	spotty		weekly or more	
CookStyle	fussy	sloppy	fussy	sloppy
Corrosion no	99	98	99,99	98
yes	1	2	0,01	2

3 Methodology

In this chapter we recapitulate some important definitions and theorems, which are useful to understand the inference about trouble shooting, diagnosis and risk assessment, which will be presented in chapter 4. Let us begin with the definition of a DAG, which is a basic tool for our analysis.

DEF. 1: DAG
Let G = (**V**, **E**) be a graph with vertex set **V**, an edge set **E**, and a mapping φ: $\mathbf{V} \times \mathbf{V} \rightarrow \mathbf{E}$. The edges are directed, and there exist no cycles or loops along any set of edges.

As a running example we use Example 1: DAG = (**V**, **E**)
with $\mathbf{V} = \{A, B, C, D, E, F\}$, and $\mathbf{E} = \{BC, AC, CD, CE, DF, EF\}$.

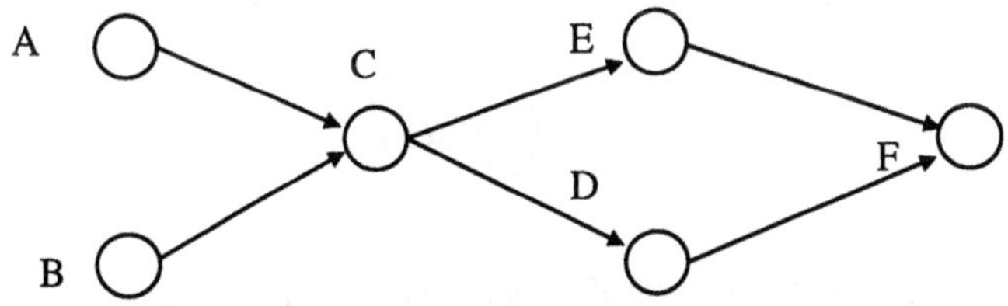

Figure 2: DAG

A joint distribution $p(\,)$ defined on product space $\mathbf{V} = A \times B \times \ldots \times F$ will be denoted by $p(A,B,C,D,E,F)$. By marginalisation we can derive from a joint distribution any conditional distribution of interest, for example

$$p(F \mid A,B) = \frac{\sum_C \sum_D \sum_E p(A,B,C,D,E,F)}{\sum_C \sum_D \sum_E \sum_F p(A,B,C,D,E,F)} \tag{1}$$

However, this "global" approach has many pitfalls, for instance:

- Formulas 'not handy'
- Joint distribution absolutely not known among human experts
- Only local (partial) expert knowledge available as marginal or low dimensional conditional distributions.

The question arises how to fuse partially overlapping knowledge of autonomous experts? The answer is given by the Bayesian point of view as Bayesian Belief Networks, cf. Pearl (1988). Inference is performed by probability propagation as will be done below. A key idea to reduce the complexity of a joint distribution is to utilise the special structure of DAGs and to apply Markov theory, Lauritzen and Spiegelhalter (1988).

$$p(\mathbf{V}) = p(v_1)p(v_2 | v_1) \quad p(v_N | v_{N-1},\dots,v_1) = \tag{2}$$
$$= \prod_i p(v_i | v_{i-1},\dots,v_1) = \prod p(v_i | \text{parents } (v_i))$$

The last equality is valid due to the Markov property of DAGs. This enables to specify a joint or global distribution by a mixture of local, i. e. marginal or conditional probabilities. This implies for example in Fig. 2 above:

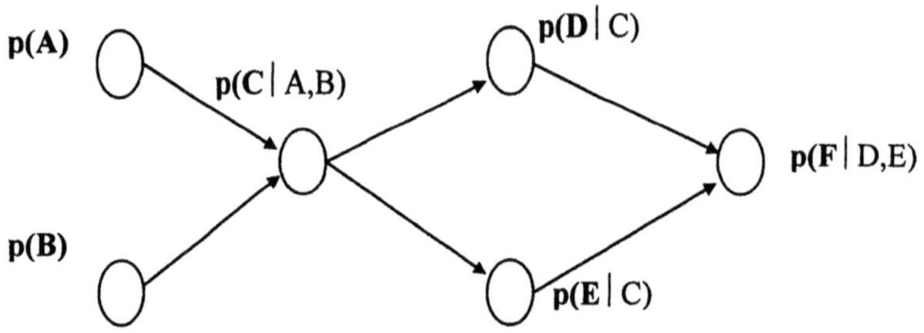

Figure 3: A DAG with allocated probability distributions

We can write

$$\begin{aligned} p(\mathbf{V}) &= \prod_i p(v_i | \text{parents } (v_i)) \\ &= p(A)p(B)p(C | A,B)p(D | C)p(E | C)p(F | D,E) \end{aligned} \tag{3}$$

The following factorisation theorem is fundamental, cf. Lauritzen (96). If a distribution p over **V** factorizes according to a DAG then it satisfies the *Global Markov Property*.
Separation (Factorisation) Theorem:

For any triple (X, Y, Z) of disjoint subsets of $\mathbf{V}$ such that Y separates X and Z in $G_{\text{moral}\,(A,B,C)}$ we have

$$X \perp Z\ Y\ .$$

Here $G_{\text{moral}\,(A,B,C)}$ denotes the so-called moral graph of the smallest ancestral set (=a set where parents of all members are in the set itself) that contain X, Y and Z. The figure below illustrates the theorem.

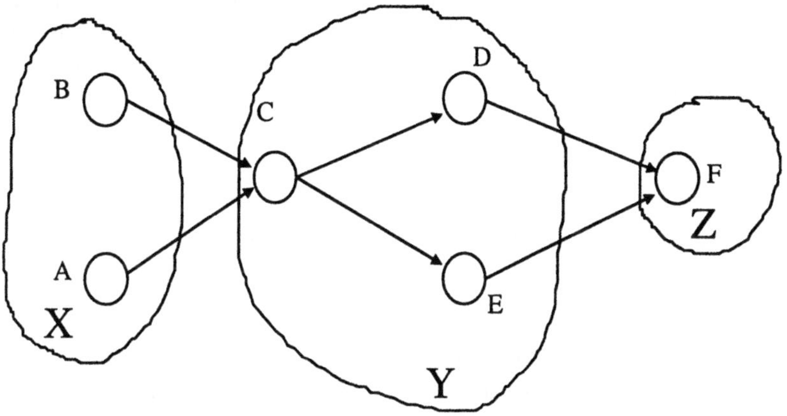

Figure 4: Separation principle illustrated

Here exist several algorithms to allow probabilistic reasoning on DAGs, i.e. probability propagation according to given evidences as part of szenarios ("what if analyses"), cf. Pearl (1988). We consider here the so-called junction-tree algorithm, cf. Lauritzen (1996), Jordan (1998). A junction tree is a clique tree of a triangulated graph. It has a reduced complexity and therefore speeds-up probability propagation.

Junction Tree Algorithm

Author: Lauritzen 1997
Purpose: Given a DAG find a clique tree for a triangulated graph (i.e. a junction tree JT).

Input: DAG, $p(\,)$ as a mixture of MPTs and CPTs

Output: JT, $\phi(\,)$ of DAG

beginJT
Step 1: Transform DAG into undirected graph G_{undir}.

Step 2: Obtain a moral graph G_{moral} of G_{undir}.

Step 3: Find all cliques C and potentials $\phi(\,)$ of G_{moral}

Step 4: Triangulate graph as G_{TG}

Step 5:Compute clique tree of triangulated graph G_{TG}, i.e. transform G_{TG} into a junction tree JT

endJT.

There exist a lot of implementations of this algorithm. One of the leading implementations with very reliable functions and an advanced graphical user interface is HUGIN from Hugin Expert A/S, Aalborg, Denmark. We use in the next chapter the licence-free version HUGIN light 6.x.

4 Diagnosis and Trouble Shooting Analysis

Let us start with some few scenarios.

Scenario I is concerned with an "average" case or "no evidence at hand" case. This means that there exist no evidences at all on baking and cooking frequencies of the household, the cooking style is not known, and there exists no specific information about the location of the gas stove. Instead of evidences we take the marginal probabilities as fixed by the experts, see MPTs above. We compute the following beliefs or risks as probabilities:

Table 5: Scenario I "No evidence at hand"

P(Corrosion = yes)	1.80 %
P(Leakage = yes)	1.57 %

Let us now change to scenario II. We only assume, that we have evidences about the location of the stove and about the cooking style of the housewife, say. Assume that the location is within a solid house, i.e. the environment of the stove is dry, and the cooking style of the housewife is ideal, i.e. fussy. This implies:

Table 6: Scenario II "Fussy wife in a solid house"

P(Corrosion = yes| CookStyle = fussy, Location = dry)	1.03 %
P(Leakage = yes| CookStyle = fussy, Location = dry)	1.33 %

The risk of corrosion reduces by about 43%, while the probability of a caused leakage decreases by about 15 %. Assume now for scenario III, that the household prefers to cook regularly, and that at least once a week the stove is used for baking. Baking has the effect of evaporating any spoiled water, which might penetrate through the inefficient insulation and is kept on the pipes inside the body of the stove. The risk reduces strongly due to interaction of all factor levels:

Table 7: Scenario III "Ideal household and location"

P(Corrosion = yes| CookStyle = fussy, CookFreq = weeklyormore, BakingFreq = weeklyormore, Location = dry)	0.01 %
P(Leakage = yes| CookStyle = fussy, CookFreq = weeklyormore, BakingFreq = weeklyormore, Location = dry)	1.00 %

In Scenario IV we assume just a complementary type of user mode and environment compared to scenario III. This means that everything is either sporadic or sloppy, and the cottage with its humidity has a further negative effect.

Table 8: Scenario IV "A real risky scenery"

P(Corrosion = yes| CookStyle = sloppy, CookFreq = spotty, BakingFreq = spotty, Location = humid)	10 %
P(Leakage = yes| CookStyle = sloppy, CookFreq = spotty, BakingFreq = spotty, Location = humid)	4.2 %

Evidently, the risk of rust corrosion increases by more than a factor 5, and the risk of a leakage by about 3 compared with the "no evidence" case. It is clear what such trouble shooting analysis can offer to a call back campaign. The most critical environments can be identified retrieving customer's addresses from dealer's files and using maps. Moreover, with some extra money, individual features of the usage mode can be learned from interviews over a phone. This will contribute to reduce costs of the campaign.

5 Sensitivity Analysis

Some further insight into the behaviour of causal relationships modelled by a BBN can be gained by a sensitivity study. Generally speaking, the sensitivity with respect to variation in the probabilities while keeping the structure of a DAG constant depends upon

- the complexity of the DAG measured by $c = n / m$ with n = number of arcs of DAG, and m = maximum number of arcs of DAG where each node $v \in \mathbf{V}$ is connected to each node $v' \in \mathbf{V}$, $v \neq v'$ if no cycles are induced,
- the structure of CPTs and
- the levels of probabilities assigned to the MPTs and CPTs.

In the case of DAGs and two-point distributed variables moderate sensitivity can be expected if variations of up to ± 20% around the midpoint of the ranges occur and one-factor at a time type of analysis ("ceteris paribus" studies) is considered only.

This is demonstrated in Fig. 4 below. We stick to the probabilities of scenario I ("No Evidence" Case) and vary the probability of CookStyle = fussy in the range from 0.25 to 0.85. Note, that the curve P(Corrosion = yes) flattens out around 0.5. In our diagnosis studies above the probability of a fussy cooking style was fixed to 55%.

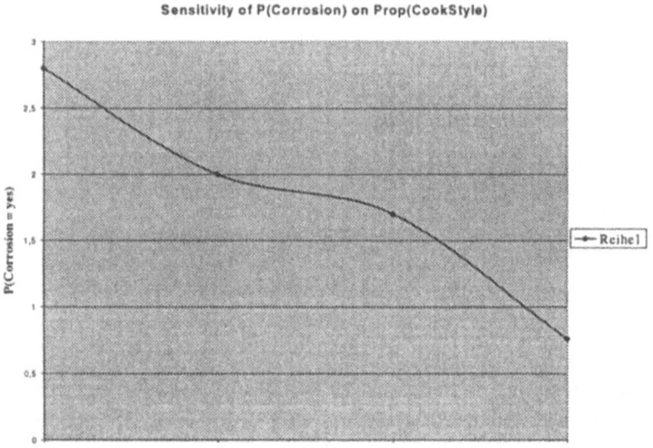

Figure 5: Scenario I with one factor variation of Prob (CookStyle = fussy) in %

The study of the full impact of variations of values from the conditional probability tables (CPTs) is tedious due to the well-known course of dimension. As is evident from above this influence depends upon the number of variables, the sizes of cardinality of the value sets of the variables and the graph complexity measure.

6 Summary

Troubleshooting may help management to assess risks of a low qualified production lot. As the lot was uniformly badly designed and produced, a global model reflecting causes and effects could be established. As there was only one sample available with a detected exploded oven, no data analysis or data mining techniques could be applied. Instead of sampling a task force of experts from various fields like engineering, marketing, accounting, dispatching, craftsmanship, statistics and law became active. They agreed on a Bayesian Belief Network and the corresponding subjective marginal and conditional probabilities.

These probabilities were used to study the sensitivity of the approach and to analyse various scenarios under different kind of evidences. Special interest was paid to an extreme scenario, which showed relative high risks of corrosion and leakage at private households. Gas fitters under special contracts got hints to such locations and to the possible cooking and baking profiles of such households. They used those hints to identify risky installations based on their experience at various sites and to inspect them according to their temporal priority.

The author is proud to report that due to such probability model based trouble-shooting no further leakage was reported.

References

1. Breese , J. and Heckerman, D. (1995) Probabilistic Case-based Reasoning. In: preliminary Papers of the Fifth workshop on Artificial Intelligence and Statistics, Fisher, D. and Lenz, H.-J. Ft. Lauderdale, 56-63
2. Girshik, M.A. and Rubin, H. (1952) A Bayes Approach to a quality control model. In: Ann. Math. Statist., 23, 114-125
3. HUGIN light 5.5 (2000) www.hugin.dl
4. Jordan, M. (1988) Learning in Graphical Models. Dordrecht, Kluver
5. Lauritzen, S.L. (1995) Statistical Methods for Inducing Models from Data, A Tutorial. In: preliminary Papers of the Fifth workshop on Artificial Intelligence and Statistics, Fisher, D. and Lenz, H.-J.. Ft. Lauderdale

6. Lauritzen, S.L. and Spiegelhalter, D.J. (1988) Local computations with probabilities on graphical structures and their application to expert systems. In: JRSS, ser. B, 50, 157-224
7. Pearl, J. (1988) Probabilistic Reasoning in Intelligent Systems. Morgan Kaufman, San Mateo
8. Shafer, G. (1996) The Art of causal Conjecture. MIT Press, Cambridge
Lauritzen, S.L. (1996) Graphical Models. Clarendon, Oxford

An Alternative Expression of the Fractional Factorial Designs for Two-level and Three-level Factors

Yoshikazu Ojima, Tomomichi Suzuki and Seiichi Yasui

Tokyo University of Science, 278-8510, Japan

Abstract. Two-level and three-level fractional factorial designs are widely and effectively applied in many practical fields for finding active factors. We introduce an operation for obtaining interaction effects in the design matrices of the experiments in this paper. The relationship between two factors and their interaction can be found by applying group theory with the operation, which is called the Hadamard product for two-level factors, and the generalized Hadamard product for three-level factors, based on the columns of the design matrix. The group structure can be applied to solve many related problems including the fold-over technique. The operation, i.e. the Hadamard product and the expression of the design matrices for three-level factorial experiment, are established by introducing the complex cubic root of 1 and some other required generalization.

1 Introduction

Two-level and three-level fractional factorial designs arc widely and effectively applied in many practical fields for finding active factors. There are many textbooks describing two-level fractional factorial experiments, however, their styles are different. In particular, the definition of main effects for two-level designs is consistent with neither the usual linear models nor three-level designs.

We propose a standard expression for the 2^k factorial designs in section 2.1. The difference from other expressions is discussed in section 2.2. The standard expression enables to introduce an operation obtaining the column of the interaction. The operation is defined as the Hadamard product of the columns of main effects in section 3.1. Some applications of the expression are described in sections 3.1 and 3.2 including the fold-over technique.

The generalization of the expression by means of the generalized Hadamard matrix to multi-level designs including three-level is described in the section 4.

2 Expressions for the 2^3 Factorial Design

2.1 A Standard Expression for the 2^3 Factorial Design

The fractional factorial designs are easily derived from the full factorial design. In this section, we mainly discuss the 2^3 factorial design, because large designs are space-consuming, however the results obtained here are valid for larger designs. We propose a standard expression for the 2^3 factorial design as shown in Table 1. The entries of the table form an 8×8 matrix, and let $\boldsymbol{H}_8$ denote the matrix.

Let $\boldsymbol{y}$ be a data vector, whose i-th element corresponds to the i-th run, and let $\boldsymbol{\theta}$ be a parameter vector whose elements are grand mean μ, main effects α and β, two-factor interaction $\alpha\beta$, another main effect γ, and interactions $\alpha\gamma$, $\beta\gamma$, and $\alpha\beta\gamma$, respectively. Using $\boldsymbol{H}_8$ as the design matrix, the data model is expressed as

$$\boldsymbol{y} = \boldsymbol{H}_8 \boldsymbol{\theta}. \tag{1}$$

Table 1. Standard Expression for the 2^3 Factorial Design Including H_8

run	**1**	***a***	***b***	***ab***	***c***	***ac***	***bc***	***abc***
1	+1	+1	+1	+1	+1	+1	+1	+1
2	+1	−1	+1	−1	+1	−1	+1	−1
3	+1	+1	−1	−1	+1	+1	−1	−1
4	+1	−1	−1	+1	+1	−1	−1	+1
5	+1	+1	+1	+1	−1	−1	−1	−1
6	+1	−1	+1	−1	−1	+1	−1	+1
7	+1	+1	−1	−1	−1	−1	+1	+1
8	+1	−1	−1	+1	−1	+1	+1	−1

The matrix $\boldsymbol{H}_8$ is one of the Hadamard matrices, and $\boldsymbol{H}_8^T \boldsymbol{H}_8 = 8\boldsymbol{I}$, where $\boldsymbol{H}_8^T$ is the transposed matrix of $\boldsymbol{H}_8$ and $\boldsymbol{I}$ is the identity matrix. The parameter vector $\boldsymbol{\theta}$ can be estimated as

$$\hat{\boldsymbol{\theta}} = \tfrac{1}{8} \boldsymbol{H}_8^T \boldsymbol{y}. \tag{2}$$

The matrix $\boldsymbol{H}_8$ can be constructed systematically from $\boldsymbol{H}_2$ which is defined in (3), as the following;

$$\boldsymbol{H}_2 = \begin{bmatrix} +1 & +1 \\ +1 & -1 \end{bmatrix} \tag{3}$$

$$H_8 = H_2 \otimes H_2 \otimes H_2 \equiv H_2^{\otimes 3} \tag{4}$$

where $A \otimes B$ is a Kronecker product of A and B (e.g. see section 1b.8 of Rao (1973)) defined as $A \otimes B = (a_{ij} B)$, where $A = (a_{ij})$.

For the case of the $2^k (\equiv n)$ factorial design, the matrix H_n is generally obtained as $H_2^{\otimes k}$

2.2 Other Expressions of 2^3 Factorial Design

2.2.1 The Expression Used in Montgomery (1997)

Table 2 is from Table 6-3 (p. 231) of Montgomery (1997). The entries + and – can obviously be replaced with +1 and –1. Let M_8 denote the matrix in Table 2. Concerning M_8, we have the following equation.

$$M_8 = \begin{bmatrix} +1 & -1 \\ +1 & +1 \end{bmatrix} \otimes \begin{bmatrix} +1 & -1 \\ +1 & +1 \end{bmatrix} \otimes \begin{bmatrix} +1 & -1 \\ +1 & +1 \end{bmatrix} \tag{5}$$

Reversing the order of rows of M_8, H_8 can be obtained as (6).

$$H_8 = \begin{bmatrix} & & & & & & & 1 \\ & O & & & & & 1 & \\ & & & & & 1 & & \\ & & & & 1 & & & \\ & & & 1 & & & & \\ & & 1 & & & O & & \\ & 1 & & & & & & \\ 1 & & & & & & & \end{bmatrix} M_8 \tag{6}$$

Based on the relationship $H_2 = \begin{bmatrix} 0 & 1 \\ 1 & 0 \end{bmatrix} M_2$ and the formula for exchanging the order of a matrix product and a Kronecker product $(A \otimes B)(C \otimes D) = (AC) \otimes (BD)$, the above relation, $H_8 = \begin{bmatrix} 0 & 1 \\ 1 & 0 \end{bmatrix}^{\otimes 3} M_8$ is easily obtained.

This expression is suitable for Yates algorithm, and similar expressions are widely used. Box, Hunter and Hunter (1978) and Wu and Hamada (2000) tabulated M_8

without the first column for the 2^3 experiment. They denoted the columns of A, B and C, as **1**,**2** and **3**, and arranged the columns from left as **1**,**2**,**3**,**12**,**13**,**23** and **123** in the table. ISO 3534-3 (1999) uses the same expression.

Table 2. The Expression of the 2^3 Factorial Design by Montgomery (1997)

Treatment	Factorial Effect							
Combination	I	A	B	AB	C	AC	BC	ABC
(1)	+	−	−	+	−	+	+	−
a	+	+	−	−	−	−	+	+
b	+	−	+	−	−	+	−	+
ab	+	+	+	+	−	−	−	−
c	+	−	−	+	+	−	−	+
ac	+	+	−	−	+	+	−	−
bc	+	−	+	−	+	−	+	−
abc	+	+	+	+	+	+	+	+

2.2.2 *The Expression Used in Taguchi (1987)*

Table 3 is from Table 11 (p.1128) in the Appendix of Taguchi (1987). The entries 1 and 2, which are the level number, can be replaced with +1 and −1. Let column 0 be a vector of $(+1, +1, +1, +1, +1, +1, +1, +1)^T$, and set the column 0 on the left side of L_8, then we have an 8×8 matrix which may be denoted as $\boldsymbol{T}_8$. Reordering the rows of $\boldsymbol{T}_8$, we have $\boldsymbol{H}_8$, as follows.

$$\boldsymbol{H}_8 = \begin{bmatrix} 1 & 0 & 0 & 0 & 0 & 0 & 0 & 0 \\ 0 & 0 & 0 & 0 & 1 & 0 & 0 & 0 \\ 0 & 0 & 1 & 0 & 0 & 0 & 0 & 0 \\ 0 & 0 & 0 & 0 & 0 & 0 & 1 & 0 \\ 0 & 1 & 0 & 0 & 0 & 0 & 0 & 0 \\ 0 & 0 & 0 & 0 & 0 & 1 & 0 & 0 \\ 0 & 0 & 0 & 1 & 0 & 0 & 0 & 0 \\ 0 & 0 & 0 & 0 & 0 & 0 & 0 & 1 \end{bmatrix} \boldsymbol{T}_8 \qquad (7)$$

It should be noted that the bottom row of Table 3 shows the group number, which is thought to reflect a group structure of the columns.

Table 3. Taguchi's $L_8(2^7)$, the 2^3 Factorial Design

	1	2	3	4	5	6	7
1	1	1	1	1	1	1	1
2	1	1	1	2	2	2	2
3	1	2	2	1	1	2	2
4	1	2	2	2	2	1	1
5	2	1	2	1	2	1	2
6	2	1	2	2	1	2	1
7	2	2	1	1	2	2	1
8	2	2	1	2	1	1	2
Gp	1	2		3			

3 Hadamard Product and Its Application

3.1 Hadamard Product and Interaction Effect

The Hadamard product (e.g. see section 1b.8 of Rao (1973)) of matrices is defined as follows.

$$\boldsymbol{A} * \boldsymbol{B} = (a_{ij}\ b_{ij}),\ \text{where}\ \boldsymbol{A} = (a_{ij}),\ \ \boldsymbol{B} = (b_{ij}) \tag{8}$$

Applying two columns of arbitrary main effects in $\boldsymbol{H}_8$, i.e. $\boldsymbol{x}$ and $\boldsymbol{y}$, the column of their interaction is obtained as the Hadamard product, $\boldsymbol{x}*\boldsymbol{y}$. The column $\boldsymbol{ab}$ is obviously the interaction of two main effects $\boldsymbol{a}$ and $\boldsymbol{b}$.

Concerning the Hadamard product, a set of columns in Table 1, $G_{ABC} = \{\mathbf{1}, \boldsymbol{a}, \boldsymbol{b}, \boldsymbol{ab}, \boldsymbol{c}, \boldsymbol{ac}, \boldsymbol{bc}, \boldsymbol{abc}\}$ is a group. Obviously, $\mathbf{1}$ is the unit element, and the inverse element of any element is the element itself, $\boldsymbol{x}^{-1} = \boldsymbol{x}$. A set of unit element, $G_1 = \{\mathbf{1}\}$ is the minimal subgroup, and sets of columns $G_A = \{\mathbf{1}, \boldsymbol{a}\}$, $G_{AB} = \{\mathbf{1}, \boldsymbol{a}, \boldsymbol{b}, \boldsymbol{ab}\}$ are also subgroups, i.e. $G_1 \subset G_A \subset G_{AB} \subset G_{ABC}$. Taguchi's GP numbers correspond to $G_A - G_1$, $G_{AB} - G_A$, and $G_{ABC} - G_{AB}$, applied to split-plot experiments.

$\boldsymbol{H}_8$ is an 8×8 matrix, and all the columns are mutually orthogonal. Further the set of all columns is a group concerning Hadamard product. Then we can find a column related to the interaction of any two main effects in $\boldsymbol{H}_8$. This situation is valid for any 2^k design whose design matrix is expressed as $\boldsymbol{H}_2^{\otimes k}$

For the case of $n = 12$, there exists an Hadamard matrix which is referred to as the Plackett-Burman design. All the columns of the matrix are mutually orthogonal, however the set of all columns is not a group concerning the Hadamard product. The column vector associated with the interaction of any two columns of main effects can be obtained as the Hadamard product of them. Because the matrix is of full rank, any interaction vector is linearly dependent on column vectors of the matrix. Letting two vectors of main effects be denoted by $\boldsymbol{x}$ and $\boldsymbol{y}$, the interaction of them is expressed as $\boldsymbol{x}*\boldsymbol{y}$. The inner products related to $\boldsymbol{x}*\boldsymbol{y}$ are $(\boldsymbol{x}*\boldsymbol{y}, \boldsymbol{x}) = 0$, $(\boldsymbol{x}*\boldsymbol{y}, \boldsymbol{y}) = 0$, however $(\boldsymbol{x}*\boldsymbol{y}, \boldsymbol{z}) = \pm 4$ where $\boldsymbol{z} \neq \boldsymbol{x}$, $\boldsymbol{z} \neq \boldsymbol{y}$. The squared norm of vectors is 12, two factor interaction can be said to be partially confounded to main effects.

3.2 Application to the Fold-Over Technique

The fold-over technique, which is described in section 4.4.3 of Wu and Hamada (2000), is a technique for planning a follow-up experiment with de-aliasing of some main effects. Let the design matrix of the original experiment be $\boldsymbol{H}_n$, a simple replication of the experiment is expressed as

$$\begin{bmatrix} H_n \\ H_n \end{bmatrix}.$$

Because the size of the whole experiment, i.e. the original and the follow-up experiments, is twice that of the original one, the design matrix of the whole can be written as,

$$\begin{bmatrix} H_n & H_n \\ H_n & -H_n \end{bmatrix} = \boldsymbol{H}_2 \otimes \boldsymbol{H}_n . \tag{9}$$

In the fold-over technique, levels of some of the main factors are exchanged in the follow-up experiment. This change can be made by moving columns from i-th to $i+n$ -th in the matrix (9). The results concerning the fold-over technique including de-aliasing are easily explained by the matrix (9) and the Hadamard product.

4 Generalization to 3^k Factorial Designs

Three-level factorial designs are described in Taguchi (1987). The minimal design matrix $\boldsymbol{H}_3$, can be expressed as

$$H_3 = \begin{bmatrix} 1 & 1 & 1 \\ 1 & \omega & \omega^2 \\ 1 & \omega^2 & \omega \end{bmatrix} \tag{10}$$

where ω is a complex cubic root of 1.

The 3^k factorial design can be obtained by a Kronecker product as $H_3^{\otimes k}$. For the case of the 3^2 factorial design, H_9 is expressed as

$$H_9 = H_3 \otimes H_3 = \begin{bmatrix} 1 & 1 & 1 & 1 & 1 & 1 & 1 & 1 & 1 \\ 1 & \omega & \omega^2 & 1 & \omega & \omega^2 & 1 & \omega & \omega^2 \\ 1 & \omega^2 & \omega & 1 & \omega^2 & \omega & 1 & \omega^2 & \omega \\ 1 & 1 & 1 & \omega & \omega & \omega & \omega^2 & \omega^2 & \omega^2 \\ 1 & \omega & \omega^2 & \omega & \omega^2 & 1 & \omega^2 & 1 & \omega \\ 1 & \omega^2 & \omega & \omega & 1 & \omega^2 & \omega^2 & \omega & 1 \\ 1 & 1 & 1 & \omega^2 & \omega^2 & \omega^2 & \omega & \omega & \omega \\ 1 & \omega & \omega^2 & \omega^2 & 1 & \omega & \omega & \omega^2 & 1 \\ 1 & \omega^2 & \omega & \omega^2 & \omega & 1 & \omega & 1 & \omega^2 \end{bmatrix} \tag{11}$$

The 2nd column of (11) is the conjugate of the 3rd column, and the 4th, 5th and 6th columns are conjugates of the 7th, 8th and 9th columns, respectively. These vectors can be labeled as **1**, $\boldsymbol{a}$, $\boldsymbol{a}^2$, $\boldsymbol{b}$, $\boldsymbol{ab}$, $\boldsymbol{a}^2\boldsymbol{b}$, $\boldsymbol{b}^2$, $\boldsymbol{ab}^2$ and $\boldsymbol{a}^2\boldsymbol{b}^2$, from the left hand side.

As the matrix H_3 is the complex, then H_3^*, the conjugate transposed matrix (the adjoint matrix) of H_3 is required for estimating effects and calculating sums of squares.

The Hadamard product is also extended in the complex sense, i.e.

$$\boldsymbol{a} * \boldsymbol{b} = (\bar{a}_i b_i), \tag{12}$$

where $\boldsymbol{a} = (a_i)$, $\boldsymbol{b} = (b_i)$, and $\bar{a}_i$ is the conjugate of a_i.

One vector $\boldsymbol{x}$ and its conjugate $\boldsymbol{x}^2$ correspond to a main effect X. The interaction of two main effects X and Y correspond to four vectors, $\boldsymbol{x}*\boldsymbol{y}$, $\boldsymbol{x}*\boldsymbol{y}^2$, $\boldsymbol{x}^2*\boldsymbol{y}$ and $\boldsymbol{x}^2*\boldsymbol{y}^2$. As $\boldsymbol{x}*\boldsymbol{y}$ is the conjugate of $\boldsymbol{x}^2*\boldsymbol{y}^2$ and $\boldsymbol{x}^2*\boldsymbol{y}$ is the conjugate of $\boldsymbol{x}*\boldsymbol{y}^2$, two other main effects can be allocated to ($\boldsymbol{x}*\boldsymbol{y}$ and $\boldsymbol{x}^2*\boldsymbol{y}^2$) and ($\boldsymbol{x}^2*\boldsymbol{y}$ and $\boldsymbol{x}*\boldsymbol{y}^2$).

Some further results for the three level factorial experiments can be obtained by direct application of the Hadamard product and the design matrices.

References

1. Box, G. E. P., Hunter, W. G., and Hunter, J. S. (1978) Statistics for Experimenters - An Introduction to Design, Data Analysis, and Model Building, John Wiley & Sons, New York
2. ISO 3534-3 (1999) Statistics - Vocabulary and symbols - Part 3: Design of experiments, ISO, Geneva
3. Montgomery, D. C. (1997) Design and Analysis of Experiments, 5th Edition, John Wiley & Sons, New York
4. Rao, C. R. (1973) Linear Statistical Inference and Its Applications, 2nd Edition, John Wiley & Sons, New York
5. Taguchi, G. (1987) System of Experimental Design - Engineering Methods to Optimize Quality and Minimize Costs, Volume 2, UNIPUB/Kraus International Publications, New York
6. Wu, C. F. J., and Hamada, M. (2000) Experiments - Planning, Analysis, and Parameter Design Optimization, John Wiley & Sons, New York

Part 4

Studies in Areas Related to Reliability

Building In Reliability. Is Statistical Process Control Statistical Reliability Control ?

James Prendergast [1] and Eamonn Murphy [2]

[1] Institute of Technology, Clash, Tralee, Co. Kerry, Ireland.
[2] University of Limerick Plassey Technology Park, Limerick, Ireland

Abstract. This paper deals with the concept of using experimental design to verify the ideals of building in reliability in an integrated circuit fabrication process. The paper focuses on the gate oxide loop, which is one of the most critical steps in the process and uses a combination of a linear L12 Taguchi Array and Central Composite Face Centred Design experimental techniques to prove the Building In Reliability concept. The work will demonstrate that from a quality perspective many nodes are controlled in the process but only certain nodes have an impact on reliability. If these nodes are carefully controlled reliability can be maintained and indeed predicted possibly by using an expert system thus leading to "Statistical Reliability Control".

1 Introduction

Perhaps nowhere more than in the semiconductor industry is the quest for reliability taken so seriously [1-3]. The industry has achieved excellent levels of quality and reliability and these levels continue to be surpassed with each new technology introduction. The I.C. industry has embraced the concepts of Building In Reliability (BIR) [4-7] and Statistical Process Control (SPC) totally and now seeks new methodologies to improve reliability without additional costs. This new approach involves identifying critical reliability parameters in the process and placing these under statistical reliability control, (SRC).

Reliability has historically been proven, by demonstrating that products tested at the end of manufacturing from the process are reliable [8]. This requires large samples sizes and equipment, which are solely dedicated to the reliability demonstration. The reliability tests themselves are destructive and the samples are destroyed on completion. The test equipment used is microprocessor controlled with expensive power supplies and temperature controlled ovens all of which add to the manufacturing costs [9]. More importantly the data gleaned from such tests comes too late in the process and corrective actions, which are implemented, come after failure analysis when lots have possibly shipped to customers making the process ineffective.

A more suitable, proactive technique is the "Building In Reliability" (BIR) [9,10] approach. This approach places the emphasis on controlling critical parameters in the process very stringently and treating these parameters as reliability influencing parameters. Using this technique, corrective actions can be taken within the processing facility as required limiting exposure to reliability risks in the field. This shortens the feedback loop making corrective actions more effective.

Field failure data indicates that the gate oxide process is critical to the reliability of the integrated circuit. The gate oxide process loop in figure 1 has many parameters, which can affect the quality but only a few which will affect the reliability. There are however other process parameters outside the gate oxide process loop, which will also affect the reliability.

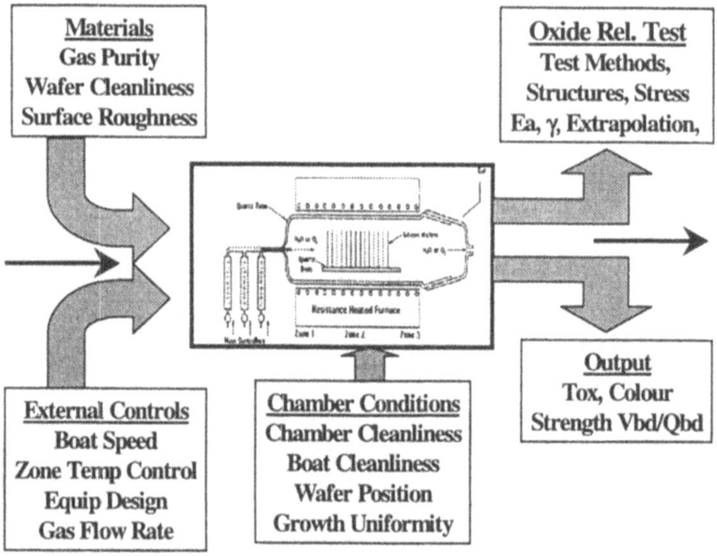

Figure 1
Gate Oxide Process Loop

The issue is to understand which parameters affect reliability only and place these under statistical control. The only logical method for understanding these parameters and their importance is experimental design.

2 Oxide Reliability Testing and Prediction

The reliability of an oxide can be tested and predicted by doing end of line reliability tests using dedicated test structures. The acceleration factors for reliability prediction are voltage and temperature. The reliability breakdown distribution for this 15nm oxide [11] is the lognormal distribution as shown in figure 2.

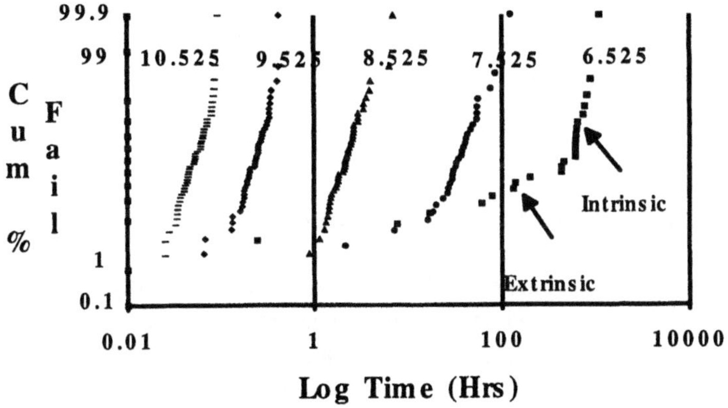

Figure 2
Typical Oxide Failure Distribution

From figure 2 both an intrinsic and extrinsic distribution can exist and analysis has shown that these distributions must be treated separately [11]. The intrinsic reliability is directly related to the strength of the oxide and the sigma values obtained for pure intrinsic reliability generally range from 0.2 to 0.5. The extrinsic distribution however is related to defects such as particles, miss processing or dislocations and is characterised by large sigma values and reduces the T0.1% failure times considerably. This reduction in T0.1% has a significant impact on the use lifetime of the oxide. For this particular oxide the model for intrinsic reliability prediction is:

$$Tf(\%)u = \{[\exp. - Ea/k(1/Tt - 1/Tu)]\exp.\gamma(Et - Eu)\}Tf(\%)t$$

where

Tf(%) is the time to failure at use and test conditions

Ea = Thermal Activation Energy

Tt and *Tu* are the test and use temperatures

Et and *Eu* are the test and use electric fields

K = Boltzman's constant

γ = field accelaration factor

Both the thermal activation energy and the field acceleration factors are defined from Time Dependent Dielectric Breakdown (TDDB) tests [11]. Using this model the time to failure for 0.1, 1.0 or 10% cumulative failures can be predicted at use conditions from the accelerated test condition.

The objective is now to relate the TDDB test results to the in line statistical process control values using experimental design and demonstrate with a degree of confidence that for specific nodes within the integrated circuit process the SPC values control the reliability.

3 Taguchi Optimisation Experiment

There are various parameters, which can affect the reliability of the oxide and using published data and engineering experience 11 critical parameters were identified for experimentation. An L12 Taguchi array was selected to refine the initial 11 parameters to 3 critical ones. It was deemed extremely important to reduce the 11 parameters to 3 to minimise the number of splits in the modelling experiment as cost was a significant factor with each experimental split costing US$ 5,000. This assumed that each parameter was independent and that there were no interactions. As this is only a screening experiment this assumption is valid and any interactions will be accounted for in the final experiment.

Table 1
Optimisation Experiment Factors and Factor Levels

Factor Description	Code	Cont	Level 1	Level 2
SAC Oxide	S	C	650 A	500 A
Gate Oxide Type	GOT	NC	Wet	Dry
G/ Poly Dep.Time	G	C	0 Hrs.	4 Hrs.
Type Of Clean A	Clean	NC	Clean A	Ohmi
Poly Dep. Temp	Temp	C	620 C	625 C
P/ Dope Method	Pdope	NC	Implant	$POCl_3$
Metal 1 Etch Time	Rf	C	15 sec	40 sec
Poly Etch Time	Pover	C	Standard	+30 sec.
RTA Temp.	R	C	1060 C	1100 C
DLM	DLM	NC	Yes	No
Etch Resist Strip	MRS	C	Std Time	2 x Std

The 11 factors selected are shown in Table 1 above and all the controllable (C) ones are varied within their statistical process control limits. There are 4 factors, which were not controllable (NC) and these were selected to generate engineering information for process improvements. As expected some of the experimental splits yielded extrinsic distributions and these allowed the ANOVA analysis to be performed on the total and the intrinsic portion of the distribution. The sample size tested for each experimental split varied from 42 to 64 units and all splits ran to full failure. The time to failure data generated from the TDDB testing at 9 MV/cm and 225 C is shown in Table 2. Table 2 indicates that for two of the splits the intrinsic and total distributions both have the same failure times due to the absence of the extrinsic population.

Table 2
Optimisation Experiment Results

Split	Total Distribution (Hrs.)				Intrinsic Distribution (Hrs.)			
	T0.1%	T1%	T10%	T50%	T0.1%	T1%	T10%	T50%
1	0	0	0	0	0	0	0	1
2	0.88	1.14	1.65	2.58	1.07	1.35	1.84	2.70
3	0.07	0.14	0.34	1.06	0.53	0.68	0.94	1.40
⋮								
11	2.27	3.22	5.18	9.31	4.58	5.65	7.52	10.73
12	0.85	1.21	1.96	3.53	2.24	2.24	2.95	4.12

3.1 Intrinsic Analysis

The ANOVA was performed on the T 0.1, T1.0, T10 and T50% failure times and the most significant factors investigated. The ANOVA table for the T0.1% cumulative failure is shown in Table 3.

Table 3
Optimisation Experiment T0.1% ANOVA Results

Factor	Pool	Df	S	V	F	S'	%rho
SAC	N	1	1.1E6	1.1E6	5.3	0.9E6	5.4
GOT	N	1	7.0E6	7.0E6	30.3	7.8E6	41.2
G	N	1	1.9E6	1.9E6	8.5	1.6E6	9.9
Clean	Y	1	0.2E6	0.2E6			
Temp	Y	1	0.1E6	0.1E6			
Pdope	N	1	1.9E6	1.9E6		1.6E6	9.4
Rf	N	1	0.6E6	0.6E6	2.2	0.3E6	1.9
Pover	Y	1	0.3E6	0.3E6			
MRS	Y	1	0.5E6	0.5E6			
R	N	1	2.3E6	2.3E6	10.4	2.1E6	12.5
DLM	N	1	0.9E6	0.9E6	4.0	0.7E6	4.3
Error (e)		4	0.9E6	0.9E6		2.6E6	15.5
Total			17.6E			17.6E	

All the times to failure were analysed in similar fashion and the six most significant factors for each time to failure are shown in Table 4.

Table 4
Intrinsic Distribution Significant Factors

%Tf	Factors					
	Most Significant					Least Significant
T50	GOT	Pdope	R	Rf	MRS	Pover
T10	GOT	Temp	Pover	Clean	S	G
T1.0	GOT	Pdope	R	G	DLM	RF
T 0.1	GOT	R	G	Pdope	S	DLM

For the total distributions the analysis was conducted in a similar fashion and the results are as shown in table 5.

Table 5
Total Distribution Significant Factors

%Tf	Factors					
	Most Significant					Least Significant
T50	GOT	Pdope	R	DML	S	Rf
T10	GOT	DML	S	G	Pdope	R
T1.0	DML	G	S	GOT	Pdope	Clean
T 0.1	DML	G	S	GOT	Pdope	Clean

The results indicate that factors such as the gate type, use of dual layer metal and poly dope methodology are important, however these are technology factors and are not controllable. While focusing on the controllable factors for both the intrinsic and combined data sets and investigating the lower percentage failures times the most significant factors, which affect the gate oxide reliability are:

A) Rapid Temperature Anneal (R)

B) Sacrificial Oxide Thickness (S)

C) Poly Gate Deposition Delay (G)

4 Modelling Experiment

The 3 critical factors listed above were subjected to a central composite face centred experimental design using RSDiscover/Explore. The experimental design in this section of the work is assumed to be non linear i.e. quadratic and the concepts of robust design were employed. Again the 3 factors were varied within their statistical process control limits to establish and model the effects of variation within the limits and comprehend any variability and interactions within the system. In this section test structures and test equipment used were the same as those in the optimisation experiment and the test conditions were 225 Deg. C and 8 MV/cm. The data sets generated for each experimental split gave an extrinsic tail to the distribution and this again allowed the analysis of both the intrinsic and the total distribution.

4.1 Experimental Worksheet and Results

The worksheet for the quadratic experiment is shown in Table 6. This shows the T50% and T0.1% cumulative failure results and the sigma values for intrinsic and total distributions. While not all the experimental splits exhibited an extrinsic population there was sufficient to analyse the factors against both the intrinsic and total distributions and compare results.

Table 6
Experiment Worksheet

Factors			Intrinsic Dist. Res.			Total Dist. Res.		
S	**G**	**R**	**T50%**	**T0.1%**	**Sigma**	**T50%**	**T0.1%**	**Sigma**
575	2	1060	36.29	12.39	0.35	35.31	10.77	0.38
575	2	1080	37.17	20.39	0.19	34.60	15.43	0.26
500	0	1100	29.65	15.30	0.21	27.50	10.82	0.30
⋮								
575	2	1080	42.13	20.33	0.22	42.13	20.33	0.22
575	2	1080	45.78	23.59	0.24	45.78	23.59	0.24

4.2 Statistical Analysis of the Intrinsic Data Set

Tables 7 and 8 indicate that the initial design model is a very good fit containing some very significant factors such as the S^2 and R^2 terms as both have a high F ratio. The model itself obeys the hierarchy with no evidence of lack of fit.

Table 7
Least Squares Component ANOVA, Final Design

Source	Df	Sum of Sq.	Mean Sq.	F-Ratio	Signif.
Constant	1	6182.3493			
S	1	1.4440	1.4440	0.24	0.6352
R	1	14.0660	14.0660	2.30	0.1534
S^2	1	26.3997	26.3997	4.31	0.0582
R^2	1	142.5766	142.5766	23.30	0.0003
Residual	8	75.5596	6.1200		

The initial model was decidedly non linear with an F Ratio of 4.15. The interaction terms had low F Ratios compared to the quadratic terms, were deemed to be insignificant and were discarded from the final model. The least squares component ANOVA for the refined and final design is shown in Table 7 and the least squares summary ANOVA is shown in Table 8.

Table 8
Least Squares Summary ANOVA, Final Design

Source	Df	Sum Of Sq.	Mean Sq.	F-Ratio	Signif.
Total (Corr.)	17	240.5748			
Regression	4	161.0152	40.2538	6.58	0.0040
Linear	2	15.5100	7.7550	1.27	0.3142
Non Linear	2	145.5052	72.7526	11.89	0.0012
Residual	13	79.5596	6.1200		
Lack Of Fit	10	72.5973	7.2597	3.13	0.1889
Pure Error	3	6.9623	2.3208		

$$T\ 0.1\% = 20.68 + 0.38S - 1.186R + 2.92S^2 - 6.78\,R^2 \qquad (1)$$

Using the time to failure model previously defined for intrinsic reliability the SPC values for the selected nodes can be translated to reliability use conditions.

4.3 Residual Analysis

Figure 3 represents a plot of the studentized residuals versus the fitted values of the response.

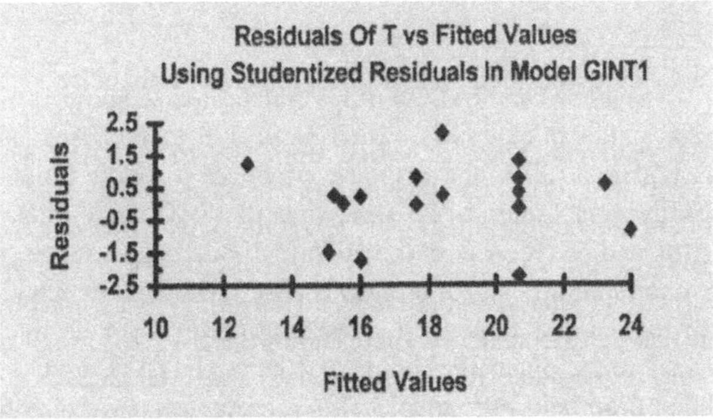

Figure 3
Plot of Residuals Of TDDB 0.1% vs. Fitted Values

The plot is random and no obvious pattern exists indicating that the model assumptions may be correct. Response curves were also generated and these indicate that the SAC term is a positive quadratic while the RTA term is negative quadratic.

4.4 Response Surface Analysis of the Intrinsic Data

The graph in figure 4 shows a contour plot which analyses the results of the experiment to maximise the TDDB (0.1%) reliability result versus the process settings. It has been seen from the analysis that the gate to poly deposition delay was not a critical factor and the factors plotted are the SAC oxide thickness versus the RTA. From the graph the competing effects of SAC and RTA are seen.

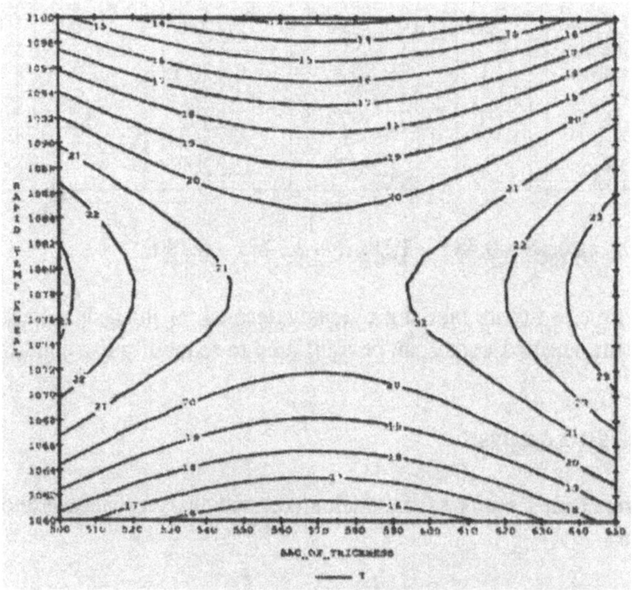

Figure 4
Contour Plot of TDDB (0.1%) Intrinsic Reliability

When the SAC oxide thickness is varied from 500 to 650 Ang. and the RTA varied from 1060 to 1100 C the response produces a saddle point. The TDDB (0.1%) results are a maximum of 23 Hrs. when the SAC oxide thickness is either 500 or 650 Ang. and the RTA is in the 1075 to 1082 Deg. C. range. At the centre point settings the reliability sits at around 20 Hrs., which is acceptable, but below the maximum that can be achieved by resetting the SAC oxide thickness. If the RTA values vary to the outer SPC limits of 1060 and 1100 Deg. C. the reliability of the oxide can drop below 17 and 13 Hrs. respectively. From the figures 4 it is recognisable that the approximate process setting of an RTA equal to 1080 Deg. C. and a SAC oxide thickness of 575 Angstroms give the most robust processing conditions.

4.5 Confirmation Run

Confidence intervals at +/- 95% were generated and a confirmation run was performed to evaluate the experimental results. These results are shown graphically in Figure 5 and indicate that all the values from the confirmation run fall within the +/- 95% confidence intervals. It is evident from the data that there is good agreement between the observed and the predicted results.

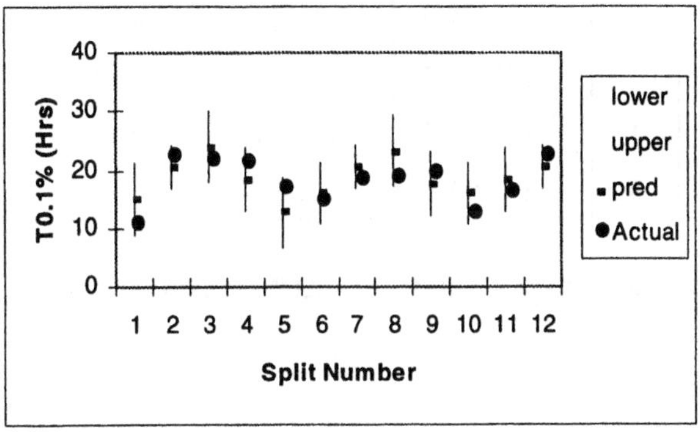

Figure 5
Confirmation Run Failure Times vs. Predicted and Confidence Intervals

4.6 Extrinsic Analysis

Table 6 discussed previously, also contains the data for the extrinsic worksheet. Once again the model was assumed to be quadratic and the initial model contained all quadratic and interaction terms. In this instance the RTA and the interaction between the Gate Delay and SAC oxide thickness were very significant. The model was refined and the least squares component ANOVA is shown in Table 9.

The inclusion of the interaction term is very important as it is the product of the gate delay and the sacrificial oxide thickness both which can contribute to the extrinsic distribution. These factors will either allow particles to accumulate on the surface of the gate oxide before the poly layer is deposited or result in a sacrificial oxide thickness for which the oxide removal process is not optimised. This would lead to further inspections and possibly rework or partial removal of damaged silicon prior to the gate deposition, contributing to the extrinsic distribution.

Table 9
Least Squares Component ANOVA, Final Model for the Total Distribution

Source	Df	Sum Of Sq.	Mean Sq.	F-Ratio	Signif.
Constant	1	4576.05			
S	1	1.3496	1.3496	0.11	0.7422
G	1	7.5169	7.5169	0.63	0.4419
G	1	49.5953	49.5953	4.17	0.0637
S*G	1	43.1521	43.1521	3.63	0.0809
R^2	1	103.9525	103.9525	8.75	0.0120
Residual	8	142.6090	11.8841		

$$TO.1\% = 18.63 + 0.37S - 0.87G - 2 - 23R - 2.32SG - 4.84R^2 \qquad (2)$$

4.7 Residual Analysis

Figure 6 shows a plot of the studentized residuals versus the fitted values. There is no apparent relationship between the residuals indicating again that the model assumptions may be correct and the model is indeed quadratic.

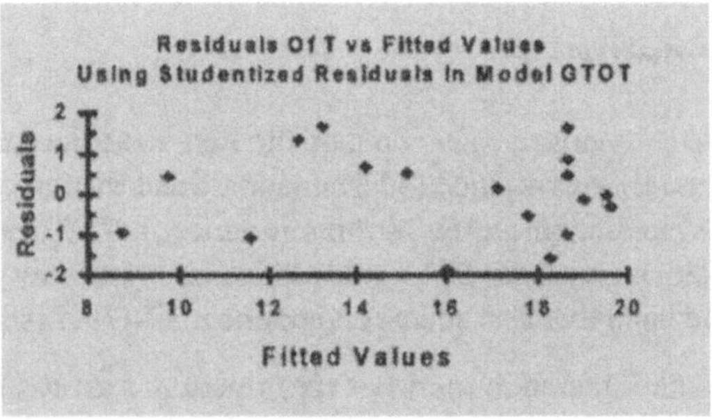

Figure 6
Residuals Plot for the Final Model, Total Distribution

4.8 Analysis of the Interaction Term

The interaction effect seen in equation 2 is very important and results from using a robust design. The interaction is shown graphically in figure 7. This shows the effect of varying the SAC oxide thickness for different gate delays ($G = 0$ and 4 Hrs.) The effect on the T0.1% is linear when each gate delay is considered individually however it is important to note that the two lines have different slopes. The slope for G equal to 0 is positive and the time to failure increases with increasing SAC oxide thickness. On the other hand the slope of the G = 4 hours is negative and as a result the time to failure decreases with increased SAC oxide thickness.

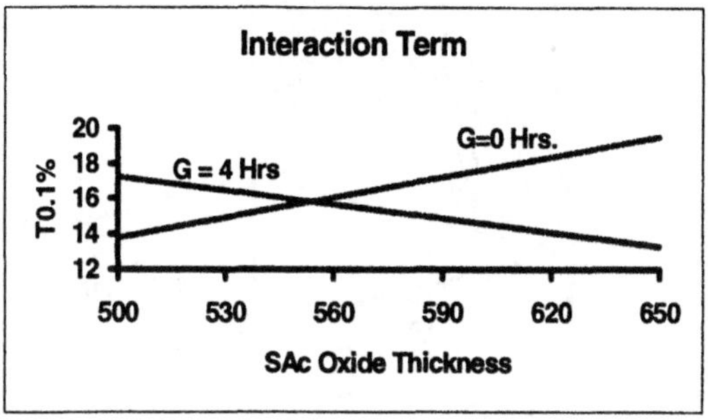

Figure 7
Interaction Effect of SAC Oxide Thickness with the Gate to Poly Deposition Delay for T0.1% Failure

For the statistical process control limits for the process the lines intersect at a SAC oxide thickness of approximately 560 Angstroms. At 650 Angstroms SAC oxide thickness the difference in failure times for an RTA setting of 1080 C is approximately 6.5 Hrs. with a gate delay of 0 Hrs. giving a higher result. At 500 Angstroms the difference in failure times is 3 Hrs. but in this case the gate poly deposition delay of 4 hours gives the higher failure time of 17 hours. At the 575 Angstrom SAC oxide thickness, which is the optimum for intrinsic reliability the delay of 0 hours is slightly better than the 4 hour delay. This result is very important as it confirms the presence of the gate delay, which was found during the Taguchi L_{12} optimisation experiment. It is noticeable that the gate delay did not appear in the intrinsic model and this is very understandable. The gate delay contributes to the presence of particles on the gate surface before the polysilicon is deposited. These particles are very small and distort the electric field across the gate oxide thus producing extrinsic failures.

4.9 Response Surface Analysis of the Total Distribution

Figures 8 and 9 are threeds showing the response surface of the gate oxide reliability as the gate poly deposition delay is varied from 0 to 4 hours. In these figures the negative quadratic effect of the RTA is clearly seen. Also obvious is the interaction effect outlined above when the poly deposition delay goes from 0 to 4 hours. When the gate poly deposition delay changes from 0 to 4 hours the T0.1% failure pattern changes and the pattern is most stable for a gate poly deposition delay of 2 hours.

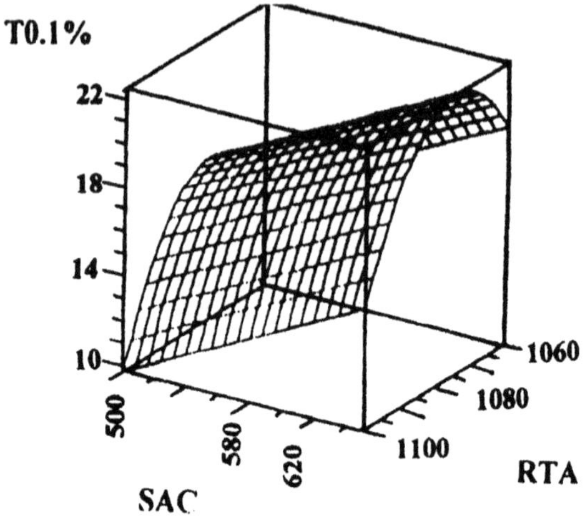

Figure 8
Threed of SAC Oxide Thickness vs. RTA
Gate Poly Deposition Delay of 0 Hours

The most robust process conditions are at 575 Angstroms SAC oxide thickness, a gate poly deposition delay of 2 hours and an RTA of 1080 C. However if the delay is not controlled to 0 hours the T0.1% cumulative failure time can vary between 10 and 22 hours which is not a very robust process if the SAC oxide thickness and the RTA are allowed to vary over the full process range.

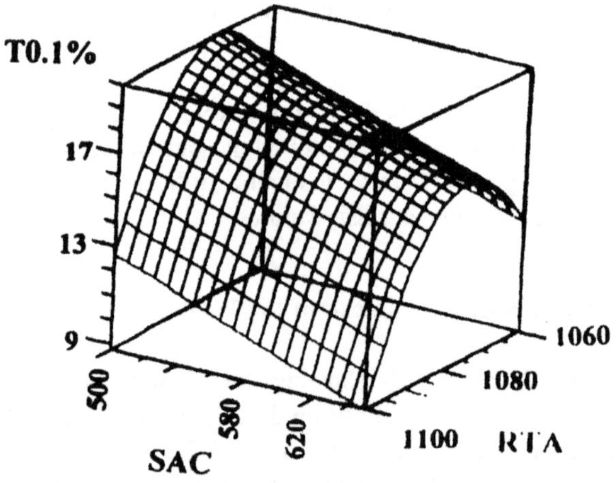

Figure 9
Threed of SAC Oxide Thickness vs. RTA
Gate Poly Deposition Delay of 4 Hours

For a gate poly delay of 4 hours and an RTA of 1080 deg. C. the T0.1% cumulative failure time can vary between 16 and 20 hours but if the RTA is not controlled the T0.1% failure time can reach as low as 9 hours. For a gate poly deposition delay of 2 hours and an RTA value of 1080 deg. C. a robust setting can be obtained by allowing the SAC oxide thickness to vary from 500 to 650 Angstroms, which results in a T0.1% cumulative failure of 19 hours. However controlling the gate to poly deposition delay to less than 2 hours may not be a feasible option in production so a preferred option may be setting the SAC oxide thickness value to 575 Angstroms, the RTA to 1080 deg. C. and allowing the gate poly deposition delay to vary between 0 and 4 hours. This will give a T0.1% cumulative failure time of between 19 and 22 hours respectively. The settings of 575 Angstroms for the sacrificial oxide thickness and 1080 deg. C. for the RTA are also the optimum settings for the intrinsic reliability.

4.10 Confirmation Run for the Total Distribution

Confidence intervals were generated at +/- 99.5 % and a confirmation run was performed to confirm the results. The results of the confirmation run are shown in Table 10. The results show that the values obtained from the confirmation run agree with the predicted values at 99.5% confidence.

Table 10
Confirmation Run Values for the Total Distribution

Factors			Predicted	Confirmation	Within
S	G	R	Value	Run Value	C.I. Values
575	2	1080	18.63	22.48	yes
650	2	1080	19.00	13.03	yes
650	0	1060	19.57	16.68	yes
575	2	1100	11.57	7.52	yes
650	0	1100	15.12	14.39	yes
500	0	1060	14.20	12.29	yes
650	4	1100	8.75	12.20	yes
650	4	1060	13.20	9.19	yes
575	2	1080	18.63	14.70	yes

5 Discussion and Conclusion

1) The use of experimental design validates the BIR process and confirms the work done by Jensen and others.
2) The work identified 3 critical nodes from the ones selected and used these in a face centred cubic array to model the response
3) A quadratic response was identified and modelled.
4) The models shorten the reliability corrective action feedback loop.
5) The RTA has a significant quadratic effect but the SAC oxide thickness is linear.
6) The results from the selected parameters allow the prediction of the purely intrinsic and total reliability of the oxide from statistical reliability control nodes.

Note 1: The full version of any truncated tables can be obtained from the authors.

References

1. Jensen, Finn. Electronic Component Reliability. Wiley
2. Nelson, Wayne. Accelerated Testing. Wiley 1990
3. Bentley, John P. Introduction to Reliability and Quality Engineering. Addison Wesley
4. Prendergast, J. Reliability Yield and Quality – Correlation for A Particular Failure Mechanism. 31st. Annual Reliability Physics Symposium 1993
5. Prendergast, J. Building In Reliability – Implementation and Benefits. International Journal of Quality And Reliability Management, Vol. 13, No.5, 1996
6. Condra, L.W. Reliability Improvements with Design of Experiments. New York: Marcel Decker
7. Prendergast, J. Reliability Yield & Quality, Correlation for A Particular Failure Mechanism. 31st. I.E.E.E International Reliability Physics Symposium, 1993
8. Ender, U. Quality And Reliability Assurance for Electronic Systems – Actions and Results. Quality And Reliability Engineering International, Vol. 2, 127-130, 1987
9. Crook, D.L. Evolution of VLSI Reliability Engineering. 2nd. European Symposium on Reliability of Electronic Devices, Failure Physics and Analysis, 1991
10. Schafft, H.A., Baglee, D.A., Kennedy, P.F. Building In Reliability – Making it Work. Proc 29th. Annual IEEE Reliability Physics Symposium 1991
11. Prendergast, J. Investigation of the Intrinsic SiO2 Area Dependence Using TDDB Testing. 1997 I.E.E.E International Reliability Workshop, Oct. 1997

Optimal Warranty Period when Sale-Price Increases with the Lower Specification Limit

Shaul P. Ladany and Haim Shore

Department of Industrial Engineering, Faculty of Engineering Science,
Ben-Gurion University of the Negev, P.O.B. 653, Beer-Sheva 84105, Israel

Abstract. The optimal warranty period (OWP) for a product with a single critical variable (like life-time of a car battery) often becomes the manufacturer's Lower Specification Limit (LSL). In this paper the OWP problem is addressed under the assumption that with the increase of the warranty period the cost incurred on non-conforming items grows but so is also the achievable sale-price. The OWP problem is first solved where the critical quality variable follows an exponential distribution, and then a flowchart is developed for the solution routine of the general case (any distribution). Employing Shore's inverse normalizing transformation, a distribution-free solution is demonstrated for several numerical cases.

1 Introduction

Warranties offered by producers of manufactured goods are integral characteristics of the quality of products. Many different warranty policies exist, and likewise a considerable number of research publications address various issues associated with the determination of optimal warranty policies (for a survey of some of these issues refer to Elsayed, 1996). While these publications commonly deal with the evaluation of the economic cost of the warranties and the best warranty policy for a given warranty period, the problem of determining optimal warranty period in relation to the manufacturer's lower specification limit rarely appears in the literature.

In this work we deal with the situation in which the manufacturer's upper specification limit (USL) is given, and it is assumed that the warranty period is determined by the lower specification limit (LSL) of the life-time of the manufactured product. Since an increased warranty period is expected to incur an increased sale-price, we assume here that the sale-price per manufactured item increases linearly with the warranty period, and therefore also with the LSL. The optimal warranty period (OWP) problem is therefore to determine an optimal LSL that would maximize the expected revenue per item. In Section 2 we investigate the case where the critical quality characteristic is exponentially distributed, and investigate the conditions for the existence of an optimal solution (LSL which provides maximum average revenue per produced item). Section 3 generalizes the analysis to any statistical distribution. Section 4 applies one of Shore's inverse normalizing transformations (Shore, 2000) to derive solutions when the distribution of the quality characteristic is unknown, however a sample of observations allows estimation of moments of second degree, at most. In Section 5 we demonstrate numerically the ease-of-use of the derived method for partially specified life-time distributions. Section 6 discusses some possible extensions of the current models, relate to some sensitivity analysis issues and offers concluding remarks.

2 Exponentially Distributed Life-Time

Assume that the manufacturer generates a revenue (or net contribution to profit) of r_s per conforming unit (a unit with life-time between the LSL and the USL). While obviously customers will not complain if life-time of the product exceeds the USL, this may constitute an undesirable outcome for the manufacturer who will sell fewer items. The effect of the reduced sales volume may be expressed in terms of lower revenue (or contribution to profit) per unit, $(r_U)\,r_s\,(0<r_U<1)$, for units with life-time above the USL, where $(1\text{-}r_U)$ represents the fractional (relative) loss per unit for these items.

If the life-time is within the warranty period (below LSL), the manufacturer incurrs a cost for realized warranty expenditures, but possibly also for some good-will loss. In this case the revenue (or contribution to profit) per unit will be $(r_L)\,r_s$ $(0<r_L<1)$, where $(1\text{-}r_L)$ is the fractional (relative) loss per unit for items with life-time below the LSL. Though r_L may vary between units according to their actual life-time, it is assumed here that r_L represents the average of these varying proportions.

Denote by $f(x)$ the density function of the life-time of an item. The expected revenue (or contribution to profit) per item, π, is:

$$\pi = r_s \cdot r_L \int_{x=\infty}^{\mathrm{LSL}} f(x)dx + r_s \int_{x=\mathrm{LSL}}^{\mathrm{USL}} f(x)dx + r_s r_U \int_{x=\mathrm{USL}}^{\infty} f(x)dx \tag{1}$$

If the manufactured product has life-time that is exponentially distributed with parameter λ, namely:

$$f(x) = \lambda\ \mathrm{Exp}(-x), x > 0 \tag{2}$$

the expected revenue per item is:

$$\pi = r_s[\ r_L + (1 - r_L)\mathrm{Exp}(-\lambda\ \mathrm{LSL}) + (r_U - 1)\mathrm{Exp}(-\lambda\ \mathrm{USL})\ ] \tag{3}$$

Let us now incorporate in the analysis the effect of the warranty period on the sale price. It is obvious that the latter may be increased with increased warranty period, and therefore r_s is expected to be a uniformly increasing function of LSL. It is assumed here that a linear relationship exists between the two:

$$r_s = a + b\text{LSL} \tag{4}$$

where the parameters, a and b, are positive. Substituting (4) into (3) we obtain:

$$\pi = (a + b\text{LSL})\ [\ r_L + (1 - r_L)\text{Exp}(-\lambda\text{LSL}) + (r_U - 1)\text{Exp}(\lambda - \ \text{USL})] \tag{5}$$

The optimal warranty period, LSL*, will maximize (5). Differentiating (5) with respect to LSL we obtain:

$$\partial\pi / \partial\text{LSL} = (1 - r_L)\ \text{Exp(-}\ \lambda\text{LSL})\ [-\lambda\ a + b - \lambda\ b\ \text{LSL}]\ +$$

$$b(r_U - 1)\text{Exp}(-\lambda\text{USL}) + br_L \tag{6}$$

Equating (6) to zero, we obtain a necessary condition for optimality:

$$\text{Exp}(-\lambda\text{LSL*})\big[1 - \lambda(a/b + LSL\,*\big] =$$

$$\big[(1 - r_U)/(1 - r_L)\big]\big[\text{Exp}(-\lambda\text{USL}) - r_L/(1 - r_U)\big] \tag{7}$$

The right-hand-side (RHS) is independnet of LSL, and it may be either positive or negative. Thus, two cases may be investigated:

Case A: $\text{Exp}(-\lambda\text{USL}) - r_L/(1 - r_U) < 0$

Case B: $\text{Exp}(-\lambda\text{USL}) - r_L/(1 - r_U) \geq 0$

For case A, eq. 7 mandates that:

$$\text{Exp}(-\lambda\text{LSL*})\big[1 - \lambda(a/b + \text{LSL*})\big] < 0 \tag{8}$$

which can be satisfied only if: $\big[1 - \lambda(a/b + \text{LSL*})\big] < 0$, i.e., when:

$$\text{LSL*} > (1/\lambda) - (a/b) \tag{9}$$

Similarly, for Case B a necessary condition for optimality is:

$$\text{LSL*} \leq (1/\lambda) - (a/b) \tag{10}$$

Note, that in the above analysis we have assumed that r_L and r_U are confined to the interval (0,1), as earlier specified. Though situations might exist when $r_U \leq 0$ and/or $r_L \leq 0$, these are highly unusual cases, and we will ignore them here.

The solution of (7) provides maximal π only if the second derivative with respect to LSL, at LSL*, is negative. From (6) we obtain for the second derivative:

$$\partial^2\pi / (\partial\text{LSL})^2 = -\lambda(1 - r_L)(b)\text{Exp}(-\lambda\text{LSL}) -$$

$$\lambda(1-r_L)(b)\text{Exp}(-\lambda LSL)\left[1-\lambda(a/b+LSL)\right] \tag{11}$$

Substituting for the second term in (11) from the RHS of (7) we get:

$$\partial^2\pi/(\partial\text{LSL})^2{}_{|\text{LSL*}} = \lambda(1-r_L)(b)\text{Exp}(-\lambda\text{LSL*}) -$$

$$\lambda(1-r_U)(b)\left[\text{Exp}(-\lambda\text{USL}) - r_L(1-r_U)\right] \tag{12}$$

For (12) to be negative we observe that:

$$\lambda(1-r_L)(b)\text{Exp}(-\lambda\text{LSL*}) > -\lambda(1-r_U)(b)\left[\text{Exp}(-\lambda\text{USL}) - r_L/(1-r_U)\right] \tag{13a}$$

which reduces to:

$$\text{Exp}(-\lambda\text{LSL*}) > -\left[(1-r_U)/(1-r_L)\right]\left[\text{Exp}(-\lambda\text{USL}) - r_L/(1-r_U)\right] \tag{13b}$$

Again introducing from (7), into the RHS of (13b), we note that:

$$\text{Exp}(-\lambda\text{LSL*}) > -\text{Exp}(-\lambda\text{LSL*})\left[1-\lambda(a/b+\text{LSL*})\right] \tag{14a}$$

which reduces to:

$$\text{LSL*} < 2/\lambda - a/b \tag{14b}$$

The necessary (but not sufficient) conditions for an optimal (maximum) solution are summarized in a flowchart in Figure 1.

Two numerical examples now follow.

Example 1

$\lambda = 0.025;\ a = 1;\ b = 0.5;\ \text{USL} = 1000;\ r_L = 0.2;\ r_U = 0.8$

$$\text{Exp}(-\lambda\text{USL}) - r_L/(1-r_U) = 1/\text{Exp}(25) - 1 = -1 \ \rightarrow \text{Case } A$$

$$38 < \text{LSL+} < 78$$

The solution of eq. 7 yields: LSL* = 78. This does **not** meet the second condition (eq. 14), which implies that the second derivative, at LSL*, is not negative, and therefore **no** optimal LSL exists that **maximizes** π. Ploting π as a function of LSL shows that π increases with LSL indefinitely.

Example 2

$\lambda = 0.025,\ a = 1;\ b = 0.05;\ \text{USL} = 40;\ r_L = 0.05;\ r_U = 0.5$

$$\text{Exp}(-\lambda\text{USL}) - r_L/(1-r_U) = 1/\text{Exp}(1) - 0.05/(1-0.5) = 0.27 \ \rightarrow \text{Case } B$$

$$\text{LSL*} \le 20$$

The solution of eq. 7 yields: LSL* = 12.3. This solution meets both conditions and generates a maximum revenue of $\pi = 0.9118$.

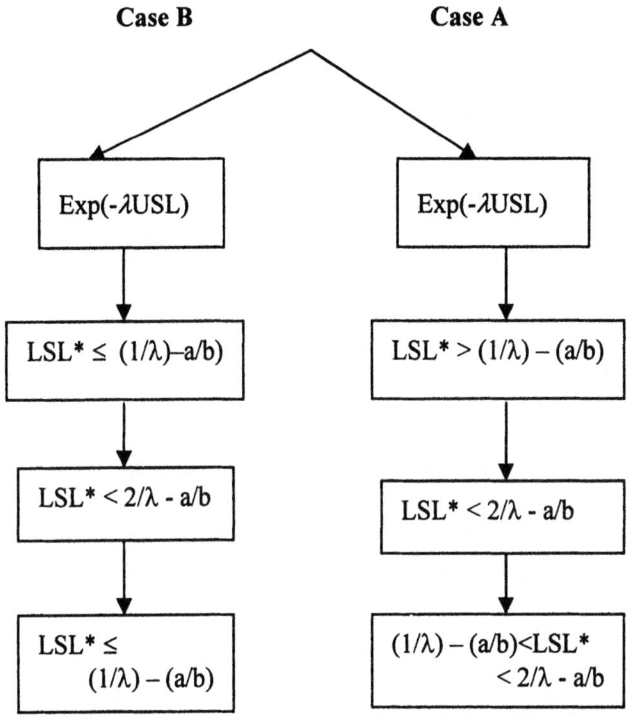

Fig. 1: A flowchart for the necessary conditions for optimality (maximal π) for the exponential case

3 Generally Distributed Life-Time

If we discard from the previous analysis the assumption about the life-time being exponentially distributed, the analysis may be repeated in more general terms. Let the cumulative density function at $X = x$ be $F(x)$.

Equation (1) can be re-written as:

$$\pi = r_s r_L F(\text{LSL}) + r_s\left[F(\text{USL}) - F(LSL)\right] + r_s r_U\left[1 - F(\text{USL})\right] \tag{15}$$

Introducing for r_s from (4) we get:

$$\pi = (a + b\text{LSL})[(r_L - 1)F(\text{LSL}) - (r_U - 1)F(\text{USL}) + r_U] \tag{16}$$

$$f(x) = \text{Exp}(-x), x > 0$$

Analogously with (7), we obtain on differentiating (16) with respect to LSL and equating to zero:

$$[(a/b)+\mathrm{LSL}^*]f(\mathrm{LSL}^*)+F(\mathrm{LSL}^*)=[(1-r_U)/(1-r_L)]\,[F(\mathrm{USL})-r_U/(r_U-1)]=C \quad (17)$$

The RHS of (17) is independent of LSL, and therefore it was denoted as a constant, C. For positive a and b, and, as before: $0 < r_L, r_U < 1$, eq. 17 implies:

$$[\,F(\mathrm{USL})-r_U/(r_U-1)]>0 \quad (18a)$$

or:

$$F(\mathrm{USL})>r_U(r_U-1) \quad (18b)$$

which is always met since the RHS is always negative.

The second derivative with respect to LSL is, from (16):

$$\partial^2\pi/(\partial\mathrm{LSL})^2=b(r_L-1)[\,2f(\mathrm{LSL})+(a/b+\mathrm{LSL})\partial f(\mathrm{LSL})/\partial\mathrm{LSL}\,] \quad (19)$$

To evaluate (19) at LSL*, we observe from (17) that:

$$a/b+\mathrm{LSL}^*=[C-F(\mathrm{LSL}^*)]/f(\mathrm{LSL}^*) \quad (20)$$

On substituting into (19) we obtain:

$$\partial^2\pi/(\partial\mathrm{LSL})^2_{|\mathrm{LSL}^*}=b(r_L-1)\{2f(\mathrm{LSL}^*)+[C-F(\mathrm{LSL}^*)]/f(\mathrm{LSL}^*)$$

$$\partial f(\mathrm{LSL}^*)/\mathrm{LSL}^*\} \quad (21a)$$

and since:

$$\{\partial f(\mathrm{LSL}^*)/\partial\mathrm{LSL}^*\}/f(\mathrm{LSL}^*)=\partial\{\log[f(\mathrm{LSL}^*)]\}/\partial\mathrm{LSL}^*$$

eq. (21a) can be expressed also as:

$$\partial^2\pi/(\partial\mathrm{LSL})^2_{|\mathrm{LSL}^*}=b(r_L-1)\{2f(\mathrm{LSL}^*)+$$

$$[C-F(\mathrm{LSL}^*)]\{\log[f(\mathrm{LSL}^*)]\}/\partial\mathrm{LSL}^* \quad (21b)$$

For LSL* to generate maximal π, eqs. 21 should be negative, namely:

$$2f(\mathrm{LSL}^*)+[C-F(\mathrm{LSL}^*)]\partial\{\log[f(\mathrm{LSL}^*)]\}/\mathrm{LSL}^*>0 \quad (22a)$$

which can also be expressed as:

$$\{[F(\mathrm{LSL}^*)-C]/f(\mathrm{LSL}^*)\}\partial\{\log[f(\mathrm{LSL}^*)]\}/\mathrm{LSL}^*<2 \quad (22b)$$

Let us investigate (22). The expressions f(LSL*), F(LSL*) and C (refer to eq. 17) are always positive. However, ∂f(LSL*) / ∂LSL* (and likewise, ∂\{log [fLSL*)]\}) / ∂LSL*) may be positive, negative or zero, dependent on the life-time distribution and on LSL*. For the exponential distribution, this derivative is always negative, but for the normal it is positive for LSL* below the mean, and negative for LSL* above the mean.

The solution procedure for the general case, taking into account condition (22), is summarized in the flowchart of Figure 2.

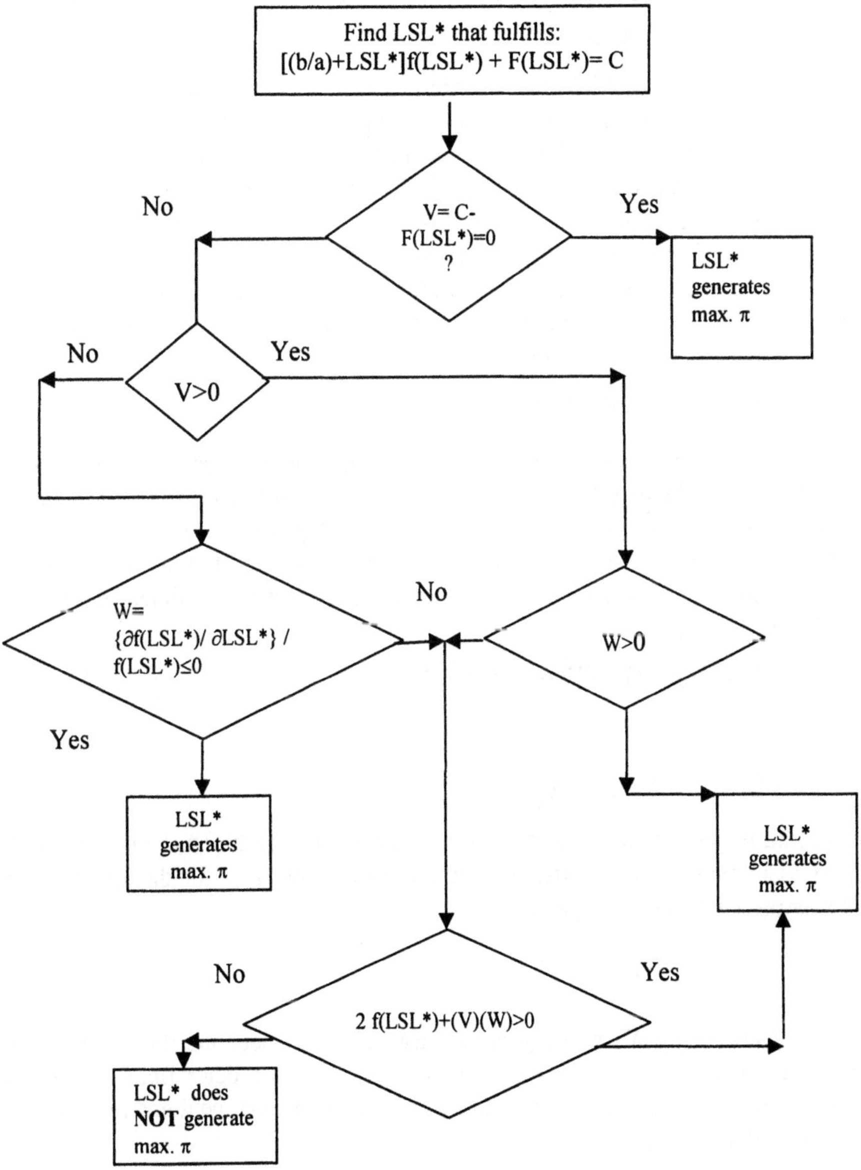

Figure 2: A flowchart for the conditions that generate/do not generate maximal π

For the specific case when life-time is exponentially distributed, eqs. 17 and 22 reduce to eqs. 7 and 12, respectively, as expected.

4 A General Solution when Life-Time Distribution is only Partially Specified

Occasionally the exact life-time distribution is not specified, and we wish to find an optimal solution when only partial distributional information, like the first few moments of the life-time distribution, is available. In this section we apply Shore's inverse normalizing transformations (INTs; Shore, 2000) to obtain a general distribution-free solution to the OWP problem. Let (x, z) be the respective p-th quantiles of X, life time, and Z, a standard normal variate, where: $F(x) = \Phi(z) = p$, and $\Phi(z)$ is the cumulative density function of Z. In Shore (2000) various INTs have been derived which allow us to express quantiles of X in terms of quantiles of Z. Here we will use the following INT:

$$x = M\ \mathrm{Exp}\{B[\mathrm{Exp}(C_z) - 1] + D_z\} \tag{23}$$

For given sample data, various moment-fitting oriented procedures have been developed in Shore (2000). These procedures require estimation of moments (partial and complete) of second degree, at most. A procedure which requires only first degree moments is presented here. Assume that the life-time median value, M, is known, and let: $\mathrm{LT} = \log(X)$ be the log-transformed life-time. The mean of LT is, from (23):

$$\mu(\mathrm{LT}) = Log(M) + B[\mathrm{Exp}(C2/2) - 1] \tag{24}$$

and therefore:

$$B = [\mu(\mathrm{LT}) - Log(M)]/[\mathrm{Exp}(C^2/2) - 1] \tag{25}$$

Substituting for B back into (23), we obtain a quantile function in terms of M, μ(LT)and two unknown parameters: C and D. Let and μ_L (X)be the upper and the lower partial means of X, namely:

$$\mu_U(X) = \int_M^{\infty} xf(x)dx \tag{26}$$

and $\mu_L(X) = \mu - \mu_U(X)$, where μ is the mean of X. By equating the partial means with those of (23), the unknown parameters may be determined. Numerical demonstration for the accuracy of this fitting procedure can be found in Table 4 of Shore (2000).

Introducing into (23) the quantile that corresponds to LSL, we obtain, in terms of the corresponding standard normal quantile, z_L :

$$\text{LSL}(z_L) = M\ \text{Exp}\{B[\text{Exp}(Cz_L) - 1] + Dz_L\} \tag{27}$$

Thus, instead of searching for LSL*, using (16), we will search for the optimal $\{z_L^*, P_L^*\}$, where: $P_L^* = \Phi(z_L^*)$. Re-formulating the objective function in terms of these decision variables, we obtain for eq. 16, in terms of $\{z_L, P_L\}$:

$$\pi = [a + b\text{LSL}(z_L)][(r_L - 1)P_L - (r_U - 1)P_U + r_U] \tag{28}$$

where P_U (= F(USL)) may be easily obtained, for given USL, from the fitted eq. 23. The expression in (28) may be simplified if we use the recently derived highly accurate approaximation for $\Phi(z)$ (Shore, 2000):

$$P = \Phi(z) \cong 1 - \text{Exp}\{-Log(2)\ \text{Exp}\{B[\text{Exp}(C_z) - 1] + D_z\}\} \tag{29}$$

where: $B = -1.81280$; $C = -0.472245$; $D = 0.294549$

For $z > 0$, the error does not exceed ± 0.00002. For $z < 0$, the error may be somewhat larger (maximally around 0.0007). Introducing for P from (29) into (28) we obtain an explicit expression in z, the optimal value of which may be easily searched.

5 Numerical Examples

The procedure detailed in the previous section was applied to various combinations of the cost parameters and life-time distributions. The results, including some sensitivity indicators, are exhibited in Table 1. P_{Le}^* is the exact CDF corresponding to the exact LSL*, namely: $P_{Le}^* = F(\text{LSL}^*)$. Analogously, P_{La}^* is the approximate probability value derived from the optimal solution of eq. 28.

We realize that highly accurate solutions are obtained. Note, however, that occasionally a local maximum is derived, and the true global optimum resides at the end points of the interval, namely: either at $P_L^* = 0$ or $P_L^* = 1$. These points should therefore be also searched to ascertain that indeed a real global optimum has been achieved.

6 Concluding Remarks

In this paper it was assumed that the relative reduction in contribution to profit for items with life-time below the LSL, $1\text{-}r_L$ is constant. Since the selling price for an item was assumed to grow linearly with LSL, while larger LSL will increase the percentage of non-confoming units (thus reducing expected total profit), an optimization problem is generated. The solution of this problem has been

Table 1: Numerical examples for five life-time distributions

	Distribution				
Distribution Parameters	**Normal** (14,2)	**Exponential** (1,5)	**Gamma** (3,2)	**Weibull** (1.125,5)	**Weibull** (0.4,5)
Mean (μ)	14	5	6	4.790	5.665
Var. (σ^2)	4	25	12	18.19	50.99
Skewness	0	2	1.155	1.676	2.815
Median (M)	14	3.466	5.348	3.610	2.000
***B* (eq. 23)**	-0.1342	-2.155	-1.676	-1.933	-5.059
***C* (eq. 23)**	-0.3887	-0.4318	-0.267	-0.4300	-0.4450
***D* (eq. 23)**	0.0905	0.2237	0.160	0.1951	0.6314
USL	20	70	90	60	20
Cost Par.:					
a	100	100	100	100	100
b	2	100	5	20	15
r_U	0.9	0.5	0.5	0.4	0.9
r_L	0.5	0.1	0.05	0.3	0.01
Solutions:					
{P_{Le}*,	.02865	.6838$^{(1)}$	.02246	.4969$^{(1)}$	.8959$^{(2)}$
P_{La}*}	.02920	.6861$^{(1)}$	.02247	.5034$^{(1)}$	.8962$^{(2)}$
{$LSL_e(P_{Le}$*),	10.20	5.751	1.187	3.581	38.51
$LSL_e(P_{La}$*)}	10.21	5.794	1.187	3.641	38.60
{$\pi_e(P_{Le}$*),	118.6	259.9	103.7	111.9	74.12
$\pi_e(P_{La}$*)}	118.6	259.9	103.7	111.9	74.12
$LSL_a(P_{La}$*)	10.19	5.804	1.188	3.641	38.86

$\pi_e(P)$ - Exact profit, given P (eq. 16); $LSL_e(P)$ - Exact LSL, given P

$LSL_a(P)$ - Approximate LSL, given P (eq. 27);

(1) Local max. Highest profit at P_L* = 1; (2) Local max. Highest profit at P_L* = 0.

developed in this paper. We will now discuss such extensions and issues associated with the new model.

First, while a constant r_L is the common case (as assumed here), other situations may be envisaged, where r_L itself depends on LSL. In such cases the contribution $r_s r_L$ may no longer be a linear function of LSL (as in this paper). These cases will be investigated in a future study. Secondly, one may wish to investigate cases other than NIB (nominal is best), which necessarily implies a double-sided specification limit. This case was pursued here since we have assumed that very high life-time may damage future sales. However, cases associated with only one-sided specification limit are also common, for example: when only an USL is given since the lower the quality characteristic the better (known as LIB, lower is better). In that case, lowering the guaranteed USL becomes a marketing advantage, which the producer may wish to optimize. A solution procedure similar to the one developed here may be easily obtained.

Another relevant issue is the sensitivity of the optimal solution to the parameters of the model. Some parameters may not be easy to estimate or quantify, like r_u. An investiagtion of how estimation errors or changes in these parameters may affect the optimal solution may deliver some insight with respect to the allowable errors in determining these parameters.

Finally, some factors not currently in the model may affect the optimal warranty policy, like market conditions or warranty policies of competitors. These factors may be incorporated into the model via the objective function or as additional constraints. In particular, warranty policies that are customary in certain industries may place restrictions on the allowable values of the optimal decision variable. For example, current warranty policies of most car manufacturers provide warranty coverage for 60*k* kilometers. Such an acceptable norm apparently places a lower bound on the allowable optimal LSL. These issues and others will be addressed in more detail by some future research work.

Acknowledgment

This research was partially supported by the Paul Ivanir Center of Robotics, Ben-Gurion University of the Negev.

References

1. Elsayed, A., E. *Reliability Engineering*. Book. Addison Wesley Longman,, Inc., Reading, Massachusetts. **1996.**
2. Ladany, Shaul P. Optimal set-up of a manufacturing process with unequal revenue from oversized and undersized items. In "*Frontiers in Statistical Quality Control 6*", H.-J Lenz and P.-Th. Wilrich (eds.). **2001.** Physica-Verlag, Heidelberg, Germany. 93-101.
3. Shore, H. General control charts for variables. *International Journal of Production Research,* 38(8), **2000**, 1875-1897.

Evaluation of Quality and Market Risk of Complex Industrial Products

Evgueni Solojentsev[1] and Elart von Collani[2]

[1] Intelligent Integrated Systems Laboratory, Institute of Mechanical Engineering Problems, Russian Academy of Sciences, St.Petersburg, Russia
[2] University of Würzburg, Sanderring 2, D-97070 Würzburg, Germany

Abstract. This paper considers the production of complex products of possibly different types with several quality characteristics including all the relevant technical, economic and market features. Each characteristic may fail or be in an operating state where it may adopt one of several levels. Accordingly, a discrete probability model is used for describing the stochastic situation and statistical methods are developed for determining the actual values of the relevant probabilities.

For evaluating a given situation with respect to market failure or market success of a product and for selecting the most profitable production settings, the market risk is defined as a function of the quality characteristics and their probability distributions and the actual market conditions, which include also the view of the consumer.

The proposed approach is illustrated by an example referring to the introduction of an automobile the success of which depends for example on design, speed, acceleration, safety, reliability, weight, comfort, ease of control, ecological characteristics, number of sales agencies, price, advertising, market conditions like stagnation, recession or boom. Further examples which have been dealt with are the manufacturing of compressors and the building of a bridge.

1 Introduction

Quality management is a difficult task and involves many decisions about products, processes and after-services and has to take into account, for example, the market conditions, the employees and, of course, the customers. In the following the decision refers to certain objects, which could be complex products.

It is of great importance that the risks of the different objects are known quantitatively thus enabling a classification of the objects which could be used as a basis for rational management decisions.

In this paper an approach is developed for solving the relevant problems which is similar to the one used in reliability theory of complex systems also known as reliability analysis of stochastic networks.

The different characteristics of the object are modelled by Bernoulli variables which may operate on one of several levels. If some initiating events occur with respect to the characteristics, the object enters a certain state called class. This paper only deals with the most simplest case that there are two final classes called operation and failure, respectively. But a generalization to more than two classes can be achieved in a straightforward manner. Initiating events leading to object failure define logic connections *AND, OR, NOT* between the characteristics. From these logic connections the so-called *L*-model of the object is derived.

If the object is a parallel system and the characteristics are components, then *AND* is the logic connection of the components. In a serial system, for instance, the logic connection is given by *OR*.

The levels of the characteristics allow for modelling different states of the characteristics which might be not linearly ordered.

One basic aim of the paper is to provide an intuitively attractive objection function for optimization. More specifically, the following tasks have to be accomplished:

1. Numerical evaluation of the failure.
2. Classification of the object or the object conditions in "good" or "bad" objects.
3. Determination of the cost of the risk.
4. Analysis of the contributions of different characteristics to the object risk.

2 The L, P-Model

Consider an object to be analyzed with respect to its quality and market loss risk (QMLR). The risk of a failure of the object depends on n characteristics given by the random vector

$$\boldsymbol{X} = \left(X_1, \ldots, X_n\right)^T \tag{1}$$

It is assumed that the characteristics X_j, $j = 1, \ldots, n$, are selected in a way that they are stochastically independent random variables.

The characteristic X_j might operate in one of N_j levels. Which of the levels is adopted is defined by the random vector

$$\boldsymbol{Z}_j = \left(Z_{j,1}, \ldots, Z_{j,N_j}\right)^T \tag{2}$$

with the following meaning:

$$Z_{j,r} = \begin{cases} 1 & \text{if } X_j \text{ has level } r \\ 0 & \text{if } X_j \text{ has not level } r \end{cases} \tag{3}$$

It follows that the image space of $\boldsymbol{Z}_j$ is given by the N_j-dimensional unit-vectors.

The overall operating level of the object can be represented by the random vector

$$\boldsymbol{Z} = \left(Z_1, \ldots, Z_n\right)^T \tag{4}$$

with

$$Z_j = \sum_{r=1}^{N_j} r Z_{j,r} \tag{5}$$

The L-model specifying the initiating events leading to object failure is given by the random variable

$$Y = \psi(X_1, \ldots, X_n) = \Psi(\boldsymbol{X}) \tag{6}$$

with

$$Y = \begin{cases} 1 & \text{if there is an object failure} \\ 0 & \text{if there is no object failure} \end{cases} \tag{7}$$

For example an object which fails if any one characteristic fails has the L-model

$$Y = \psi(X_1, \ldots, X_n) = X_1 \vee \ldots \vee X_n \tag{8}$$

Let

$$P_j = P_{X_j}(\{1\}) = 1 - Q_j \quad \text{for } j = 1, \ldots, n \tag{9}$$

the failure probability of X_j. The P-model is defined by the probability of object failure:

$$P = P_Y\left(\{1\}\right) = \phi(P_1, \ldots, P_n) \tag{10}$$

The function ϕ depends on the L-model and the arguments are the failure probabilities P_j of the characteristics.

For example, the P-model of an object with L-model given by (8) adopts the form:

$$P = P_1 + P_2Q_1 + P_3Q_1Q_2 + \ldots \tag{11}$$

So far the levels of the characteristics have not been taken into account. The probability of failure of X_j depends on its operating level. Thus, the probability of failure of an object depends on the overall operating level $\boldsymbol{Z}$:

$$P = P_{Y|\boldsymbol{Z}}\left(\{1\}\right) = \phi\left(P_{X_1|Z_1}, \ldots, P_{X_n|Z_n}\right) \tag{12}$$

The level probabilities are defined by

$$W_{j,r} = P_{Z_{j,r}}(\{1\}) \quad \text{for } j = 1, \ldots, n \text{ and } r = 1, \ldots, N_j \tag{13}$$

with

$$\sum_{r=1}^{N_j} W_{j,r} = 1 \quad \text{for } j = 1, \ldots, n \tag{14}$$

The conditional failure probabilities of the characteristics are denoted by

$$P_{j,r} = P_{X_j|Z_{j,r}=1}(\{1\}) \quad \text{for } j = 1, \ldots, n \text{ and } r = 1, \ldots, N_j \tag{15}$$

The unconditional failure probabilities of the characteristics, which can be looked upon as average failure probabilities, are denoted by $P_j^{(m)}$ and given by

$$P_j^{(m)} = P_{X_j}(\{1\}) = \sum_{r=1}^{N_j} P_{j,r} W_{j,r} \quad \text{for } j = 1, \ldots, n \tag{16}$$

Moreover, the conditional level probabilities are defined by

$$P1_{j,r} = P_{Z_{j,r}|X_j=1}(\{1\}) \quad \text{for } j = 1, \ldots, n \text{ and } r = 1, \ldots, N_j \tag{17}$$

with

$$\sum_{r=1}^{N_j} P1_{j,r} = 1 \quad \text{for } j = 1, \ldots, n \tag{18}$$

There are some more quantities of importance for the risk analysis.

- The admitted risk denoted by $P^{(ad)}$ separates the "bad" from the "good" items:

$$\begin{aligned} P < P^{(ad)} &\Leftrightarrow \text{object is good} \\ P \geq P^{(ad)} &\Leftrightarrow \text{object is bad} \end{aligned} \tag{19}$$

- The average risk of an object

$$P^{(av)} = \sum_{(z_1,\ldots,z_n)} P_{Y|(z_1,\ldots,z_n)}(\{1\}) P_Z(\{(z_1, \ldots, z_m)\}) \tag{20}$$

- The average gradation probability for characteristic X_j

$$W_j^{(m)} = \frac{1}{N_j} \sum_{r=1}^{N_j} W_{j,r} = \frac{1}{N_j} \quad \text{for } j = 1, \ldots, N_j \tag{21}$$

The problem is to develop a method by means of which the above probabilities may be determined.

3 Object Classification

If the probabilities $P_{j,r}$ are known for $i = 1, \ldots, n$ and $r = 1, \ldots, N_j$ then it is possible to calculate the risks for any possible object given by a realization $\boldsymbol{z} = (z_1, \ldots, z_n)^T$ of $\boldsymbol{Z}$ by inserting the resulting probabilities $P_{X_j|z_j}$, $j = 1, \ldots, n$ into the P-model specified by ϕ. Thus we obtain the risk for an object being in the overall operating level $\boldsymbol{z}$:

$$P(\boldsymbol{z}) = \phi\Big(P_{X_1|z_1}, \ldots, P_{X_n|z_n}\Big) \tag{22}$$

A comparison of $P(\boldsymbol{z})$ with the admitted risk $P^{(ad)}$ according to (19) yields the classification into bad and good items.

Clearly, the risks $P(\boldsymbol{z})$ define an order in the set of all possible objects.

Assume that the probabilities $W_{j,r}$, $j = 1, \ldots, n$ and $r = 1, \ldots, N_j$, are known, and consider the object with the given overall operating level $\boldsymbol{z}_i$ and risk $P(\boldsymbol{z}_i)$. Then, by means of $W_{j,r}$ it is possible to calculate the probabilities

$$a_i = \text{probability that an object has a risk smaller than } P(\boldsymbol{z}_i) \tag{23}$$
$$b_i = \text{probability that an object has a risk larger than } P(\boldsymbol{z}_i) \tag{24}$$

Another measure which can be used are the probabilities that the risk is larger than a given one among the good items (with risk smaller than the admitted risk $P^{(ad)}$ and among the bad items (with risk larger than or equal to $P^{(ad)}$.

The quantification of how good or how bad an object given by $\boldsymbol{z}_i$ is can be based on the distance to the admitted risk:

$$d_i = |P(\boldsymbol{z}_i) - P^{(ad)}| \tag{25}$$

The measure d_i may finally be used for defining the cost of different risks.

4 Identification of the P-Model

The identification of the probabilities defined previously shall be done on the basis of T given items of the object of interest, for which the overall operating levels $\boldsymbol{z}_k$, $k = 1, \ldots, T$, and the states with respect to the characteristics are known. Moreover, the number of good items T_g with a satisfactory performance and the number of bad items T_b with an unsatisfactory performance are also assumed to be known and we have

$$T = T_g + T_b \tag{26}$$

The problem is that generally the necessary probabilities cannot be estimated as they are not identifiable. Therefore, an optimization problem is developed and solved which yields an "optimal" P-model.

The optimization problem consists of determining the probabilities $P_{j,r}$, $j = 1, \ldots, n$ and $r = 1, \ldots, N_j$ in a way that the function

$$F = T_{gs} + T_{bs} \stackrel{!}{=} \max \tag{27}$$

is maximum, where T_{gs} is the number of items among the given T items which are correctly classified as good by the assumed P-model and T_{bs} is the number items among the given T items which are correctly classified as bad by the assumed P-model.

The optimization problem has the following restrictions:

1. The probabilities $P_{j,r}$ satisfy the condition

$$0 < P_{j,r} < 1, \quad j = 1, \ldots, n \text{ and } r = 1, \ldots, N_j \tag{28}$$

2. The average risk $P^{(av)}$ meets the condition

$$P^{(av)} = \frac{1}{T} \sum_{k=1}^{T} P(z_k) = \frac{T_b}{T} \tag{29}$$

3. The admitted risk $P^{(ad)}$ must provide the given relation for wrongly classified good and bad objects

$$E_{gb} = \frac{T_g - T_{gs}}{T_b - T_{bs}} \tag{30}$$

In order to solve the optimization problem various existing relations between the different probabilities are utilized and simplified by replacing the probabilities $W_{j,r}$ by relative frequencies or by the known average probabilities $W_j^{(m)}$.

The task of identifying the risk P-model is characterized by the following features:

- Optimization has to be performed over a large number of variables, as there are $\sum_{j=1}^{n} N_j - n$ probabilities $P_{j,r}$ to be determined.
- The average initial values of probabilities P_{jr} may be determined easily.
- The target function is integer-valued and, of course, bounded.

- It is not possible to calculate derivatives of the target function F with respect to the probabilities $P_{j,r}$;
- Proceeding from the L, P-model structure, the target function has local extremes.
- It is not possible to search for the optimum F_{max} by means of positive or negative increments to all parameters P_{jr}, because the condition of the known average risk (29) must be met.

In order to evaluate the quality of the P-model the following relative errors may be used:

$$E_g = \frac{T_g - T_{gs}}{T_g} \tag{31}$$

$$E_b = \frac{T_b - T_{bs}}{T_b} \tag{32}$$

$$E_m = \frac{T - F}{T} \tag{33}$$

5 The Optimization Algorithm

The optimization problem is rather complex and solved by an iterative algorithm, which generates $P1_{j,r}$ maximizing F and leads to suitable values for the number of the good N_{gs} and bad $N_{bs} = N - N_{gs}$ items. This algorithm is briefly outlined below, where the iterations are consecutively numbered by $0, 1, 2, \ldots, v, \ldots, N_{opt}$.

Initial values: .

1. Initialize $P_{j,r}^{(0)}$ and $P1_{j,r}^{(0)}$ for $j = 1, \ldots, n$ and $r = 1, \ldots, N_j$ which meet condition (29).
2. Calculate ${}^{(0)}P_j^{(m)}$ and ${}^{(0)}P1_j^{(m)} = \sum_{r=1}^{N_j} P1_{j,r}^{(0)} W_{j,r}$.
3. Set the target functions F_{max} to a sufficiently small number, for instance $\frac{T}{2}$.
4. Calculate

$$K_j = \frac{{}^{(0)}P_j^{(m)}}{{}^{(0)}P1_j^{(m)}} \tag{34}$$

Optimization:

Maximize the target function F by an iterative process until the target function F increases N_{opt} times. The optimization steps under the condition

$$F > F_{max} \tag{35}$$

are given below, where F, F_{max} are the current and the earlier achieved values of the target function.

5. Generate small increments $dP1_{j,r}^{(v)}$, $j = 1, \ldots, n$ and $r = 1, \ldots, N_j$.
6. Calculate $P1_{j,r}^{(v)}$

$$P1_{j,r}^{(v)} = \frac{P1_{j,r} + dP_{j,r}^{(v)}}{\sum_{r=1}^{N_j} \left(P1_{j,r} + dP1_{jr}^{(v)}\right)} \tag{36}$$

7. Calculate $P_{j,r}^{(v)}$

$$P_{j,r}^{(v)} = K_j \cdot P1_{j,r} \tag{37}$$

8. Calculate the risks of all items

$$P^{(v)}(\boldsymbol{z}_k) = P_{Y|\boldsymbol{z}_k}\left(\{1\}\right) \quad \text{for } k = 1, \ldots, T \tag{38}$$

   by using the new probabilities $P_{j,r}^{(v)}$.
9. Calculate the object average risk ${}^{(v)}P^{(av)}$.
10. Determine $P^{(ad)}$ yielding N_{gs} and N_{bs}.
11. Calculate $F^{(v)}$.
12. If $F^{(v)} > F_{max}$, then

$$F_{max} = F^{(v)}, \; P1_{j,r} = P1_{j,r}^{(v)}, \; P_{j,r} = P_{j,r}^{(v)}. \tag{39}$$

$$K_j = \frac{{}^{(v)}P_j^{(m)}}{{}^{(v)}P1_j^{(m)}} \tag{40}$$

13. If the target function did not increase during the given number N_{mc} of trials, then the algorithm fixes the impasse. The actual value F_{max} is reduced by dF (approximately 2-4 units) and with this new value the maximization process is restarted:

$$F = F_{max} - dF, \tag{41}$$

It can be shown that the optimization algorithm has a polynomial complexity of n^3, where n is the number of characteristics in the QMLR object description.

6 Example: QMLR for an Automobile

The car is described by $n = 20$ characteristics, where each characteristic has at least 2 and at most 11 levels, with a total of $\sum_{j=1}^{n} N_j = 94$ levels. The characteristics may be classified into

1. technical and ecological ones,
2. sale and service ones,
3. economical ones,
4. seasonal factors,
5. market conditions.

The simple L-model (8) was used for the analysis.

There were information available about $T = 1000$ cars with respect to the characteristics and the gradations yielding an average object risk of $P^{(av)} = 0.3$.

Applying the optimization algorithm yielded the following results:

Number of Levels	F_{max}	T_{gs}	T_{bs}	T_g	$P^{(ad)}$	E_{gb}	E_m
94	798	554	244	610	0.307	2.61	0.202

Table 1: Overall results of the optimization procedure

The next table contains the average risks for the 20 characteristics used to describe the automobile.

Characteristic X_j	Average Risk $P_j^{(m)}$	Number of Levels N_j
X_1	0.018467	4
X_2	0.013939	10
X_3	0.017141	5
X_4	0.021644	11
X_5	0.021861	10
X_6	0.019017	5
X_7	0.017536	5
X_8	0.015764	4
X_9	0.017549	4
X_{10}	0.019574	3
X_{11}	0.017730	4
X_{12}	0.014789	4
X_{13}	0.015687	5
X_{14}	0.016288	3
X_{15}	0.019961	3
X_{16}	0.013191	4
X_{17}	0.019209	4
X_{18}	0.016756	2
X_{19}	0.017482	2
X_{20}	0.020789	2

Table 2: The average risks of the characteristics

7 Conclusions

The proposed risk analysis by means of *LP*-models and the developed identification procedure is rather general and may be applied to various evaluation problems involving the issue of quality. Besides the above given results, it may be used for evaluating the risk contributions of the different characteristics and, thereby, make a rational quality improvement policy possible.

References

1. Solojentsev E.D., Karasev V.V., Solojentsev V.E. (1999): *Logic and probabilistic models of risk in banks, business and quality.* Ed. E. Solojentsev, Nauka, St.Petersburg.
2. Taguchi G. (1986): *Introduction to quality engineering. Designing quality into products and processes.* Asian Productivity Organization.- Tokyo.
3. Adler Yu.P., Shper V.L. (1998): The new tendencies of development of the theory of quality of machines and devices. In Preprint 132 *New in the theory of accuracy and quality machines and devices.* Issue 6., St.Petersburg: IPMASH of RAS, 2-5.
4. Ryabinin I.A. (1976): *Reliability of engineering systems. Principles and analysis.* Moscow: Mir Publishers, 532 p.
5. Ryabinin I.A. (1994): A suggestion of a new measure of system components importance by means of boolean difference. *Microelectronics and Reliability* 34, 603-613.
6. Henley E.I., Kumamoto H. (1985): *Designing for reliability and safety control.* Prentice-Hall, Inc., Englewood Chiffs.
7. *The theory and information technology of modeling of safety complex(difficult) systems.* Eds. I.Ryabinin, E.Solojentsev, St.Petersburg: IPMASH of RAS. - 1994-95.- Issues 1-5, 400 p.
8. *Account of accuracy of machines and devices.* Eds.V.P. Bulatov and I.G. Friedlender. St.Petersburg: Politekhnika, 1993. 495 p.
9. Seitz J., Stickel E. (1992): Consumer loan analysis using neural network. Adaptive intelligent systems. *Proc.of the Bankai Workshop*, Brussels, 14-19 October, 177-189.
10. Solojentsev E.D., Karasev V.V. (1999): The logic and probabilistic method of assessment of credit risks and identification of risk models. Paper: 1043-017 in: *Proc. of International ICSC Congress, Computational Intelligence: Methods and Applications* (CIMA'99), June 22-25, 1999, RIT, Rochester, N.Y., USA.
11. Karasev V.V., Solojentsev E.D. (1998): Logic and probabilistic models of risk of loss of quality (QLR) and their training. *Proc. of 3-th International School "Modern fundamental problems and applied tasks of the theory of accuracy and quality of machines, devices, systems".* Eds. V.P.Bulatov, A.S. Zakharzhevskiy. June 29 - July 1, 1998, St.Petersburg,79-93.
12. Solojentsev E.D., Karasev V.V. (1999): Adaptive optimisation by slight increments of parameters for identification of the risk logical and probabilistic models. *Proc. 6th Saint Petersburg Symposium On Adaptive Systems Theory* (dedicated to the memory of Ya. Z.Tsypkin). September 7-9 1999, Saint Petersburg, Russia.

GPSR Compliance

The European Union's (EU) General Product Safety Regulation (GPSR) is a set of rules that requires consumer products to be safe and our obligations to ensure this.

If you have any concerns about our products, you can contact us on ProductSafety@springernature.com

In case Publisher is established outside the EU, the EU authorized representative is:

Springer Nature Customer Service Center GmbH
Europaplatz 3
69115 Heidelberg, Germany

Batch number: 10725973

Printed by Printforce, the Netherlands